Technical Editing

THE ALLYN AND BACON SERIES IN TECHNICAL COMMUNICATION

SERIES EDITOR: SAM DRAGGA, TEXAS TECH UNIVERSITY

Thomas T. Barker
Writing Software Documentation:
A Task-Oriented Approach

Dan Jones
Technical Writing Style

Charles Kostelnick and David D. Roberts
Designing Visual Language: Strategies for
Professional Communicators

Carolyn D. Rude
Technical Editing, Second Edition

Second Edition

Technical Editing

Carolyn D. Rude
Texas Tech University

Allyn and Bacon
Boston • London • Toronto • Sydney • Tokyo • Singapore

Vice President: Eben W. Ludlow
Editorial Assistant: Linda M. D'Angelo
Marketing Manager: Lisa Kimball
Editorial Production Service: Omegatype Typography, Inc.
Manufacturing Buyer: Suzanne Lareau
Cover Administrator: Suzanne Harbison

Copyright © 1998 by Allyn & Bacon
A Viacom Company
160 Gould Street
Needham Heights, MA 02194

Internet: www.abacon.com
America Online: keyword: College Online

Library of Congress Cataloging-in-Publication Data

Rude, Carolyn D.
  Technical editing / Carolyn D. Rude. — 2nd ed.
    p.   cm.
  Includes bibliographical references and index.
  ISBN 0-205-20032-X
  1. Technical editing.   I. Title.
T11.4.R83   1998
808'.02—dc21   97-24302
                                          CIP

Printed in the United States of America
10 9 8 7 6 5 4   02 01 00

Photo Credit: Thomas M. Schaefges

To Don and Jonathan

Brief Contents

PART 1   CONCEPTS AND METHODS   1
    1   Editing: The Big Picture   3
    2   Readers, Users, Browsers, Problem Solvers...   19
    3   Copymarking: Hard Copy   34
    4   Electronic Markup and Online Editing   49

PART 2   BASIC COPYEDITING   63
    5   Basic Copyediting: An Introduction   65
    6   Copyediting for Consistency   82
    7   Spelling, Capitalization, and Abbreviations   98
    8   Grammar and Usage   112
    9   Punctuation   133
    10   Quantitative and Technical Material   158
    11   Proofreading   176

PART 3   COMPREHENSIVE EDITING   191
    12   Comprehensive Editing: Definition and Process   193
    13   Style: Definition and Sentence Structures   212
    14   Style: Verbs and Other Words   228
    15   Style: The Social and Global Contexts   243
    16   Organization   253
    17   Visual Design   273
    18   Illustrations   295
    19   Editing Online Documents   321

PART 4   MANAGEMENT AND PRODUCTION   339
    20   Collaborating with Writers   341
    21   Legal and Ethical Issues in Editing   356
    22   Type and Production   370
    23   Management   387

Contents

Foreword by the Series Editor   xxi

Preface   xxiii

PART 1   CONCEPTS AND METHODS   1

    1   Editing: The Big Picture   3
    Scenario One: Print Document, Inhouse Editor   4
        The Product Team   4
        Project Definition and Planning   4
        Writing and Editing   5
        Publication   7
        Collaboration of Writer and Editor   7
    Scenario Two: Hypertext Tutorial, Contract Company   7
        The Product Team   8
        Project Definition and Planning: Content, Structure   9
        Planning for Design and Production   9
        Editorial Review   10
        Customer Review   11
    Comment: Editing at BMC Software and SAIC   11
    Editorial Functions and Responsibilities   12
        Preparing Documents for Publication   12
        Text Editing   12
    The Document Production Process Summarized   14
    The Technical Part of Technical Editing   15
        Technical Subject Matter and Method   15
        Technical Genres   16
        Inhouse or Contract Setting   16
    Qualifications for Technical Editing   16
    Summary   17

Further Reading   17
Discussion and Application   17

2   Readers, Users, Browsers, Problem Solvers...   19
Texts and Contexts   20
    Origins and Impacts: The Problem and Solution   20
    Readers and Use   22
    Culture and Expectations   22
    Constraints on Development and Production   23
    Using Context to Improve the Text   23
The Interactive Nature of Reading   24
    Creating Meaning   24
    Reading Selectively   24
Readers and Uses of Online Documents   25
Reading to Comprehend Information: Content, Signals, Noise   25
    Content   26
    Signals   27
    Undesirable Signals: Noise   28
Designing Documents for Use   29
Summary   30
Notes   30
Further Reading   30
Discussion and Application   30

3   Copymarking: Hard Copy   34
The Symbols of Copymarking   34
    Placing the Marks On the Page   36
    Marking Consistently   38
    Distinguishing Marginal Notes from Text Emendations   38
Special Problems of Copymarking   40
    Punctuation   41
    Hyphens and Dashes   41
    Ambiguous Letters and Symbols; Unusual Spellings   42
    Headings, Tables, References, and Lists   42
    Illustrations   43
Marks for Graphic Design   43
Queries to Writers   44
Summary   46
Further Reading   46
Discussion and Application   47

4   Electronic Markup and Online Editing   49
Styles and Templates   50
SGML   52
    Descriptive versus Procedural Markup   53
    SGML Tags and Document Type Definition (DTD)   53

SGML Documents as Databases   54
SGML Coding and Editing   54
HTML   *54*
Online Editing: Policies and Methods   *55*
Version Control   58
Tracking Changes   59
Electronic Copymarking   60
Comparison of Hard Copy and Online Editing   60
Summary   *61*
Note   *61*
Further Reading   *61*
Discussion and Application   *61*

PART 2   **BASIC COPYEDITING**   **63**

   5   **Basic Copyediting: An Introduction**   **65**
Document Correctness and Consistency   *66*
Document Accuracy   *67*
Completeness: Books, Manuals, Long Reports   *68*
Parts of a Book, Manual, or Long Report   68
Copyediting Illustrations   *71*
Parts of Illustrations   72
Placement of Illustrations in the Text   75
Quality of Reproduction   76
Copyediting Hypertext   *77*
How To Copyedit   *77*
Summary   *79*
Further Reading   *79*
Discussion and Application   *80*

   6   **Copyediting for Consistency**   **82**
Document Consistency   *82*
Verbal Consistency   83
Visual Consistency   84
Consistency of Mechanics   85
Content Consistency   88
A Foolish Consistency...   88
Style Manuals   *89*
Comprehensive Style Manuals   90
Discipline Style Manuals   90
House or Organization Style Manuals   92
Document Style Sheet   94
Using the Computer to Establish Consistency   *95*
Summary   *96*

Further Reading   97
Discussion and Application   97

7   Spelling, Capitalization, and Abbreviations   98
Spelling   99
    Guidelines and Tools   99
    Frequently Misused Words   102
    International Variations   105
Capitalization   105
Abbreviations   106
    Identifying Abbreviations   107
    Periods and Spaces with Abbreviations   107
    Latin Terms   107
    Measurement and Scientific Symbols   108
    States   109
Summary   109
Further Reading   109
Discussion and Application   110

8   Grammar and Usage   112
Definitions   113
Sentence Structure   113
    Verbs and Sentence Patterns   113
    Adjectives, Adverbs, and Modifying Phrases   117
Relationships of Words in Sentences   118
    Subjects and Predicates   118
    Verb Tense and Sequence   121
Modifiers   122
    Misplaced Modifiers   122
    Dangling Modifiers   123
    Pronouns   125
Conventions of Usage   127
Guidelines for Editing for Grammar   128
Summary   129
Further Reading   129
Discussion and Application   130

9   Punctuation   133
The Value of Punctuation   133
Clauses, Conjunctions, and Relative Pronouns   134
    Independent and Dependent Clauses   134
    Conjunctions   135
    Relative Pronouns   136

Sentence Types and Punctuation   137
    Punctuating Simple Sentences: *Don't Separate the Subject
        and Verb with a Single Comma*   138
    Punctuating Compound Sentences: *Determine Whether
        There Is a Coordinating Conjunction*   138
    Punctuating Complex Sentences   140
    Punctuating Compound-Complex Sentences   142
Punctuating Phrases   143
    Series Comma and Semicolon   143
    Commas with a Series of Adjectives (Coordinate
        Adjectives)   144
    Parallelism   144
    Introductory and Interrupting Phrases   146
Punctuation Within Words   147
    The Apostrophe   147
    The Hyphen   148
Marks of Punctuation   150
    Quotation Marks   151
    Dash   151
    Colon   152
    Ellipsis Points   152
    Typing Marks of Punctuation to
        Emulate Typesetting   152
Method of Editing for Punctuation   153
Summary   153
Further Reading   153
Discussion and Application   153

10   Quantitative and Technical Material   **158**
Using Numbers   158
Measurement   159
Marking Mathematical Material   161
    Fractions   161
    Equations   162
    Copymarking for Typesetting   164
Statistics   165
Tables   167
    General Guidelines   167
    Application: Copyediting A Table   168
Standards and Specifications   172
Summary   173
Note   173
Further Reading   173
Discussion and Application   174

11 Proofreading   176
The Value and Goals of Proofreading   177
Proofreading Marks and Placement on the Page   179
Computers and Proofreading   184
Strategies for Effective Proofreading   185
Summary   187
Further Reading   187
Discussion and Application   187

PART 3   COMPREHENSIVE EDITING   191

12 Comprehensive Editing: Definition and Process   193
Example: Copyediting vs. Comprehensive Editing   194
The Process of Comprehensive Editing   198
Analyze the Document's Readers, Purpose, and Uses   199
Evaluate the Document   200
Establish Editing Objectives   201
Review Your Editing Plans with the Writer   201
Complete the Editing   201
Evaluate the Outcome   202
Review the Edited Document with the Writer or Product Team   202
Application: The Service Call Memo   203
Analysis   203
Evaluation   203
Establishment of Editing Goals   204
Evaluation of the Outcome   204
Determining Whether Comprehensive Editing Is Warranted   206
The Computer in Comprehensive Editing   208
Summary   209
Further Reading   209
Discussion and Application   209

13 Style: Definition and Sentence Structures   212
Definition of Style   213
Writer's Persona and Tone   213
Style and Comprehension   213
Example: Analysis of Style   214
Guidelines for Editing for Style   216
Context: Make Style Serve Readers and Purpose   216
Sentence Structures: Use Structure to Reinforce Meaning   216
Place the Main Idea of the Sentence in the Structural Core   217
Use Subordinate Structures for Subordinate Ideas   218

Use Parallel Structure for Parallel Items   219
Sentence Arrangement   *220*
Place the Subject and Verb Near the Beginning of the Sentence   220
Arrange Sentences for End Focus and Cohesion   220
Prefer S-V-O or S-V-C Word Order   222
Sentence Length   *222*
Adjust Sentence Length to Increase Readability   223
Use People as Agents when Possible   223
Prefer Positive Constructions   224
Summary   *225*
Further Reading   *226*
Discussion and Application   *226*

14   Style: Verbs and Other Words   228
Verbs: Convey the Action in the Sentence Accurately   *228*
Build Sentences around Action Verbs   228
Choose Strong Verbs   229
Avoid Nominalizations   230
Prefer Active Voice   231
Use Concrete, Accurate Nouns   *233*
Prefer Single Words to Phrases or Pairs
and Simple to Complex Words   234
Application: Editing for Style   *236*
Analysis   236
Evaluation and Review   238
Summary   *240*
Further Reading   *240*
Discussion and Application   *241*

15   Style: The Social and Global Contexts   243
The Language of Discrimination   *243*
Application: Discriminatory Language   244
Editing for a Nonsexist Style   246
Style for a Global Workplace   *247*
International Correspondence   247
Globalization   248
Translation   248
Localization   249
Researching Social and Cultural Information   *250*
Summary   *251*
Further Reading   *251*
Discussion and Application   *251*

16 Organization   253

Organization for Performance: Task-Based Order   253

Organization for Comprehension: Content-Based Order   254

  The Schema Theory of Learning   254

  Organization and Comprehension   255

  Analysis as a Means of Understanding   256

Principles of Organization   257

  Follow Pre-Established Document Structures   257

  Anticipate Reader Questions and Needs   258

  Arrange from General to Specific and Familiar to New   258

  Use Conventional Patterns of Organization: Match Structure
      to Meaning   259

  Group Related Material   260

  Use Parallel Structure for Parallel Sections   263

Paragraph Organization   263

  Linking Sentences   263

  Repetitions and Variations   264

Application: The Problem Statement for a Research Proposal   264

Summary   269

Further Reading   269

Discussion and Application   270

17 Visual Design   273

Terms Related to Visual Design   274

Visual Design Options   275

Functions of Visual Design   278

  Comprehension   278

  Usability   279

  Motivation   282

  Ongoing Use   282

Headings   283

  Wording   283

  Levels of Headings   283

  Frequency of Headings   284

Application: Radar Target Classification Program   285

Guidelines for Editing for Visual Design   291

Summary   292

Further Reading   293

Discussion and Application   293

18 Illustrations   295

Reasons to Use Illustrations   296

  Convey Information   296

  Support The Text   297

Enable Action   297
Types of Illustrations   298
Comprehensive Editing of Illustrations   298
Appropriateness and Number   301
Match of Form, Content, and Purpose   301
Arrangement: Sequential and Spatial   302
Emphasis and Detail   302
Perspective, Size, and Scale   304
Relationship of Text and Illustrations   305
Discriminatory Language and Good Taste   307
Maximizing Data Ink   307
Application: Cassette Instructions   308
Preparing Illustrations for Print   312
Editing Illustrations with Computers   314
Summary   316
Further Reading   316
Discussion and Application   316

19   Editing Online Documents   321
Understanding Hypertext   322
Hypertext Links   323
Navigation   324
Principles of Design for Online Documents   325
Content and Information Design   325
Organization and Navigation   326
Style and Ease of Reading   329
Screen Design and Color   329
Grammar and Verbal Consistency   331
Graphics and Other Media   331
Political, Ethical, and Legal Issues   332
Process of Editing Online Documents   332
Summary   336
Further Reading   336
World Wide Web Resources   336
Discussion and Application   336

PART 4   MANAGEMENT AND PRODUCTION   339

20   Collaborating with Writers   341
The Editor–Writer Relationship   341
Strategies for Working with Writers   343
Edit Effectively   343
Manage Efficiently   344

Develop an Attitude of Professionalism   346
The Editor–Writer Conference   *347*
    Conference Organization   348
    Review of the Edited Typescript or Hypertext   348
    The Language of Good Relationships   349
    Furniture Arrangement   351
Correspondence with Writers   *352*
    Letters of Transmittal   352
    Email Correspondence   353
Summary   *354*
Further Reading   *354*
Discussion and Application   *354*

21 Legal and Ethical Issues in Editing   356
Legal Issues in Editing   *356*
    Intellectual Property: Copyright, Trademarks, Patents,
        Trade Secrets   357
    Copyright   357
    Permissions and "Fair Use"   358
    Copyright and Online Publication   359
    Trademarks, Patents, and Trade Secrets   359
    Product Safety and Liability   361
    Libel, Fraud, Misrepresentation   361
Ethical Issues in Editing   *362*
    Professional Codes of Conduct   362
    Misrepresentation of Content or Risks   365
    Environmental Ethics   365
    Bases for Ethical Decisions   366
Establishing Policies for Legal and Ethical Conduct   *366*
Summary   *367*
Further Reading   *368*
Discussion and Application   *368*

22 Type and Production   370
Type Fundamentals   *371*
    Typefaces   371
    Type Size   372
    Leading, Letterspacing, Wordspacing, and Line Length   373
Paper   *376*
The Production Process for Print Documents   *378*
    Typesetting and Page Makeup   378
    Illustrations   379
    Imposition, Stripping, and Platemaking   379
    Printing: Offset Lithography   380

Color Printing   382
Binding   383
Design Tips for Beginning Designers   *384*
Summary   *385*
Further Reading   *386*
Discussion and Application   *386*

23 Management   387
The Case for Managing the Document Development Process   *387*
The Life–Cycle Model of Publications Development   *388*
Planning   *389*
Estimating Time   *389*
Classification of Editorial Tasks and Responsibilities   389
Record Keeping   393
Sampling   394
Setting Priorities   394
Document Scheduling and Tracking   *395*
Management Plans   395
Tracking   397
Soliciting Bids   *398*
Setting Policy   *398*
Computers in Project Management   *399*
Summary   *399*
Notes   *400*
Further Reading   *400*
Discussion and Application   *400*

Glossary   402

Index   414

Foreword by the Series Editor

The Allyn and Bacon Series in Technical Communication is designed for the growing number of students enrolled in undergraduate and graduate programs in technical communication. Such programs offer a wide variety of courses beyond the introductory technical writing course—advanced courses for which fully satisfactory and appropriately focused textbooks have often been impossible to locate. This series will also serve the continuing education needs of professional technical communicators, both those who desire to upgrade or update their own communication abilities as well as those who train or supervise writers, editors, and artists within their organization.

The chief characteristic of the books in this series is their consistent effort to integrate theory and practice. The books offer both research-based and experience-based instruction, describing not only what to do and how to do it but explaining why. The instructors who teach advanced courses and the students who enroll in these courses are looking for more than rigid rules and ad hoc guidelines. They want books that demonstrate theoretical sophistication and a solid foundation in the research of the field as well as pragmatic advice and perceptive applications. Instructors and students will also find these books filled with activities and assignments adaptable to the classroom and to the self-guided learning processes of professional technical communicators.

To operate effectively in the field of technical communication, today's students require extensive training in the creation, analysis, and design of information for both domestic and international audiences, for both paper and electronic environments. The books in the Allyn and Bacon Series address those subjects that are most frequently taught at the undergraduate and graduate levels as a direct response to both the educational needs of students and the practical demands of business and industry. Additional books will be developed for the series in order to satisfy or anticipate changes in writing technologies, academic curricula, and the profession of technical communication.

Sam Dragga
Texas Tech University

Preface

Concepts and Goals

Technical Editing aims to prepare editors who function as information designers. To design information, editors need expertise in traditional areas of editorial responsibility—grammar, punctuation, and style. But they also need principles for structuring and displaying information, for making information easy to locate, and for choosing when to represent information visually rather than verbally. They must imagine the document in use—the text in its context—and make textual choices that facilitate use and comprehension. They also need to understand the process of document development and how to work effectively on teams that also include subject matter experts, writers, and graphic designers. This book extends the definition of editor from checker and "clean-up" specialist to decision maker and designer. It also asserts that effective editing begins early in a project.

Technical Editing helps editors develop concepts as well as skills so that they can make informed decisions. Because the book claims that analysis, evaluation, imagination, and judgment are primary editorial functions, it explains and illustrates principles and strategies of information design. It does not merely rehearse rules, but it does review principles of grammar, punctuation, style, organization, and visual design so that editors will understand how all of these text features may be used to clarify meaning, to persuade, to make information easy to locate, or to encourage the accurate completion of a task.

The book conveys an attitude of respect for novice editors and of editors for writers. It encourages professionalism through such means as using the vocabulary of the field, making choices based on principles rather than preference, and managing work to respect deadlines.

Revisions for the Second Edition

The second edition has been updated to incorporate changes in technology and the global marketplace. New chapters on editing online documents and using computers in editing redefine editing as an online as well as a paper-based task. A new chapter on editing for the global marketplace expands awareness of audience and culturally bound language choices. A separate chapter emphasizes legal and ethical issues. References in the "Further Reading" section of each chapter have been updated.

Whereas the first edition's theoretical approach to editing was primarily textual and cognitive, this edition adds the rhetorical and social perspectives. The social perspective comes out in numerous ways, mainly in the insistence that editors consider the context as well as the text in determining what is "good" and that they always consider the document in use.

Two significant changes in terminology reflect the social perspective. "Substantive" editing is now "comprehensive" editing, and "format" is now "visual design." "Substantive" and "format" suggest the text as it is bound by the page, whereas "comprehensive" and "design" point to editorial choices that enable use beyond the page. Also "readers" from Chapter 2 have now become "readers, users, browsers, problem solvers...," reflecting different reasons for using a technical document.

This edition should be easier to use than its predecessor. Some long chapters in the first edition have been broken into shorter chapters to present information in more manageable chunks. Revisions at the sentence level have shortened most chapters. Tables have been added for easy reference, especially in the grammar and punctuation chapters. Additional discussion and application projects have been added, and others have been grouped to teach specific concepts and skills.

Audience, Pedagogical Methods, and Structure

This book is written for students who have completed at least one college course in technical communication and for practicing editors with some experience in technical genres. It presumes that readers have been introduced previously to such terms as *style, noun, line graph,* and *instruction manual,* and it presumes some competence in technical writing. Chapters on spelling, grammar, and punctuation review concepts readers have learned before but do not substitute for a basic textbook or handbook. The chapters refresh students' vocabularies so that they can talk about their editorial choices using the language of the profession. The glossary reinforces the premise that professional technical communicators master the vocabulary of their discipline.

Scenarios, examples, illustrations, discussion questions, and applications complement explanations to enhance learning. Tables summarizing key points in chapters on copyediting enable ready reference. The instructor's manual includes workplace documents correlated to chapters so that students can apply the principles that the chapters discuss without the distraction of other editorial needs. Chapter activities and assignments at the Allyn and Bacon Web site (abacon.com/rude) support teaching in computer classrooms and online editing.

The book is organized to parallel the typical career path of editors. Just as editors prove themselves as copyeditors before they accept responsibility for comprehensive editing, so does this book teach copyediting before the more complex and less structured principles of comprehensive editing and management. This structure also facilitates the use of the book in two quarters or semesters, with the first term devoted to copyediting and the second to comprehensive editing. The modular character of the book, however, enables a top-down approach, with issues of comprehensive editing preceding the review of copyediting.

Part 1, Concepts and Methods, includes introductory material. The first chapter illustrates the breadth and diversity of editorial responsibilities through scenarios and discussion. The second chapter explains what readers and users do with documents—the basis for editorial choices beyond adherence to rules. Chapter 3 immerses students in copymarking on paper, and Chapter 4 introduces online editing and marking.

Part 2, Basic Copyediting, covers editorial choices that make a document conform to language standards, including grammar, punctuation, and consistency. It explains those standards in the context of a reader's need to understand, locate, and act.

Part 3, Comprehensive Editing, offers an analytical process and principles for evaluating style, organization, and the visual features of a text, whether printed or online. Editors learn to look at whole documents and imagine their use by readers so that they can guide decisions about these high-level features of documents. Teachers who prefer the comprehensive approach to editing or whose students are skilled in basic copyediting can assign Part 3 directly after Part 1, using Part 2 for reference as needed.

Part 4, Management and Production, takes students into the workplace to consider relationships with writers, legal and ethical issues of publication, and methods of production and management.

Acknowledgments

I appreciate the assistance of typescript reviewers, users of the first edition, editors in the workplace, and my students and colleagues at Texas Tech University, all of whom have contributed immeasurably to my understanding of editing and of effective pedagogy and who have suggested useful revisions. The reviewers were Laura Gurak, University of Minnesota; Dan Jones, University of Central Florida; M. Jimmie Killingsworth, Texas A&M University; and Katherine Staples, Austin Community College. Stuart Selber consulted with me on Chapter 19. My former student, Paula Green, shared her workplace experience as a technical communicator and influenced the first chapter. Kathy Klimpel and Charlene Strickland, featured in the scenarios in the first chapter, provided insight into editorial practice and also helped to edit the chapter. Steven Auerbach, Gerard Bensberg, Kae Hentges, Ken Morgan, Lane Mayon, Carlos Orozco-Castillo, Ellen Peffley, Tony Santangelo, and William Stolgitis helped to locate examples. Special thanks to my students, who always teach me more than I teach them. Love and appreciation to my husband and son, Don and Jonathan, who patiently supported me while a book took priority for time.

C. D. R.

P a r t **1**

Concepts and Methods

Chapter 1
Editing: The Big Picture

Chapter 2
Readers, Users, Browsers, Problem Solvers . . .

Chapter 3
Copymarking: Hard Copy

Chapter 4
Electronic Markup and Online Editing

Chapter *1*

Editing: The Big Picture

If you are coming to the study of editing without prior experience, you may think of it as cleanup work after a document is written. Editors correct errors in spelling and punctuation.

Editing does require establishing standards of language use, but you will find through study and experience that this work is a small part of what technical editors do. Technical editing, like writing, requires information design—creating documents that work for the people who will use them. Functional documents require more than correctness. And editors who help to create these document must be able to imagine documents in use by particular readers, to use good judgment as well as handbooks of grammar, to manage long-term projects, and to collaborate with others.

In this chapter, you will see specific editing responsibilities in the context of the whole process of conceiving, writing, reviewing, and publishing documents—the big picture of editing. Let us begin with some definitions—specifically, with answers to these questions: What does an editor do and why? How does a *technical* editor differ from any other editor?

To help answer these questions, we will review the history of two editors in different settings and their work on two different documents: a printed software manual, and a hypertext tutorial. One editor, Kathy Klimpel, works for a software company. Her manual was developed inhouse—within the company that needed the manual to help customers use its software. The second editor, Charlene Strickland, works for a consulting company that contracts for documentation projects from other companies. These two settings represent typical employment for technical editors.

The editors' roles in both projects were comprehensive:

- helping to define the need, purpose, and scope for the document
- working with writers and subject matter experts
- reviewing the text for completeness, accuracy, visual design, and overall effectiveness

Scenario One: Print Document, Inhouse Editor

Kathy Klimpel works as an editor for BMC Software, Inc., in Houston. BMC Software develops and maintains mainframe and open systems software. Kathy works with mainframe IMS products. (IMS is IBM's Information Management System, a database management system used in many major data processing centers around the world.) Before a BMC Software product can be sold, it must have a manual that describes how to use the product. BMC Software does not provide training, so the reference manual must be thorough. Customers may call a toll-free number to talk with a product support representative if they need help, but BMC Software hopes the manual will be clear enough to minimize the number of calls. The scope of the manual can vary depending on how complicated the product is, how much background information users need, and time constraints.

The Product Team

BMC Software uses a team concept to develop its products, and the writers and editors belong to these teams. The team members have these responsibilities:

- The product manager leads the group.
- The product author (usually an engineer) designs the product and oversees its development.
- The developers (programmers) write the code.
- The quality assurance representatives test the product and the documentation.
- The product support representatives answer customer questions and recreate customer problems for team members. This recreation identifies the exact nature of the problem and the conditions when it occurs. This information helps the team locate and fix the problem and test the product to make sure the problem is solved.
- The writer writes and maintains the documentation.
- The editor works with the writer.
- The publications manager assigns writers and editors to the product groups as needed. The writers and editors report to the publications manager.

The publications manager and editor work with multiple product groups. Thus, the editor works with several writers simultaneously on manuals for different products.

In other companies the publications group is separate from the product development group, but the BMC team arrangement helps writers and editors learn the product thoroughly by integrating them with the engineers and programmers.

Project Definition and Planning

Kathy works with Paula Green, a technical writer, on various projects. When a product idea is approved, the entire product team meets a few times to discuss the scope of the project and to decide how long it will take to complete. Kathy and

Paula attend the meetings and learn what expectations the team has for the product and the manual.

Paula writes the initial outline for the manual, and Kathy reviews the outline and makes suggestions for revision, mostly about content and organization. They then develop a schedule that allows enough time to complete the manual by the deadline for product completion. All products change during the course of development, and Paula and Kathy make corresponding changes to the structure of the manual and to the schedule. The initial planning and scheduling are necessary to ensure that the manual will be ready when the product is ready to sell.

Writing and Editing

As the product team begins developing the product, Paula begins writing the manual. She uses FrameMaker templates that BMC Software has developed to make the documentation for all of their products look as much alike as possible. (FrameMaker is publishing software, and templates are a set of instructions for margins and different types of text, such as headings and paragraphs.) The templates save the writers and editors time by providing a standard page design for the manuals. The templates also help users of multiple BMC Software products by making manuals for various products look familiar. Paula also refers to the comprehensive BMC Software style and design guides that tell how a document should look and give guidelines for writing, such as capitalization of terms and use of numbers.

Kathy creates a style sheet for each product. The style sheet includes spelling and capitalization and other choices for terms specific to the product. By keeping a record of these choices, Paula and Kathy can be consistent throughout the manual. The style sheet remains important even after the product is released. When there are product upgrades and revisions in the manual, the writer and editor can refer to the style sheet. This style sheet for the specific product supplements the comprehensive guide.

Paula submits each chapter to Kathy as she writes it. If the chapter includes complex or technical material, Paula first submits it to one or two members of the product team for a review of content. Kathy reviews the chapters comprehensively. Her overall goal is to help Paula produce a readable and usable manual. "Readable" and "usable" define whether readers can find and understand what they need to know to use the product. Kathy also reviews the drafts for these qualities:

- adherence to BMC Software standards as defined in the comprehensive style and design guide
- content, organization: completeness and logic; use of examples to explain concepts; logical arrangement of sections
- style: short but not choppy sentences; conciseness; ability to be translated into a foreign language
- visual features: graphics; page design; appropriate use of the templates
- grammar and punctuation
- mechanics and consistency: spelling, capitalization, hyphenation

An example illustrates the way Kathy may influence the content of the manual and its usability. On one reference manual, Kathy and Paula changed the outline after Kathy reviewed it. The manual described three related products that allow a user to restart computer programs that have failed. The three products accomplish the same goal, but each one works in a different environment.

Paula wanted to write about all of the cross-product tasks together (for example, all implementation information in one chapter with subheadings for each environment). Kathy, who tries to take the reader's point of view, suggested that the book might be more usable if the subjects were grouped by environment rather than by task. If a customer buys the product for the IMS environment, that customer probably doesn't care about the other two environments the product works in. Most computer users want to know what they need to know, when they need to know it. Therefore, it will probably be easier for them to use a book if all the information they need is in one place and is not cluttered by other, irrelevant information.

Paula rearranged the topics according to environment. Information that applied to the product in more than one environment was centralized, and Paula used cross-references to point to information as needed.

When she edits, Kathy marks hard (paper) copy of the drafts because users see the book on paper. She prefers a purple pen, avoiding red because she doesn't want to suggest to Paula that she is grading her draft, like a teacher, instead of editing.

Kathy tries to return an edited chapter to Paula in a day or two. Paula incorporates the edits, but if she questions them or if Kathy has written "let's discuss," they talk together about alternatives. Often they arrive at a better solution together.

As she edits, Kathy considers the product as well as manual design to determine whether the product could be easier to use. If the writing for a procedure seems complex, the reason may be an unnecessarily complex program.

After Kathy edits a chapter, Paula sends the chapter to the entire product team for a technical review (for accuracy and completeness of content). This process of multiple reviews during the course of product development achieves technical accuracy as well as effectiveness of the writing. It also keeps the writer and editor in touch with the programmers in case there are changes in the original plan for the product. Because of what the writer and editor observe from documenting the product, they may be able to advise the programmers on how to modify the product to increase efficiency or ease of use. Team members all benefit from the interaction throughout product development.

After she reviews each chapter individually, Kathy edits the entire manual, including all front and back matter (such as the title page and appendixes). She tries to finish this part of the editing task in two to five days, depending on length. The chapters and overall structure of the manual are established by now. She mainly edits for completeness and the smooth connection of parts.

Developing a product and its manual can take six to nine months. During this time the writer can get caught up in the details. The editor tries to keep the perspective of the user and an eye on the big picture.

Publication

BMC Software produces all its manuals inhouse on copiers. The writer orders covers in a standard design 30 days before the product will be available. After the product team, Paula, and Kathy agree that the manual is ready for publication, Paula sends the manual to be reproduced. When the manual and product are complete, the product team meets with the senior vice president of research and development to turn over the product. The sales department begins selling the product immediately—and Paula and Kathy get to work on the manual for the next product. However, they maintain (revise and update) the manual they have just finished if the product is upgraded into a new version, if the product support representatives observe from customer requests for help that information is missing or needs clarification, or if changes in other software or hardware that the product supports require changes in this product.

Collaboration of Writer and Editor

Because Paula and Kathy work together from the product idea to completion of the manual, they have a close working relationship. Paula has more technical knowledge than Kathy and has taught her how IMS works. When Kathy inadvertently changes the content as she edits, Paula explains why a different revision may make more sense. Kathy helps Paula tune specific writing skills. Paula likes to write long sentences, and Kathy helps her make them shorter.

Kathy says, "Paula and I work together very well, mainly because we have the same goal in mind—a clear, correct, concise manual—and because we have mutual respect. I enjoy working with Paula because she teaches me so much. She has helped me become a better editor."

Paula says, "I trust Kathy to see our documents from the reader's perspective, from the big picture to the tiny details. I rely on her intelligence, common sense, sharp eyes, and sound advice to perfect my work. She makes the difference between a good document and an excellent one."

Scenario Two: Hypertext Tutorial, Contract Company

Charlene Strickland works in Information Technology for Science Applications International Corporation (SAIC) in Albuquerque. This technology firm works on contract for companies that need computer programs and documentation but do not have the staff or the time to complete the work themselves. The companies can achieve high quality at a fair price by hiring software specialists to do the work.

Charlene was editor on a project to develop a computer security tutorial on an intranet within the World Wide Web for a national laboratory. (Access is limited on an internal network to people within a company.) Like typical workplace documents, the tutorial was created because a problem needed to be solved. The laboratory is required by U.S. law to offer computer security training each year to each of its 7,000 employees. Each employee has to be recertified each year by passing a test over the material. Offering the training in a traditional classroom was expensive and time-consuming. It was a logistical nightmare as well to train 7,000 employees with different levels of experience in computer use and knowledge of security procedures. The training was further complicated by the fact that three different types of computer workstations were used in various parts of the laboratory.

The laboratory contracted with Charlene's company to produce the training as a hypertext document. (A hypertext document is electronic. Its sections are linked electronically, enabling readers to choose topics in the order that suits them, not necessarily in linear order. *Hypertext* is used broadly here to refer to online documents with graphics and other media as well as text.) Hypertext training could solve the scheduling, computer platform, and expertise problems while reducing the cost of training. World Wide Web pages allow users on different types of computers to interact. The design could incorporate testing as well as instruction. Employees could complete the tutorial at their own pace. They would open the tutorial through their computers, read the material, answer questions and get feedback, and then take a test to earn certification in computer security.

The Product Team

The team that produced the tutorial included thirteen people with different types of expertise and responsibilities. Several of them were working on multiple projects simultaneously and did not devote all their time to this project.

- The project manager met with the client and was responsible for ensuring that the project met the client's specifications and was delivered on schedule.
- Three writers provided expertise in computer security and created the content for the tutorial.
- Four instructional designers organized information for maximum comprehension and retention.
- Three programmers made the pages work technically. They coded the text and graphics for each page and created the links between subjects and modules.
- One graphic artist created the color schemes, illustrations, and screen design for the instructional modules.
- One editor reviewed content development and sentence structure and maintained coherence in the parts produced by multiple writers and instructional designers.

All of these team members had specialized roles, but the editor was a generalist in addition to being a specialist. Charlene was the one person on the team with whom all the other team members interacted. Thus, she was responsible not just for the effectiveness of language but also for coordinating the efforts of the other team members and for ensuring the completeness and consistency of the information produced collaboratively through division of labor.

Project Definition and Planning: Content, Structure

The division of labor, in which different specialists were working simultaneously on different parts of the final product, required initial planning. The team used a software development process of planning, analysis, design, development, and implementation. Analysis required the team to learn the customer's expectations and to become familiar with the content of the training. Because the laboratory had previously delivered the computer security training in a classroom setting, some teaching materials were available to SAIC for developing the content. These materials, called legacy documents, were an important source of information for the training. Yet the materials contained inconsistencies and needed to be updated. The client provided this updated information in response to queries from the writers and editor.

The team planned six instructional modules plus a test. This plan for the content to be covered in each training module was necessary before the writers and programmers could begin working on the modules assigned to them. They displayed the plans on a storyboard, a poster-sized visual and verbal outline of the structure and contents of the tutorial. The storyboard included scripts for the components of the tutorial. Scripts contained both visual and textual elements, chunked by the sequence of screens that comprised each module. A template for the scripts, written as a Microsoft Word file, contained the scripts' elements in a three-column table format. Columns included the title of the electronic file, the action of the user, and the text that would appear as a result of the action.

The editor completed these tasks:

- contributed to information gathering and planning
- developed the template for the scripts
- drafted a style sheet of terms
- compared the scripts to the source document for accuracy

Planning for Design and Production

Design and production decisions affect writing and editing decisions from the start. The team had to find out what kind of equipment and software the laboratory would use for the tutorial. Software and hardware affect what the screen will display and how fast the text and images will load. The team planned screen design to encourage reading, comprehension, and ease of moving among the screens (navigation). A consistent layout of screens helps users of hypertext documents to navigate because they can find similar types of information in the same place on every screen. Menu bars let users choose lessons. The team had to decide whether screens would require users to scroll or whether each one would be complete and how to create transitions from one screen to the next. They decided as well to use a minimalist style, with as few words as possible, to avoid filling the screens with endless blocks of text. They planned as much interaction as possible within the tutorial so that the laboratory employees would not merely read. These plans followed from analysis of the laboratory's needs and available options for meeting the needs as well as from principles of instructional design and good writing.

Planning for design was part of the planning for content and organization because design needed to support the learning goals, not substitute for them. The SAIC team wanted to avoid a trendy, eye-catching tutorial. Instead they aimed for one that would accomplish the goal of teaching computer security. Each design decision had to support the instructional goals.

All of these plans at the beginning of the project gave the editor some measures for reviewing the drafts of the modules. They functioned in the same way that the comprehensive style and design guide does at BMC Software. The editor's participation in planning meant that she would not come in at the end of the project with a different idea and try to make substantial changes. She and other team members pursued the same goals from the start.

Editorial Review

Charlene used the legacy materials plus the team's plans for content and design as guidelines for editing. She edited each module separately. Her top priority in editing was to make the content accurate, complete, and readable.

For example, she reviewed the grammar for correct usage. She made changes in subject-verb agreement and the use of conjunctions (such as confusing *since* with *because* and *between* with *among*). Because online documents are often more informal in voice than material on the printed page, she also used contractions and did not spell out any numbers.

The modules were written separately, so the editor aimed to change the voice of the writers' words to read as if written by a single author. She looked for variations in tone (such as changing from second to third person). She also made terms consistent across the modules with regard to spelling, abbreviations, and capitalization. The major problem was the name of the laboratory. Throughout the modules, it had to be named by its abbreviation, not the first word in its name.

Another inconsistency was the use of bulleted lists. Charlene followed the format of capitalizing the first letter of each item, and ending each item with no punctuation. Items in each list had to be parallel. The phrase preceding each list had to end with a colon.

The writers submitted their drafts as computer files, and Charlene edited them on the computer, without printing. Editing on the computer helped her see the screens as the laboratory employees would and thus get a reader's perspective. Because she knew the subject matter and was in close contact with the writers, she could edit sentences or reorder paragraphs with confidence that she was not distorting the meaning. Still, it is easy to introduce new errors while editing, and the work required continual proofreading. Sharing files required a careful version control process, so that writers and editor would always work on the most recent version.

Editing the tutorial as a whole was a necessary step to prevent inconsistency, contradiction, and redundancy. One of the editorial strategies was to use the tutorial as a laboratory employee would, choosing the different lessons through the menus, responding to questions, and navigating through the screens.

Customer Review

The customers reviewed printed pages of each module, in the form of the script. Then, the customers viewed each module screen by screen, with a programmer and the editor presenting and discussing each screen. The customers made suggestions, and the programmer changed the screen so the customers saw and approved changes. The editor proofread these on-the-fly changes, and she noted these on her copy of the printed pages.

Comment: Editing at BMC Software and SAIC

At BMC Software and SAIC, the editorial procedures are comparable, though different products, settings, and personnel change some details.

Both editors edit comprehensively—not just for grammar, punctuation, and consistency but also for content, organization, and design. They see the document whole, as users will use it, not just at the sentence level.

Both editors are part of the product teams from the start. Integrating the editors into product development is more efficient and effective than bringing them in at the end to correct a document when it is almost ready for publication. At that point there is not time for an editor to rescue a poorly conceived document but only time to repair superficial problems. Writers, thinking they have finished a document, are also discouraged to find out that the editor wants to change it. When editors work at the front end of document development, they can prevent problems. They contribute to the vision, not just to the *re*vision, of the document. They share responsibility for information design.

Both editors see themselves as collaborators with the writers and with other team members. They have good working relationships with other team members because they share the same goal of developing a document that will work for the users.

The differences in their jobs reflect different documents, settings, and personal strengths. Kathy edits on paper while Charlene edits on the computer. The difference in part reflects the fact that Kathy's document is printed on paper while Charlene's is published on a computer screen. The editors need to see the documents as the readers will see them in order to edit well. But whether the document will be printed or transmitted electronically, both editors are concerned with content (completeness, accuracy of details), organization (relation of parts to each other and to the whole), style, page or screen design, usability (ease of use and effectiveness in enabling comprehension and performance), and correctness of grammar and punctuation. All of these qualities determine whether the documents will enable readers to use the product they have purchased or learn about computer security well enough to pass the test and to protect their files.

The comprehensiveness of their tasks, integration into the product teams, and collaboration describe optimal situations for editing. Both BMC Software and SAIC demonstrate good management practices based on understanding of the editorial function.

Editorial Functions and Responsibilities

Why are there editors? What does an editor do? For technical editors, these questions can be answered by summarizing their two primary functions: preparing documents for publication and text editing.

Preparing Documents for Publication

The editor is the link between the writing and the publishing of a document. Writers may lack the means for publishing and distributing their materials. They may have little interest in preparing the document for publication. Their purpose is to develop the content. The editor coordinates production.

The editor provides instructions for developing the writer's draft into the form in which it will be reproduced and distributed. These instructions cover editorial emendations and graphic design. Graphic design includes choices about type for the body copy and display type (headings, titles); illustrations; page size and margins; weight, color, type of paper for printed documents; and color. In traditional publishing, when a company contracts with a typesetter and commercial printer, the editor marks the draft of the document to show how the final document should look (a process called copymarking). In desktop publishing or online documents, these choices may be made at the beginning of document development and incorporated into the files. The editor reviews the copy that is ready for reproduction or distribution to be sure that the design decisions have been implemented and that the document is complete.

A production editor may also obtain bids for production based on length, number of copies, illustrations, color, and other features. If the cost estimate exceeds the budget, the editor will have to either modify some decisions about the publication, such as choosing cheaper paper for a printed document, or negotiate for a larger budget if the choices made seem necessary. Unless the work will be reproduced inhouse, the editor contracts with vendors for production based on cost, scheduling, and reputation for quality work.

Text Editing

Editing the text means making it complete, accurate, comprehensible, and usable as well as correct. An editor, whose specialty is language and document design, can suggest ways to make the document easier for readers to understand and use. The editor knows how to use style, organization, and visual design to achieve specific goals. Even when a writer is sophisticated in the use of language, an editor can bring objectivity to the reading that the writer may lose by knowing the subject too well. The editor works with the text from the perspective of the reader. The editor serves as a readers' advocate.

Not all documents receive the comprehensive editing that has been described so far. The amount of editing depends on the importance of the document and on time constraints. A company newsletter that won't be distributed publicly may be photocopied without much editing. The editor may simply check that all the pages are present and correctly numbered. For another document, the editor may check spelling, grammar, and consistency but not completeness and accuracy of information, organization, or visual design. The editor's supervisor establishes the expectations and limits of the job.

Text editing responsibilities may be classified as *comprehensive editing,* when the editor works with the content, organization, and design of the text as well as with grammar and punctuation, and *basic copyediting,* when the editor works with grammar, punctuation, spelling, mechanics, and labeling of illustrations.

Comprehensive Editing

When the editor shares responsibility with the writer for document content and usability, the editor is editing comprehensively. Other terms for this type of editing are *developmental editing, macro editing,* and *substantive* (sub´ • stan • tive) *editing. Developmental* suggests that the editor works from the start with the writer in developing the content and organization, and the term emphasizes the process. *Macro* distinguishes comprehensive text features from "micro" features such as punctuation. *Substantive* suggests "substance," and the term emphasizes document content. *Comprehensive* is used in this textbook because it suggests both the process and the focus of editing. Comprehensive editing includes analyzing the document purposes and making decisions about the best ways to meet these purpose. As content is developed, comprehensive editing includes adding and deleting material and evaluating the reasoning and evidence. It also includes reviewing organization, visual design, style, and use of illustrations, each of which affects readers' ability to find the information they need and to comprehend it.

Comprehensive editing may occur at various stages of document development. Ideally, the editor helps to conceive the project before the first draft is written, so that writer and editor can develop a shared concept of the document and its readers and purpose. Reviews of early drafts may enable necessary reshaping of the document before too much time has been invested in its development. If the editor helps with development of the document content and structure, the work may best be called developmental editing. The editor may also edit for substance after the document is almost complete. Substantive editing sometimes requires rewriting of sections as well as sentences, but the editor may also advise the writer about revisions.

A comprehensive editor understands how different text characteristics, such as organization, visual design, and style, affect whether the reader can understand the document and use it easily. The editor reads analytically and emends the text with specific goals in mind rather than according to personal preference. Because comprehensive editing addresses the content, the editor must know something about the subject matter.

Basic Copyediting

Copyeditors check for correct spelling, punctuation, and grammar; for consistency in mechanics and from one part to the next; and for document accuracy and completeness. The copyeditor may mark the document for typesetting or desktop publishing to indicate typeface and type size, column width, and page length. Basic copyediting assumes that content, organization, visual design, and style are already established.

A good copyeditor has an eye for detail as well as a command of language. The editor refers to handbooks, style guides, and other printed or online sources and queries the writer or a technical expert to resolve inconsistencies or other text questions.

The Document Production Process Summarized

Technical editing is part of the process of developing documents that solve problems or to enable readers to use products. Editing requires knowledge of language and procedures of marking documents, but good editorial decisions also require knowledge of how those decisions affect the rest of the process and the ultimate effectiveness of the document as readers will use it. Editing is a kind of quality control to increase the chances that a document will work effectively.

In spite of good editorial judgment, the editor may miss some needs for improvement in the document. Technical reviews, inspections, and usability tests provide additional information. A technical review is a review for content by a subject matter expert, such as a computer programmer or medical researcher. The reviewer looks for accuracy and completeness of information. An inspection is a kind of technical review, but after their independent reviews, various experts meet together as a group to discuss the results of their review and to make recommendations to the editor. They resolve together any conflicts from their own reviews.

Editors review the document from the perspective of the user, but representative users can provide direct evidence about whether the document works. A usability test involves representative readers using the document under the intended conditions of use. They may try to use software by following the instructions in the manual. If they can't figure out how to do something, the testers know that the manual (or the software) requires revision. Informal usability tests for this textbook took place in classrooms, with college students as the evaluators. The students had some ideas for the book and found some confusing places in it that the writer and editor had missed.

The flowchart in Figure 1.1 summarizes the document development and production process and substeps. The process is not entirely linear, as the two-way arrows between steps 2 and 3 imply. Writers draw on responses from editors, suggestions from subject matter experts, and results of usability tests in their revisions. As the document develops, the project definition may change. Some documents develop more simply, without technical reviews and usability tests.

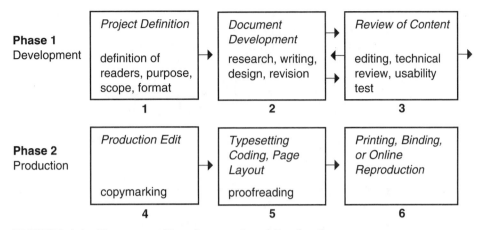

FIGURE 1.1   Document Development and Production

The production process varies depending on whether the document will be printed or distributed in some other medium. In addition, the print process depends on whether a company produces its documents by desktop publishing or by contracting with typesetters and commercial printers. If a company uses desktop publishing, type and page layout are determined at the beginning of document planning.

The Technical Part of Technical Editing

The management and text editing responsibilities are common to all kinds of editors: magazine and newspaper editors, academic journal editors, and the editors who work in commercial publishing houses on novels, trade books, and textbooks as well as to technical editors. All editors share some responsibilities for helping to make writing effective and for arranging for the publication and distribution of documents. The adjective *technical* does, however, distinguish some defining characteristics of the specific type of editing that this textbook teaches.

Technical Subject Matter and Method

Technical editors work on documents with technical subjects. *Technical* connotes technology, and typical subjects are computer science and engineering. However, technical editors also edit in medicine, science, government, agriculture, education, and business. A technical editor may be employed in any field for which the documents aim to help readers solve problems or gain information. Because of the specialized subject matter, editors ideally have technical (subject matter) knowledge as well as language expertise. A technical editor working for a software firm would know some programming, while a medical editor would know biology, chemistry, anatomy, and physiology.

Technical suggests not only the subject matter but also the method of working with the subject matter—to analyze, explain, interpret, inform, or instruct. The word derives from the Greek word, *technikos,* meaning "art and skill," which is also the source of *technique.* The art and skill of editing require specialized knowledge of the use of language and methods of making sense of information.

Technical Genres

Technical editors typically work with the document genres (or types) that permit the transfer of information or that enable readers to act by making a decision or by following instructions. Examples of such genres include instruction manuals and online help, proposals, feasibility studies, research reports, and World Wide Web sites. The documents may be produced on paper, but technical editors increasingly edit online documentation for a computer program, a slide show that is part of an oral proposal for a grant, or a World Wide Web site.

Inhouse or Contract Setting

Only a small percentage of technical editors work for large commercial or academic publishing houses—that is, places whose primary function is the publication of documents. Rather, technical editors are likely to work in a company in which the primary function is software applications, engineering, scientific research, or business. In these settings, publishing is a secondary business function. Alternatively, editors may work for companies that accept assignments on contract.

Qualifications for Technical Editing

In small companies, editors complete the full range of editing and management tasks. In larger companies, several people share the responsibilities. One editor may be responsible for development, a second for copyediting, and a third for production, while a publications manager coordinates the process. A beginning editor will probably be assigned copyediting and manuscript coordination tasks. The editor who demonstrates competence at the beginning level can advance to greater responsibility for the text and for production. As in all professions, expertise and responsibility grow with experience.

The primary qualification for basic copyediting is to understand language and know its rules. Editors also must be able to read carefully and focus on details. They need to have some knowledge of the visual characteristics of text, such as spacing and type.

Editors with responsibility for document content and development also need to analyze and evaluate the subject matter. They can imagine readers using the documents, and they know something about how readers comprehend information and use documents. They know options for format and media and reasons for making choices among the options. They are visually, as well as verbally, sophisticated. They have some understanding of the subject matter of the documents they edit. They accept responsibility for the quality and ethical integrity of the document.

Editors with management responsibilities must be organized and well disciplined. They encourage top performance from the people who work for them.

People who enjoy editing collaborate well with people and respect the contributions of people in different jobs. They set high standards for themselves, but when there isn't time to be perfect at everything, they set priorities and remain flexible.

Summary

Technical editors have two primary functions: to prepare the document for production and to edit the writing of a subject matter specialist in order to increase its effectiveness for the readers. This textbook focuses on editing to make writing more effective, but it introduces production because the good editor will make editorial decisions understanding the conception, development process, production, and use of a document.

Further Reading

Wallace Clements and Robert G. Waite. 1983. *Guide for Beginning Technical Editors.* Washington, DC: Society for Technical Communication.

Thomas M. Duffy. 1995. "Designing Tools to Aid Technical Editors: A Needs Analysis." *Technical Communication* 42.2: 262–277. Defines the editing process and tasks.

Arthur Plotnik. 1982. *The Elements of Editing: A Modern Guide for Editors and Journalists.* New York: Macmillan. See especially Chapter 2.

Judith A. Tarutz. 1992. *Technical Editing: The Practical Guide for Editors and Writers.* Reading, MA: Addison-Wesley.

Discussion and Application

1. Using the flowchart in Figure 1.1, relate the development and production of the SAIC hypertext tutorial to the different steps. What, for example, constituted research in the document preparation phase? What steps, if any, were omitted, and why?

2. Write three questions or comments you have as a result of reading this chapter. Be prepared to share them in class.

3. Some terms in this chapter may be new to you. Make a list of terms that you need to understand better. Use the glossary at the back of this book to find definitions and mark them as terms to review. These terms may be unfamiliar: *template, hypertext, desktop publishing, typesetting, copymarking, inhouse, contract, comprehensive editing, version control.*

4. Editors may receive the document to edit after it is written (at the end of the project), or they may participate in the planning and edit while the document is being developed.

   - In your own words, explain how these different procedures might affect the editor's responsibilities. Which of the procedures will invite more superficial editing, and why?
   - Speculate on how the different procedures might affect the editor's relationship with the writer.
   - How might the practice of involving the editor only at the end of the project provide an advantage in terms of what the editor is able to perceive about the work?

Chapter 2

Readers, Users, Browsers, Problem Solvers...

The title of this chapter purposely suggests multiple roles for the people who will use the documents that you edit. These people will read, but reading is not their main purpose. Rather, they will *use* the document to find specific information so that they can complete a task. They may browse as much as they read, especially if the document is a hypertext document whose links invite exploration. Reading, using, and browsing are means to the end of problem solving or decision making or operating equipment or getting information.

Just as writers begin their writing by considering who will read or use the document, in what setting, and for what purpose, editors make the best editing decisions if they consider why someone needs the document and how it will be used. Editorial decisions follow from this knowledge as much as from knowledge of handbooks and style manuals. Awareness of how the document will be used influences all good editorial decisions, including choosing the type of document (manual, poster, memo, Web site), choosing the medium (print, video, hypertext), organizing and formatting information, and correcting typos. In turn, all editorial decisions affect the reader's ability to use the document.

This chapter appears early in this book to establish the context for editorial decisions. Successful editing reflects the editor's ability to imagine and interpret the situation outside the text as well as the ability to make appropriate textual choices based on the conventions in grammar handbooks.

This chapter suggests links between textual choices and the ways people use documents. It places texts in their contexts to explain reasons for textual choices. It also reviews research on reading and comprehension and the implications for designing documents. It offers a theoretical framework for making decisions at all levels of editing by reviewing reading patterns, comprehension strategies and devices, access devices in documents, and the implications of all these for designing documents for use.

Texts and Contexts

The text of a document is its words as they are arranged in sentences, paragraphs, and sections. Editors are textual experts: they know principles of grammar, style, organization, and document design and can use this knowledge in helping writers. Yet a document can be perfect in a textual sense and still fail to achieve its purpose of enabling a person to make a decision or use equipment or find information. Usually the reason is that the text somehow does not match or meet the needs of the situation in which it exists. This situation is the document's context. As the prefix of this term suggests, the *con*text is *with* or *around* the text. The text never exists in isolation from people, places, values, and needs. Figure 2.1 illustrates the relationship between text and context.

The context includes origins and impact of the document, readers and conditions of use, the culture in which it is used, and constraints on development and production. Table 2.1 identifies some attributes of the text and the context. The rest of this section elaborates on the attributes.

Origins and Impact: The Problem and Solution

Context includes the reason for which the text is created. In technical and professional writing, this reason can probably be described as a problem to solve. A problem does not necessarily mean something bad, but it does imply some gap between an actual state and an ideal state. The document bridges this gap. For example, the purchaser of a new laptop computer may not know how to install the printer. The problem is the gap between current knowledge and the goal of successful installation; the user's manual or online help is the bridge between problem and solution because it tells the purchaser how to install the printer. Another problem could be a company's wish to inform employees of company initiatives, policies, and benefits and to develop loyalty to the company. An employee newsletter may enable achievement of those goals. Figure 2.2 illustrates the text bridge between problem and solution.

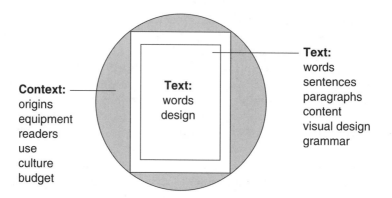

Context:
origins
equipment
readers
use
culture
budget

Text:
words
design

Text:
words
sentences
paragraphs
content
visual design
grammar

FIGURE 2.1   The Relationship of Text and Context

TABLE 2.1   Attributes of Text and Context

Text

content	headings	grammar	paper
organization	index	mechanics	size
format	appendixes	punctuation	binding
illustrations	page numbers	spelling	color
style	hypertext links	typography	screen resolution

Context	**Examples**
origins and impact	problem, need for information; solution
readers and use	prior knowledge, interests, abilities
	work tasks
	conditions of use (low light, outdoor setting)
	equipment and software available
	storage and disposal
	reading pattern (for reference, tutorial)
culture and expectations	social values (attitudes toward work, time, gender, space)
	discourse conventions (formality, patterns of organization, directness)
	language
	metaphors and cultural values
	national symbols and historical events
constraints on development and production	staff: abilities, time available (competing projects)
	budgets
	standards, policies, laws to which the document must comply
	equipment available
	related documents
	copyrights

The impact refers to what happens as a result of the document. A proposal for funds to support research succeeds if the grant is awarded, and a letter of application succeeds if a candidate wins an interview. Although textual features influence the impact, the measure of success is outside the text.

Editors aim not just to make a document correct but also functional and efficient in helping users solve problems. Recognition that documents are not ends in

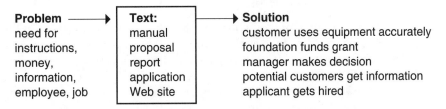

FIGURE 2.2   Text as Bridge Between Problem and Solution

themselves but means to other ends helps explain why measures of document success are currently expressed in terms of usability and persuasiveness more than readability. *Readability* refers to textual features such as sentence length and the impact of these features on comprehension. A document may be readable but still miss its purpose.

Readers and Use

A document's users focus on the outcome of reading, but reading is a necessary first step. Readers want to solve the problem—install the printer, make the decision, learn about the new health insurance policy so they can sign up for it. Although they might read novels and magazines for fun, they read technical documents in order to learn or to act. The reading is closely tied to work, whether it is work for a salary or work in the sense of completing a household task.

Because reading of technical documents is linked to work, editors can assume that readers are busy, that they have other responsibilities competing for their time, and that they are more interested in the task than in the document. They may be impatient with delays and distractions caused by reading. Unnecessary information, difficult words, clumsy sentence patterns, unusual structures or style, or difficulty in finding information divert readers from the content and task. On the other hand, lack of information can also make the text hard to understand. Thus editors and writers assess what readers probably know and what expectations they have for the document.

The anticipated interaction of the reader with the document will help establish what physical features will work best. For example, low light requires large type, and outdoor reading requires weatherproofing. Availability of equipment (such as a slide projector) and software (such as a browser for the Internet) determines choices of media and color. The need to keep a document flat on a desk, on a shelf, in a shirt pocket, or in a file cabinet influences choice of binding and paper size.

Because of the variety of contexts and uses of a document, it is impossible to create universal templates and formulas to define what a document should be, though conventions for document structure and design will help editors determine what textual choices are likely to work. A good editor at least imagines the document as readers are interacting with it to anticipate the likely uses and needs of readers. Research may be necessary to determine how and where readers will use documents.

Culture and Expectations

All readers take some of their cues for reading from their cultures. *Culture* may refer to national culture, such as the culture of Japan, to a professional culture, such as medicine, or a corporate culture, such as the work environment at Texas Instruments or Microsoft. Culture can even be defined by products, such as computers that use the Windows operating system. Readers may be parts of different cultures simultaneously—for example, programmers (profession) working at an American office (nation) of Texas Instruments (corporation).

Just as different cultures and organizations have dress codes (explicit or understood), they have codes for communication. Readers develop expectations for the structure, format, and style of documents. Because of implicit and explicit standards and conventions for science writing, the reader of a scientific research report will anticipate that the article define the research problem, the methods of investigation, and the results of the experiment, and discuss the significance. Furthermore, that reader will anticipate references to other related research reports, a formal writing style, and the author-date documentation style. A writer who fails to meet these expectations probably will not be published, and even if the publication reaches print, it will have to be extremely significant to overcome the liability of unconventional patterns of argument and presentation. Readers become aware of these conventions both deliberately (as in writing classes) and by experience.

The Asian cultures expect business correspondence to begin with courtesies regarding weather and the correspondent's health whereas correspondence from Western cultures conventionally gets straight to business. The various international cultures interpret colors in different ways, and words may have nuances of meaning that do not translate easily. Documents for international cultures may need to be more visual than verbal to minimize problems with translation. Translated documents may be shorter or longer than the originals in English. For example, German words and sentences are longer in translation than their English originals. Writers and editors make style and page design choices to accommodate these differences.

Variations from expectations may distract readers from the meaning or diminish trust in the writer and information. Therefore, editors help to ensure that the document will be appropriate for the values and expectations of the culture.

Constraints on Development and Production

Context also includes any constraints on development and production of a document. Budgets will determine whether color printing is an option, and product liability laws require warnings if safety is involved. Related documents may establish an optimum page size and layout so that all the documents will look like parts of the same set, which will help readers identify them. These external constraints may make the document better for readers (as in the case of safety warnings), or an editor may have to negotiate on behalf of readers to override constraints that cripple quality.

Using Context to Improve the Text

Good editors include in their vision not just the document as it is confined to the page or screen but also the context of use, especially of the reader using the document. This vision gives editors reasons for making textual choices and understanding of why these choices matter.

The rest of the chapter focuses on ways in which readers use the text to get the information they need or to solve the problem that brought them to the text to begin with.

The Interactive Nature of Reading

Designing usable texts requires knowledge of ways in which readers will respond to texts. Research in reading and cognitive psychology reveals that reading is much more complex than passive reception of information. Instead readers are busy making their own meanings. They depend on their prior knowledge, interests, and attitudes as well as on the text. A reader's understanding will not necessarily match the writer's intended meaning, even if the sentences and organization are clear.

In the midst of all this uncertainty about how readers will respond, writers and editors can be certain of two things: readers create their own meanings, and they read selectively. Good design, including selection of content, organization, and format, accommodates these predictable habits of reading.

Creating Meaning

The meaning comes not just from the words and other symbols on the page but from the knowledge that readers bring to the text and the way readers relate this knowledge to the information in the document. Two readers with different experiences and different memories of facts and concepts may create different meanings from the same text.

Readers also differ in their attitudes toward the material and the task, their emotional states at the time of reading, and their reading environments. They may be distracted from their reading by other thoughts, ideas, or tasks. They may skip around, miss some information, and fill in the gaps with their own creations. Furthermore, their attitudes and environments influence their responses to the text.

Nevertheless, the document will shape the reader's interpretation. Precise terms, analogies, and background information can help readers make connections between familiar ideas and new ones. Placing key concepts in prominent places in the document (such as the first sentence in a paragraph or the independent clause of a sentence) will reinforce the importance of these concepts. The document structure will give the reader an idea about the information itself. The writer's interpretation of data will influence the reader's.

Reading Selectively

Because readers of a technical document are oriented to the task or problem that the document addresses, they look for shortcuts in reading. Or they may be interested in only a portion of the document. An experienced computer user may skip over introductory material in a manual and look up the directions for importing a graphic file. A manager may read only the summary and recommendations of a feasibility study. Thus, they read selectively, skipping what they think they do not need. Writers and editors can help them find what they need with good access devices such as indexes, cross-references, and headings. If readers will perform a task while they read, editors will make it easy to get from the text to the task and back—

perhaps by using numbers to identify steps in the task or white space on the page to set off the steps.

Both print and online documents require navigational devices. A table of contents, index, chapter titles, and headings aid readers in finding specific sections. Color and tabs on pages are more expensive options. Numbering and use of space or rules between sections are visual signals about meaning, but they also aid in access. The first screen of an online document provides information about content and its organization to guide readers in locating what they need. Each screen marks location and provides directions for moving forward or back.

Placement of material is another device to make the most useful information the most accessible. Within a section, such as paragraphs grouped under a heading, important information usually appears early so that readers who read only the beginning of the section will be sure to read the important material. Information of secondary importance may be placed in an appendix. Such placement decreases the odds that the material will be read, but it makes the information of primary importance more accessible by removing the clutter of secondary information.

Readers and Uses of Online Documents

Like print documents, effective online documents are designed to enable comprehension and use. They depend, like print documents, on organization, visual design, and style to convey information and enable users to find the information they need. In selecting content, writers and editors anticipate readers' needs and knowledge. Good organization and good signals to organization in a table of contents or home page define the document and identify its contents. Menus and navigation buttons help readers keep their place in complex documents.

Many online documents use hypertext, with embedded links to related online documents. Hypertext encourages associative and selective reading. A reader does not have to read in a linear way but rather is likely to click on a link to find information on a particular topic. Although the contexts and reading patterns may differ in online and print documents, both kinds of documents share textual features of organization, style, and design, and both require similar kinds of editorial attention. As with printed documents, imagining the online document in use is a good editorial strategy.

Reading to Comprehend Information: Content, Signals, Noise

Whatever the practical outcome of reading, one purpose will be to get information. In order to act, readers will have to know something. Information will empower the reader to act. This information may be conceptual; that is, readers may need to understand and remember concepts and relationships between ideas. These readers read in order to learn. Or the information may be factual; readers may merely need to find a switch on a machine. These readers read in order to do.[1] Knowing that one broad purpose of reading is to get information, editors aim to make documents comprehensible. They make sure that terms are accurate, that conceptual information precedes detailed information, and that the level of detail is right for the readers. All levels of editing, from developmental planning to checking for correctness, work toward the goal of comprehension.

Documents give two types of information to readers: the content and the signals that help readers interpret that content. The content may be a review of research related to an experiment, a description of procedures or of a mechanism, a recommendation to make a purchase, or a parts list. Signals can relate either to the content or to the document itself. Verbal signals, such as the phrase *in conclusion,* show one's place in the document as well as the relationship of ideas. Structural signals, such as the table of contents and placement of important information at the beginnings of sentences and sections, orient readers to the document and help them recognize what is important. Visual signals, such as boldface or italic type or numbering of steps, indicate emphasis as well as structure. All of these signals help readers use the document and understand its content.

Content

The document's content enables readers to solve the problem or take action—assuming readers understand and interpret it accurately. Content must be complete and at the appropriate technical level for the readers. Careful organization facilitates accurate interpretation. Writers and editors cannot know for certain how much information to include and at what level, but they will make better decisions if they imagine how readers will respond to the document as a whole and at each point of reading.

Anticipating Readers' Questions
The content must be complete enough to make sense in light of the readers' previous experience and learning, as well as complete enough to enable the readers to do the task that motivates the reading. Editors have to anticipate what readers know and want. Imagining readers' questions is a good strategy for evaluating completeness. Readers ask predictable questions such as "What is it?" "How does it work?" "How does part A relate to part B?"

Linking New with Familiar Information
Meaningful content relates to a reader's prior experience and knowledge. Learning takes place when readers associate the new information with remembered information. Readers will struggle with comprehension if the content is unfamiliar, too much prior knowledge is assumed, and terms are unfamiliar. Analogy, background description, and reviews of literature are explicit ways of making the new information relate to something familiar. As new information is presented in the document, it becomes familiar or "old" once readers have absorbed and interpreted it. Then additional new information can be linked to information previously presented.[2]

Organizing Information

Knowledge is stored in memory in *schemata,* or patterns. Readers do not memorize separate facts but arrange them into structures. These structures distinguish concepts from details and show the relationship of parts. Content will be easier to comprehend if readers can sense its structure. The information in an organized document is easier to learn than that in a disorganized one because the structure of the document suggests the structure of the content.

Good organization reveals a hierarchy of information—the most important points and the supporting details. It also indicates how pieces of information relate, as in a cause–effect relationship, a temporal (chronological) one, or a spatial one. (See Chapter 16 for a more detailed discussion of organization.)

Signals

Signals, as well as the content itself, communicate information about the content and the document. These signals can be verbal, structural, or visual. Readers use the information communicated by the signals to help interpret the content. Editors use signals to point readers to the correct meaning.

Verbal Signals

Verbal signals are phrases or sentences that create a framework for the content or provide clues about how to interpret it. Introductions and overviews establish a framework for the new information that will follow. A forecasting statement or information map in an introduction indicates the way the document will develop. This forecasting statement may be just a list of topics that will be developed in the document. (The introduction to each chapter in this book ends with a forecasting statement.) It helps readers form a mental outline of the key issues.

The writer can signal the hierarchy of ideas with phrases such as "the most important fact is" or "the significance is." Verbal signals also include transitional words to indicate the relationship of ideas. Examples of such words are *however,* which signals contrasting information; *thus,* which signals a conclusion; and *then,* which signals a time relationship. Such signal words help readers understand the content by revealing the relationship of facts and ideas.

Structural Signals

Structure refers to the arrangement of words into sentences, sentences into paragraphs, paragraphs into sections, and sections into whole documents. Accurate structural signals help readers interpret the content by showing the hierarchy of ideas and distinguishing main ideas from supporting details. Some explicit structural signals are the table of contents, which shows the main divisions of the document, and headings, which reveal major divisions as well as subdivisions. In hypertext, a table of contents shows the way information is grouped just as in a print document. The links to other texts mark key concepts and terms and indicate development of ideas.

The arrangement of words and sentences implicitly cues readers about the relative importance of the words and sentences. The first item in a list or first words in a paragraph will get more attention because of their positions, and readers look for key information at the beginnings of sections.

The sequence of content items shows their relationships as well as hierarchy. Items may be ordered chronologically, spatially (such as top to bottom), in order of importance, or from general to specific. The sequence, along with the words themselves, helps readers interpret the material by revealing the way the parts are related. Even when the sequence the reader follows cannot be predicted, as in hypertext, the top level menu items suggest the hierarchy. This hierarchy plus categories of information revealed by menu items bind the parts into meaningful relationships.

Visual Signals

Visual signals can be seen on the page or on the screen. A common visual signal is indentation to signify a new paragraph, which, in turn, leads readers to expect a new concept or piece of information. Numbers and headings marked by variations in typography can also be used to identify sections of a document and to mark them visually as well as structurally. Boldface or italics indicate words of particular importance. Visual signals in graphs, such as an incline or decline in a line graph, guide the interpretation of the data. Even the paper and the type quality are visual signals about the content. High-quality paper and typesetting signal importance and encourage attention and respect. Neat work encourages readers to take the same care in reading that the writer and editor have taken in production.

The signals in documents to be read on a computer screen are often more visual (or auditory) than verbal: icons direct procedures, and color identifies different types of information, such as instructions for navigating the document and hypertext links to other texts.

Undesirable Signals: Noise

A document sometimes includes verbal, structural, or visual signals that interfere with comprehension of the content or that create a negative response to it. These signals are noise in the document, in the way that static in a radio broadcast is noise that partly covers the music or talk.[3] Just as listeners are annoyed and distracted by static, readers are annoyed and distracted by document noise. If the noise becomes too great, listeners will turn off the radio, and readers will stop reading. Noise interferes with comprehension by distracting readers. It makes interpretation more difficult and increases the chance of errors in interpretation.

Verbal noise can be misspelled words and grammar errors, which distract readers from the content. Inconsistencies in the use of terms or in capitalization represent noise if readers have to interrupt their reading in order to interpret whether the inconsistency signals a distinction in meaning. This book, for example, tries to be consistent in using the term *readers* rather than mixing the term with *audience* lest some readers wonder whether a distinction is intended between the two terms. A medical columnist used the terms *heart beat, heart rate,* and *pulse* interchangeably, but a reader, presuming the terms referred to three different things, queried the author for clarification. The variation gave a misleading signal about meaning.

Noise can also consist of irrelevant information, such as a digression from the main theme, unnecessary background information, or too many definitions. In addition, an inappropriate writer voice, or persona, can cause distracting noise in a document. For example, if the writer offends readers by seeming to be prejudiced or uninformed, readers may react negatively to the content, or they may not hear it at all.

Structural noise can result from the arrangement of content in a text that inaccurately reflects its actual structure (such as placing steps in a task out of chronological order). Structural noise results in hypertext when items seem randomly linked, making it hard for readers to find their place. Visual noise could be smudges on the page or an exaggerated mixture of typefaces that calls so much attention to itself that readers see only the chaos, not the content. It could be a busy screen pattern that diminishes contrast of words and background.

Although verbal, structural, and visual signals can help readers interpret and find information, too many signals can create noise. Not every sentence needs a transition word, nor does every paragraph need a heading. If readers are constantly processing signals rather than content, they will soon come to focus on the signals themselves. Signals should be relatively unobtrusive—integrated logically into the document and recognizable only when readers consciously seek them. When noise becomes so great that it dominates the content, the noise *is* the content, and the document has failed.

Designing Documents for Use

The more writers and editors know about how people will use a document and about the readers' prior knowledge and expectations, the more likely they will be to design useful products. Just as architects design buildings and engineers design roads, tools, and machines, technical writers and editors design documents. Design is more than a visual or aesthetic concept: a good designer considers the people who will use the building or the road or the document and plans the design accordingly. A conception of the functional whole leads to choices about specific features, such as materials and space. Document design parallels architectural and engineering design in its scope and purpose.

Document designers (writers, editors) begin their work with an awareness of how the document will be used. Even if you edit a document that has been completed when you receive it, a good editorial practice is to think of the document in use before you begin making specific emendations. This conception of the whole helps editors choose and modify specific components as they design the document for use.

A document designer's tools are words (including organization, style, grammar, and mechanics); visuals; design and typography; color, paper, size, and binding; and (in the case of online documents) sound and motion. Good design results in documents that readers can use efficiently and successfully.

Summary

The context in which a document will be used determines some of the editor's textual choices. Editors imagine the document in use in order to make choices about content, organization, format, and style. Documents enable readers to solve problems, complete tasks, or add to their knowledge. Reading is a means to an end. Readers interact with the document by using their prior knowledge and established reading patterns and by seeking specific information. The editor, as readers' advocate, language expert, and designer, is an essential member of the publication team. The rest of this book offers information and advice for editorial choices as measured in part by the contexts in which the texts will be used.

Notes

1. T. Sticht. 1985. "Understanding Readers and Their Uses of Text." In *Designing Usable Texts*, ed. Thomas M. Duffy and R. Waller. Orlando, FL: Academic Press: 315–340.

2. Herbert H. Clark and Susan E. Haviland. 1977. "Comprehension and the Given-New Contract." In *Discourse Production and Comprehension*, ed. R. O. Freedle. Norwood, NJ: Ablex.

3. Claude Shannon and Warren Weaver. 1964. *The Mathematical Theory of Communication*. Urbana: Illinois UP.

Further Reading

Ann Hill Duin. 1988. "How People Read: Implications for Writers." *The Technical Writing Teacher* XV: 185–193.

Barbara Mirel. 1992. "Analyzing Audiences for Software Manuals: A Survey of Instructional Needs for 'Real World Tasks.'" *Technical Communication Quarterly* 1: 13–38.

Barbara Mirel. 1993. "Beyond the Monkey House: Audience Analyses in Computerized Workplaces." In *Writing in the Workplace: New Research Perspectives*, ed. Rachel Spilka. Carbondale: Southern Illinois University Press, 21–40.

Mary Elizabeth Raven and Alicia Flanders. 1996. "Using Contextual Inquiry To Learn About Your Audiences." *Asterisk* 20.1: 1–13.

Janice C. Redish. 1988. "Reading to Learn to Do." *The Technical Writing Teacher* XV: 223–233.

Discussion and Application

1. Use a document that you have at hand to explain the difference between *text* and *context* in your own words. Imagine some details of the context of your document. Did the person or people who developed the document reveal a good sense of the context of use?

2. Locate a document of a technical nature, such as a textbook, proposal, or user's manual. Find examples in it that illustrate the writer's and editor's awareness of one or more of the concepts discussed in Chapter 2. Depending on your instructor's directions, either elaborate on how the document illustrates one of the principles or show how the document's designers were aware of several design principles. If your document fails to ac-

knowledge one or more of these principles, show where and how it fails. Bring your examples to class for sharing.

3. Compare the home pages of three or four universities on the World Wide Web. What types of information do the home pages include? What assumptions about Web surfers do the home pages suggest—for example, purposes for seeking the site, prior knowledge, needs for information? Assess how easily each one enables surfers to complete these specific tasks (typical reasons for checking a university Web site): get information on degrees offered, apply for admission, find the phone number or email address of a student, move easily between links at the site. What textual features (words, arrangement, color, visuals and graphics, typography, symbols, menus) facilitate the completion of these tasks? What features, if any, interfere with these tasks? As an editor, could you recommend improvements? If your instructor directs you to, write up the results of your analysis in a brief report.

4. Check the introductory paragraphs of the chapters in a textbook for forecasting statements or information maps—statements that tell what the chapter will cover and in what order. How will students use these verbal signals?

5. Verbal signals indicate relationships of pieces of information. Some categories of relationships are listed below, with an example for each category in italics. Extend the list of examples.

Contrast	*on the other hand*
Time	*after*
Space	*above*
Continuation	*furthermore*
Frequency	*sometimes*
Example	*for example*
Conclusion	*in conclusion*
Cause-effect	*because*
Explanation	*that is*

6. How might a table and a bar graph visually invite readers to compare?

7. Vocabulary: Check definitions of these terms if you are unsure of their meanings as they relate to documents and their readers or users: *context, design, signals, noise, selective reading.*

8. The material shown in Figure 2.3 is from *User's Guide: Microsoft Word.* Discuss how it reflects one or more of the following concepts from Chapter 2.

   a. Readers of technical documents read in order to act. (What specifically in the example encourages or enables action?)

   b. Readers read selectively. (What are the aids to selective reading?)

   c. Verbal signals help readers understand the hierarchy of ideas. (What verbal signals give clues to the hierarchy of ideas?)

   d. Visual signals help readers understand information as well as locate it. (What are the visual signals, and how do they help readers locate and understand information?)

   e. Document noise may interfere with the message of the text. (Does this document contain noise? If so, how does it interfere with the message?)

Managing Multiple Documents

With the Find File command on the File menu, you can search for, open, copy, print or delete several documents at the same time. For more information, see Chapter 22, "Locating and Managing Documents."

Saving a Document

When you open a document, Word copies it from the disk and displays it on your screen. Changes you make to the document are stored temporarily in the computer's memory. To keep the changes permanently, you must save the document on a disk. You should also make backup copies of important documents

Word can periodically save documents as you work. For more information about the Automatic Save feature, see "Controlling How Word Saves Documents," later in this chapter.

Saving Documents

When you save a document, it remains open on your screen so that you can continue working. To clear a document from your screen, choose the Close command from the File menu. For more information about closing a document, see "Closing a Document and Quitting Word," later in this chapter.

You can set an option so that when you first save a document, Word prompts you for summary information, such as a descriptive title and keywords suggesting the content of the document. If you fill in this information, see "Adding Summary Information," later in this chapter.

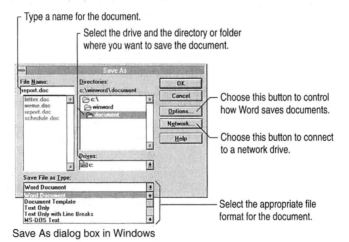

Save As dialog box in Windows

FIGURE 2.3   Sample Page from *User's Guide: Microsoft Word*.

9. The following paragraphs are from a series of guidelines on public relations (PR) for volunteer associations. Underline the verbal signals, and classify the type of relationship (contrast, cause-and-effect, sequence, etc.) indicated. Trace your own mental processes as you read a verbal signal. That is, what expectations do the signals create? Why did the writer use two paragraphs for the first point rather than running paragraphs 1 and 2 together? Paragraphing is a structural signal that may shape your interpretation of content.

The first principle of effective PR is that there is no "general public." Rather, there are a variety of publics defined by combinations of factors such as income, interests, profession, and geographical location. Most people belong to several publics. The crucial first step in any PR strategy is to select a specific audience—teachers, students, parents, doctors, agency personnel, legislators, potential employers—and then to design a message that will interest this group. Effective communications aim at people's self-interest. That is, you analyze your audience and point your message at the things they consider important. Only when you have selected a specific audience can you tailor your PR message to their interests.

This selective approach is clearly contrary to the shotgun approach of blanketing all media with the same messages in an effort to reach as many people as possible. These shotgun approaches will not be nearly as effective as a carefully planned message delivered to a specific audience. Use mass media efforts only to complement specific, targeted PR activities and not as the core of your PR program.

The second principle of PR is that personal contact is more effective than impersonal communication. While you should use many kinds of communication channels, try to include personal channels (face-to-face discussions, speeches, questions and answers with groups, conventions, telephone, letters) whenever possible. Use other media (newspaper, radio, television, billboards) to complement personal efforts. Also learn what members of the group you're trying to reach have the most influence in the group. Gain the support of these "opinion leaders."

Chapter 3

Copymarking: Hard Copy

Editors mark documents with instructions for revision of the text and for type and page design. These instructions tell the next person who works on the copy how to incorporate the editing. The person who uses the editor's marks may be the writer (in revision) or compositor (desktop publishing specialist or typesetter). The editor's instructions, marked on the document itself, are written with a special set of symbols. Marking the document with these instructions in symbolic form is known as *markup* or *copymarking*.

Marks for the writer may include suggestions for revision and queries to clarify meaning. Marks for a compositor show where changes need to be made to establish correctness, consistency, accuracy, and completeness. In addition, copymarking includes type specifications (face, size, and style) for headings, paragraphs, and other parts of the text; line length and page depth; and placement of illustrations.

When transmission of the text from writer to editor to printer is electronic (via disks or a network), the editor may make the changes or insert the suggestions for revision electronically. However, even with the possibilities for online editing and transmission of text, many editors prefer to work with hard copy at some point in production. Thus, all editors should know the accepted symbols of copymarking.

This chapter gives examples of copymarking symbols for copyediting and for graphic design. This chapter covers only copymarking on hard copy; Chapter 4, "Electronic Markup and Online Editing," describes electronic markup.

The Symbols of Copymarking

Editors, compositors, and graphic designers all understand a special set of symbols to indicate corrections and design choices. Some of the symbols and methods of giving instructions may seem cryptic to you at first, and you may be tempted to write out fuller instructions. However, such variations in conventions will confuse rather than help designers and compositors. Table 3.1 shows the symbols for

TABLE 3.1   Copymarking Symbols: Words, Letters

Symbol/Meaning	Example	Result	Comment
delete	deleete	delete	Use the closeup mark, too, if the word could be spelled as two
delete, close	proof reading	proofreading	
delete a word	in the ~~the~~ back	in the back	
insert	inert	insert	Place the caret beneath the line. Write what is to be inserted above the line.
insert space OR /#	inserspace	insert space	Usually the line alone will suffice; use the space symbol if there could be a question
	markup a text	mark up a text	
transpose	transpose	transpose	If multiple transpositions in a word make the edited version difficult to read, delete the whole word and print the correction above it
	Australia	Australia	
close up	close	close	
capital letters	ohio; ibm	Ohio; IBM	
small caps	6 a.m.	6 A.M.	Since not all fonts include small caps, make sure they are available before you mark them.
lower case	Federal	federal	
lower case, whole word	FEDERAL	federal	
initial cap	FEDERAL	Federal	
italics	Star Wars	*Star Wars*	Underline to change the type style from roman to italic or vice versa.
roman type or (rom)	*Star Wars*	Star Wars	Roman type is the opposite of italic, with straight rather than slanted vertical lines.
	Star Wars rom	Star Wars	Underline to convert from italic to roman, just as you do to convert roman to italic.
boldface	emphasis	**emphasis**	
superscript	Masters degree	Master's degree	Use the superscript sign to identify apostrophes, quotation marks, or exponents.
	A2	A^2	
subscript	H2O	H_2O	
delete an underline	revelry	revelry	OR: White out the line. Be careful not to cover up descenders or punctuation.
spell out an abbreviation or number	(2)	two	Circle an abbreviation or number you want spelled out. Spell the word as well as circling if the spelling may be in question.
	(Assn.)	Association	
	(hp)	horsepower	
"let it stand"; or .... ignore the editing	precede	precede	If you have edited in error or changed your mind, direct the compositor to set the copy in its original unedited form.

indicating changes in letters, spacing, and type style of words. Table 3.2 shows symbols for copymarking punctuation. Table 3.3 shows how to mark for spacing.

Placing the Marks on the Page

Copyediting marks appear within the lines of the text. Typically, typescripts are double spaced to leave room within the lines. Interlinear marks help the person making the changes in the text files because they appear right where the change must be made.

Instructions to the compositor other than for corrections appear in the margin. These directions cover line length, justification (whether the margins are to be aligned on the right or left or both), and typeface. They may also cover math symbols, extra space, special design material, instructions for handwritten material—anything not

TABLE 3.2   Copymarking Symbols: Punctuation

Symbol/Meaning		Example	Result	Comment
⊙	period	...forever⊙	...forever.	Circle the period to call the compositor's attention to this small mark. Do not circle other punctuation.
⩘	comma	copper iron and silver	copper, iron, and silver	Place an inverted caret over the comma. Do not place it over other punctuation.
(:)	colon	following(:)	following:	The oval makes the colon more clear.
;	semicolon	following; following; following;	following; following; following;	To create a semicolon from a comma or colon, draw in the dot or tail and place an oval around it. Otherwise, simply insert the semicolon.
⦅ ⦆	parentheses	(1986)	(1986)	The lines in the parentheses won't be typeset, but they do reinforce your intent to include parentheses rather than other lines.
⟮ ⟯	brackets	⟮word⟯	[word]	Be sure to square the lines if the writer has used parentheses.
= ⩗	hyphen	light‿emitting diode computer‿ assisted	light-emitting diode computer- assisted	The underline or checking of the hyphen reinforces your intent to include a hyphen at that point. Mark end-of-line hyphens for clarity.
(eq)	equal sign	a = b	a = b	Since the equal sign can look like the underlined hyphen, write *eq.* by the mark and circle it to show that the information is an instruction.
⊥/M or M	em dash	a pejorative ‿ᴹ disparaging‿ᴹ word	a pejorative— disparaging—word	An em dash is as wide as the base of the capital letter *M* in the typesize and typeface used. It is used to set off parenthetical material or a break in thought.
⊥/N or N	en dash	1996 ‿ 97	1996–97	An en dash is as wide as the base of the capital letter *N* in the typeface and typesize used. Its primary use is in numbers expressed as a range.

TABLE 3.3   Copymarking Symbols: Spacing, Position

Mark	Meaning	Example	Result
⊄	begin a new paragraph	...other design features. The editor's...	...other design features. The editor's...
⊃ (Z)	begin a new line	numbers; abbreviations;	numbers; abbreviations;
ᒣ	run together (do not create a new paragraph)	...other design features. The editor's...	other design features. The editor's...
⌐ or (fl)	flush left or justify left	The editor's choice...	The editor's choice...
⌐	justify right	Book Title	Book Title
] [	center	⅃――Book Title――⊏	――Book Title――
(RR)	ragged right [Lines do not align on the right margin.]		
⊣	fill out the line [Use for right justification or an extremely ragged margin]	...form your marks. It is not the time to express your...	...form your marks. It is not the time to express your...
‖	align		
□	indent one em		
⊓ or ②	indent two ems		
②	indent the whole block of text 2 ems		
ᔕ	transpose a group of words	transpose of words a group	transpose a group of words
(	close up vertical space (as when an extra space has been skipped between paragraphs)	...too many lines skipped. Close up vertical space.	too many lines skipped. Close up vertical space.
⁀	set as a paragraph rather than as a list	numbers; abbreviations; and spelling	numbers; abbreviations; and spelling.

covered by standard copymarking symbols. Such directions often apply to whole blocks of text rather than to single words or phrases, which is why they are marginal rather than interlinear.

Compositors appreciate marks that are neat and easy to read. Marks should not be too small to see nor so big that they make the whole page look messy. In addition, you should train yourself to form your marks in conventional ways. Copymarking is not the time to express your personality in handwriting with such quirks as small circles substituting for dots over the letter *i*. Be careful when you mark not to obscure the correct type on the page. Clean copy will increase the chances of error-free copy at the next stage of production.

You can help the compositor locate specific changes by using a bright-color pencil, such as red or green. Faint pencil marks are difficult to read, and they suggest timidity or lack of self-confidence. Especially if you are marking the hard copy of an electronic file, make the marks noticeable. The person who corrects the file will not be reading line by line and needs to be able to find the changes to make.

There may be more than one way to mark a change. For example, to mark the misspelling of "electornic," an editor could transpose "or" or delete the "o" where it appears and insert it after the "r." The choice depends on which marking is more clear and on the way the compositor will keyboard the change. In this case, the transposition is the simpler mark, and the compositor is likely to think of the change as one step, not the two steps of deletion and insertion.

Corrections will normally be incorporated into the electronic copy before the document goes to production. Clean copy is less confusing than heavily marked copy. Copymarking on corrected hard copy that goes to a compositor is likely to be limited to information in the margins about graphic design.

Figure 3.1 shows a marked typescript, and Figure 3.2 shows how the same typescript looks after being typeset.

Marking Consistently

Generally you will mark each occurrence of change rather than depending on the compositor to remember what you have done on previous pages. If, for example, you are deleting the hyphen in "on-line" throughout, mark each instance where the word occurs. Compositors are taught to type what they see rather than to edit. The editing is your job. Furthermore, more than one compositor may work on the job. Prepare the typescript as though the next person to read it will begin reading only at the point of the mark you are making, rather than at the beginning.

Mark each heading to identify whether it is a level one, two, or three heading, and provide the type specifications for each level. If paragraph indentations are not clear in the typescript, mark each change. Marking is especially important with spacing and graphic design marks if the design itself includes variations. Some designers, for example, specify indentation for all paragraphs except those that follow headings. You would check all paragraph indentation and mark any places where the pattern varies.

Distinguishing Marginal Notes from Text Emendations

Marginal notes may be necessary to clarify your marks on the text. For example, if you want an equal sign but the marked text looks just like a hyphen with an underscore, you could write "equal sign" in the margin.

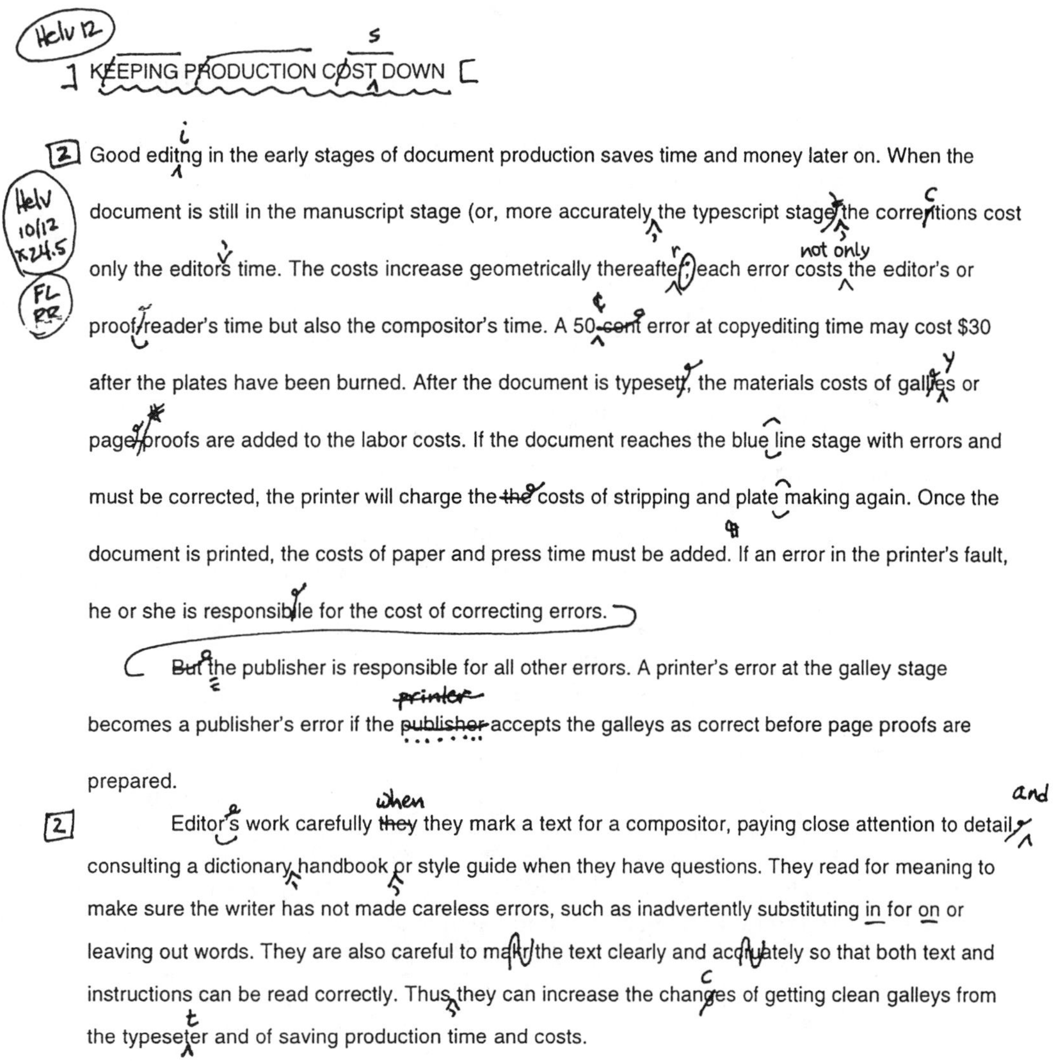

KEEPING PRODUCTION COST DOWN

Good editing in the early stages of document production saves time and money later on. When the document is still in the manuscript stage (or, more accurately, the typescript stage the corrections cost only the editor's time. The costs increase geometrically thereafter each error costs the editor's or proofreader's time but also the compositor's time. A 50-cent error at copyediting time may cost $30 after the plates have been burned. After the document is typeset, the materials costs of galleys or page proofs are added to the labor costs. If the document reaches the blue line stage with errors and must be corrected, the printer will charge the the costs of stripping and plate making again. Once the document is printed, the costs of paper and press time must be added. If an error in the printer's fault, he or she is responsible for the cost of correcting errors.

But the publisher is responsible for all other errors. A printer's error at the galley stage becomes a publisher's error if the publisher accepts the galleys as correct before page proofs are prepared.

Editor's work carefully they they mark a text for a compositor, paying close attention to detail, consulting a dictionary handbook or style guide when they have questions. They read for meaning to make sure the writer has not made careless errors, such as inadvertently substituting in for on or leaving out words. They are also careful to mark the text clearly and accurately so that both text and instructions can be read correctly. Thus they can increase the changes of getting clean galleys from the typesetter and of saving production time and costs.

FIGURE 3.1 Marked Typescript

To distinguish instructions from text insertions, circle the note. If you need to include marginal messages to both writer and compositor, you may distinguish these by using different colors of pencil for each category of message. Or you may preface the note with a label identifying the audience for the message—usually "au" for author and "comp" for compositor.

Pierce's philosophy   *au: correct?*

Influences on Darwin's *Origin of Species*   *comp: set rom*

Keeping Production Costs Down

Good editing in the early stages of document production saves time and money later on. When the document is still in the manuscript stage (or, more accurately, the typescript stage), the corrections cost only the editor's time. The costs increase geometrically thereafter; each error costs not only the editor's or proofreader's time but also the compositor's time. A 50¢ error at copyediting time may cost $30 after the plates have been burned. After the document is typeset, the materials costs of galleys or page proofs are added to the labor costs. If the document reaches the blueline stage with errors and must be corrected, the printer will charge the costs of stripping and platemaking again. Once the document is printed, the costs of paper and press time must be added.

If an error is the printer's fault, he or she is responsible for the cost of correcting errors. The publisher is responsible for all other errors. A printer's error at the galley stage becomes a publisher's error if the publisher accepts the galleys as correct before page proofs are prepared.

Editors work carefully when they mark a text for a compositor, paying close attention to detail and consulting a dictionary, handbook, or style guide when they have questions. They read for meaning to make sure the writer has not made careless errors, such as inadvertently substituting *in* for *on* or leaving out words. They are also careful to mark the text clearly and accurately so that both text and instructions can be read correctly. Thus, they can increase the chances of getting clean galleys from the typesetter and of saving production time and costs.

FIGURE 3.2   Copy Set as Marked in Figure 3.1

Marginal notes may also give instructions for the placement of illustrations if those instructions are not clear in the text.

Special Problems of Copymarking

Though the copymarking symbols will be clear in most situations, marks or letters that could be interpreted in different ways require special care. You may need to insert additional instructions when marking punctuation, distinguishing between hyphens and dashes, clarifying ambiguous letters and symbols or unusual spellings, and when marking headings, reference lists, and illustrations.

Punctuation

Because punctuation marks are so tiny, copyeditors add additional information to clarify which mark is intended. The conventions are these:

period	circle	$\odot$
comma	inverted caret	$\stackrel{\wedge}{,}$
colon or semicolon	oval	$\odot$

These marks are part of the information that identifies the punctuation; they are not interchangeable. An inverted caret means "comma," not period, and a circle is part of the information that means "period."

Hyphens and Dashes

Although hyphens differ in use and size from dashes, the distinction between hyphens and dashes is not always clear on a typescript and thus should be marked. Hyphens are the mark for combination words. They also appear in words that are broken at the end of one line and continued on the next. A line inserted under a hyphen during copymarking or a check over it indicates that the hyphen should be set as marked. If the hyphen is already typed correctly, however, you don't need to underline it—unless the word is hyphenated at the end of a line. Then, you need to clarify whether the hyphen should be retained if the word breaks differently in typeset copy. Underline an end-of-line hyphen that should be retained. Likewise, if the word is not normally hyphenated but conceivably could be, use the close-up mark with the hyphen at the end of the line on a typescript to show that the word should be set closed.

> Marking the document with instructions is called copy⊃
> marking. The copyeditor marks with the assumption that
> the compositor will enter text exactly as it is
> marked, letter for letter and mark for mark. End-of‗
> line hyphens are particularly confusing and should be marked.

You can minimize the confusion of end-of-line hyphens by preparing typescripts without hyphens except for words that are always hyphenated. Instruct writers to turn off the hyphenation on their word processors; then no hyphens will appear at the end of the line. In the preceding example, "copymarking" could have been typed on the second line without a hyphen, saving copymarking time.

Em dashes and en dashes are longer than hyphens and have different meanings. Em dashes separate words or phrases from the rest of the sentence; they function like parentheses in casual style. They are about the length of a capital letter *M* in the typeface in which they are set. Some people type two hyphens to create em dashes (a remnant of typewriter days). En dashes, which are the length of a capital letter *N* in the relevant typeface, are used in numbers, to show a range. If there can be ambiguity in interpreting which dash is intended, you should mark each occurrence.

Place two lead weights⎯each weighing 4ᴺ7 grams⎯on the model car body between the rear wheels.

Hyphens and dashes are set without space on either side. If a typist has typed spaces around them, mark the copy to close up the space.

Dashes⌒marks of punctuation used to set off parenthetical material⌒are longer marks than hyphens, and may be indicated with two hyphens. If a typist uses just one hyphen, the copyeditor must mark the em dash.

Ambiguous Letters and Symbols; Unusual Spellings

Some letters and symbols, such as the numeral *1* and the letter *l*, look similar. The editor should clarify anything about which the compositor may have to make a judgment. If context establishes the meaning, the editor does not need to clarify. For example, the compositor will recognize the insertion of letter *l* in "galey" to spell "galley" and will not insert the numeral 1. But in the following example, the compositor may think "numeral" after typing the 2 and not recognize the letter *l* as an abbreviation for liter. The note clarifies the editor's intent. The circle indicates information for a compositor rather than text to insert.

Evaporate a 2-l sample.   (ℓ)

The letter O and the numeral 0 may also be confused. You may need to write "zero" by the number and circle the word to show that you mean for a zero to be typed in that place. In equations, the x (indicating a variable or an unknown) must be shown to differ from the multiplication sign $\times$. (See Chapter 10 on editing mathematical material for more information on copymarking equations.)

If spellings are unusual, you may write and circle "stet" next to the unusual spelling to indicate that the unusual spelling is intentional. *Stet* is the Latin term for "let it stand."

Headings, Tables, References, and Lists

Copyeditors are likely to read the paragraphs of a document more carefully than other types of text, such as the headings, tables, and list of references. Yet errors occur in these parts of the text. It is easier to make content and typing errors in a reference list or in a table than in paragraphs. Thus, you must check for the accuracy and completeness of the information in these parts of the text. Be sure also to mark such text carefully, checking details such as type style (italics, roman, bold), accuracy of the numbers, and spacing. Also, pay close attention to lists, watching for incorrect end-of-line punctuation and indentation as well as for spelling, grammar, and punctuation errors.

Illustrations

As editor, you are responsible not just for the text but for the entire document, including tables and figures. The symbols used to mark illustrations are the same as those for marking text. You may need to correct spelling by deleting or inserting letters, to adjust spacing, or to request alignment of numerals on their decimal points. Headings, labels, and titles need to be marked for correctness and consistency in capitalization and type style.

If the illustrations are attached to the end of the typescript for insertion at the time of page layout, mark the place where they are to be inserted. You can do this in the margin if the text does not already indicate the location. Simply write <insert figure 1 about here> on a separate line, enclosing the words in square or angle brackets. You can specify only an approximate location because the page may not have enough room for the illustrations at the exact point you have marked.

Marks for Graphic Design

The editor or graphic designer or the printer's staff may mark the document for its graphic design—that is, the face, style, and size of type, the spacing, and the line length. If you have some training in graphic design, you may make the decisions about design and mark them too. Or you may place marks on the document according to a graphic designer's instructions. You can mark boldface, italics, and capitalization using the marks displayed in Table 3.1.

The marks for typeface, type size, and line length will make more sense to you once you are familiar with typeface names and with the printer's measures of points and picas (see Chapter 22). But the following example illustrates how you will mark such information from the graphic designer. The instructions direct the compositor to set type of a particular size and face on a line of the specified length:

$$\text{set } 10/12 \times 30 \text{ Times}$$

Here is what this note means:

set	=	set type
10/12	=	10-point type on a line 12 points deep (there will be some extra space between the lines of type)
× 30	=	the line length—30 picas
Times	=	the typeface—Times Roman

These instructions would produce type just like what you are reading here. The instructions are circled to clarify that they are not part of the text. The note appears in the left margin so the compositor will see them before typing the letters.

Whenever there is a change in the text, as from a heading to a paragraph, mark the change. You can do this either by marking the part, such as a level-one heading, or by providing the type specifications, such as size and style, each time. If you mark the part, include a type specification sheet with the marked copy.

Queries to Writers

You may not be able to make some copyediting decisions until you contact the writer for further information. Or you may wish to confirm that the marks are correct or to explain a change that will not be obvious. A question to a writer is called a *query.* The term is also used to refer to all comments from the copyeditor to the writer. Marginal notes can work for simple queries, but some questions and explanations are too elaborate to be phrased as marginal notes on the text. For these—or for all your queries, even simple ones—you can attach query slips to the typescript.

Sticky notes work well as query slips. They can be attached to the edge of the typescript, with the sticky part on the back, and folded over the edge of the page. They should be attached to the typescript at the place where the question arises. When the writer opens the slip, he or she reads the query and responds by revising the text or confirming that the editing is correct. The slip remains in place until the copyeditor checks the response and makes the necessary changes. Then the slip can be removed before the typescript is forwarded for typesetting. The query slips prevent the clutter of notes on the typescript itself that could distract the compositor. Figure 3.3 shows a marked page with a query slip attached.

Not every mark you place on the page requires a question or comment, and queries over obvious information will be annoying. Queries let you acquire information that you need to edit or mark correctly. They also let you explain marks that may puzzle a writer. In phrasing your queries, write directly and courteously, and avoid evaluative statements, especially when the evaluations are negative. For example, instead of writing "unclear" or giving the vague directions to "rewrite" or "clarify," tell the writer exactly what you need to know. Some examples follow:

Requests for information

to check a discrepancy in an in-text citation and a reference list	Page 16 cites the date of 1986 while the reference list cites 1987. Please check the date and indicate the correct one.
to verify an unclear use of quotation marks and capitalization	May I assume that the quote marks signal a quotation and that ABC should be capitalized?
to clarify the reference for a pronoun	I'm not sure whether "it" refers to the program or to the previous step. Please clarify.
to verify format	Do you have any special instructions for this figure—e.g., single or double space, paragraph indentations or flush left? Please advise.

Explanations

to explain why a numbered list has been converted to text with headings	Other numbered lists in this book present very short discussions for each item. The importance and development of each of these topics warrant the use of headings. The headings will emphasize each topic more than the numbered list does. OK?
to explain changes in headings	I expanded the main heading and deleted the sub-headings to parallel the pattern in other chapters. OK?

au: Because I frequently mark inclusive numbers in typescripts, I suggest adding that item to your list of choices for numbers. See the marked list below. OK?

Ⓒ **Abbreviation** ⌐ Abbreviation choices include whether to abbreviate (some terms are better known by their abbreviations than by the spelled-out version), what abbreviation to use if there are alternatives, and how to identify the abbreviations for readers (parenthetically or in a list of abbreviations). Abbreviations such as the following will require consistent editorial choices.

(LT)

(MCL)

a.m.	A.M.	A.M.
inches	in.	"
Pennsylvania	PA	Pa.

Comp: inches symbol (double prime)

Ⓒ **Numbers** ⌐ The various possibilities for expressing numbers require editorial choices.

(LT)

(MCL)

Commas	1000	1,000
Numerals or words	five	5
Dates	12 April 1998	April 12, 1998
Inclusive numbers	411–414	411–14
Time	eight o'clock	8 a.m.
Equation numbers	eq. 2.2	equation 2
Illustration numbers	figure I.I	figure I

411–4
8 A.M.
Figure 1

Comp: set as MCL w/ first col ital, ⊡ indent. If MCL won't fit w/ text meas, use turnovers in lefthand col

FIGURE 3.3   Copyedited Typescript Showing Notes to a Compositor and Marks for Graphic Design

Ask the writer to indicate that he or she has considered the query by initialing or checking the slip or by a verbal response.

If you note the typescript page number on the query slip, you will know where it belongs if it is accidentally detached. If you are writing queries to a designer or production editor as well as to the writer, also note on each slip who should read it. The note "au/24" identifies a query to the author on page 24 of the typescript. Some copyeditors attach the slips for the writer on the right side of the page and slips for the designer on the left or use different colors of notes for different readers. Then the designer and writer know which slips to read and which they can ignore.

Figure 3.3 illustrates a copyedited typescript page with marginal notes for the compositor, alignment marks on the tabular material, and marks to identify typeface, type size, and type style. The flag at the top of the page is a query from the copyeditor to the writer. It asks for the writer's approval of a possible change in the examples if small capital letters and italics are not available in the compositor's typewriter font.

The straight vertical lines and the circled letters at the left indicate specific design elements. BL means "bulleted list," LT means "list text," TYP means to use typewriter font for the examples, MCL means "multicolumn list," UL means "unnumbered list," and AC means "art callout." All of these types of text have design specifications: directions about font, spacing, type style, indentation, and so forth. The abbreviations request that those particular specifications be applied. The circled note by the abbreviation for inches directs the compositor to use the symbol for inches, not for quotation marks. The note in the right margin tells the compositor to use the text font for the items in the left column and the typewriter font for the other columns (the examples).

Turnovers are phrases in the first column spilling onto a second line. The irregular vertical lines within the columns specify alignment for the three columns of examples. You can see the results of these marks by looking at the section on consistency of mechanics in Chapter 6.

Summary

A version of a document that follows the typescript can only be as good as the copymarking. If the copymarking is incomplete or ambiguous, errors will appear in the proof copy. Conversely, thorough and accurate copymarking should result in clean proof copy, and the production of the document will continue on schedule.

Further Reading

Mary Stoughton. 1989. *Substance & Style: Instruction and Practice in Copyediting.* Alexandria, VA: Editorial Experts.

Discussion and Application

1. Mark the words in the first column so that they will be printed like the words in the second column.

developement	development
interogation	interrogation
emphasis	*emphasis*
italic	italic
½	one-half
three	3
teh	the
ambivalance	ambivalence
on going	ongoing
tabletennis	table tennis
2mm	2 mm
semi-colon	semicolon
CHAPTER TITLE	Chapter Title
Cpr	CPR
Research Laboratory	research laboratory
testing confidential	confidential testing
Tavist D	Tavist-D
referees decision	referee's decision
m2	m^2
M2	M$_2$
We finished quickly - we had more errands to complete.	We finished quickly—we had more errands to complete.
end of sentence	end of sentence.
end of clause,	end of clause;
introduction	introduction:
quote	"quote"

2. Your job is copyeditor—to make the document correct, consistent, accurate, and complete. You do not alter word choices or organization. Yet you know enough about style to object to the use of the passive voice in the second sentence of the paragraph in exercise 2. You argue to yourself that readers would identify more with the instruction and be more likely to follow it if the sentence read, "wash thoroughly with soap and water, and change soiled clothing as soon as practical." What can you do?

3. Circling an abbreviation instructs the compositor to spell it out. Assume you want the abbreviation *STC* spelled out. Why might circling be inadequate? What should you do instead?

4. You personally prefer to spell *proofread* as a hyphenated compound *(proof-read)* rather than solid. Do you have the choice of spelling it according to preference if it appears in a document you are editing? Why or why not?

5. Vocabulary: These terms should now have meaning for you in the context of editing: *copymarking, markup, query, compositor.* If you are uncertain of their meaning, check the glossary or review the chapter.

6. Mark the first paragraph so that it will be printed like the second paragraph. The second paragraph is set in the typeface Helvetica, and the measures are 10/12 × 23, flush left, ragged right. The title is Helvetica 12, bold.

FIRST AID

It is always wise to be cautious and aware of infectoin control

measures when asisting trauma victims. If contact with human

blood, urine, feces or other body secretions occur, through washing

withsoap and water is importent; and soiled clothing should be

changed as soon as practicle. We know, for example that the Aids

virus

    is readily killed by soap and water and by common

    disinfectants. You should avoid touching your mouth or eyes with

your hands or any items contaminated by glood, feces, or other body

secretions, personal with wounds or abrasions on exposed body

surfaces, such as the hands or face, should try to protect those

areas from contact with blood or secretions when emergency treatment

is being given. It is good practice to wear disposable gloves while

handling items contaminated, this is especially important for

personnel with wounds or abrasions on the hands.

First Aid

    It is always wise to be cautious and aware of infection control measures when assisting trauma victims. If contact with human blood, urine, feces, or other body secretions occurs, thorough washing with soap and water is important, and soiled clothing should be changed as soon as practical. We know, for example, that the AIDS virus is readily killed by soap and water and by common disinfectants. You should avoid touching your mouth or eyes with your hands or any items contaminated by blood, feces, or other body secretions. Personnel with wounds or abrasions on exposed body surfaces, such as the hands or face, should try to protect those areas from contact with blood or secretions when emergency treatment is being given. It is good practice to wear disposable gloves while handling contaminated items; this is especially important for personnel with wounds or abrasions on the hands.

Source: Theodore M. Hammett. 1987. *AIDS in Correctional Facilities: Issues and Options.* 2nd ed. U.S. Department of Justice, 112.

Chapter 4

Electronic Markup and Online Editing

Traditional copymarking, as described in Chapter 3, gives instructions to someone else to implement—perhaps the writer, or a desktop publishing specialist, or a typesetter in fullscale printing. The marks on paper direct corrections plus choices of type and spacing that affect the appearance of the document. Copymarking occurs near the end of document development.

Desktop computers enable editors to work directly in a computer file both to emend the text and to format it. Marking paper copy may be unnecessary. The editor may correct errors, rearrange paragraphs, and modify style at the keyboard. With desktop publishing, decisions about appearance are likely to be made early, at the beginning of document development. These decisions are coded in a document template that defines the choices of type and spacing for each text element, including headings and paragraphs. The writer applies the template using pull-down menus. The editor's job, then, is not to mark the design choices but rather to verify that the writer has selected the right ones (or to apply the template if the writer has not and the editor is serving as the desktop publishing specialist).

Other electronic markup goes beyond the use of templates. Electronic coding is necessary for documents that will be published in more than one medium (perhaps print plus hypertext) or that will be published only in an online version. Companies with substantial amounts of information to manage also find that coding the document parts aids in information retrieval. Standard Generalized Markup Language (SGML) is designed for publication in multiple media and for information retrieval. Hypertext Markup Language (HTML) formats hypertext.

This chapter discusses three ways to mark documents electronically: use of styles and templates, SGML, and HTML. All three of these methods are concerned with the structure and output of the document, not its content. The chapter also discusses policies and methods for editorial intervention in the text (online editing) to make it correct, consistent, accurate, and complete as well as usable and comprehensible.

Styles and Templates

Making choices about typography and page design before the document is written simplifies the process of graphic design and markup. Instead of the writer creating a double-spaced typescript that is later marked and converted to its published look, writers compose using the type and margins of the final version. This method creates typographical consistency from the start and saves a step in production.

Instructions for graphic design can be created in the word processing or page layout program. Each different text element, such as level-one headings, bulleted lists, and paragraphs, is defined as a *style*. A style can include specifications for typeface, type size and style, indentation, margins, tabs, columns, space before and after, borders, related styles, and the style to follow after a return.

A collection of styles can form a *template*. Table 4.1 shows some of the styles in a template used for the draft of this textbook. The template was attached to the file for each chapter so that the appropriate styles could be selected. By simply choosing "Chapter Title" from the styles menu, the writer applies all the features of the style to selected text. Later, if one of the styles seems inappropriate, the writer can modify the style. All the text that is marked with the style will change.

TABLE 4.1   Template for the Draft of *Technical Editing*

Style name	Specifications
chapter title	Normal + Font: Helvetica, 18 pt, bold, space after 36 pt, tab stop: 0.25"
example	Paragraph + Font: Helvetica, 9 pt, indent: left 1.625", space before 3 pt after 3 pt
example gloss	Example + Font: 9 pt, indent: left 1.88"
example cell	Example + Indent: left 0.19"
heading level one	Normal + Font: Helvetica, 14 pt, bold, kern at 14 pt, space before 12 pt after 3 pt, keep with next
heading level two	Normal + Font: Helvetica, bold, italic, space before 12 pt after 3 pt, keep with next
list level one	Normal + indent: left 1.5" hanging 0.19", space after 4 pt
list level two	List level one + indent: left 1.69"
normal	Font: Times, 10 pt, Widow/Orphan control
paragraph	Normal + indent: left 1.375" First 1.625"
paragraph cell	Paragraph + indent: left 0"
table title	Normal + bold, indent: hanging 1", space after 3 pt, border: bottom
table column head	Normal + bold, border: bottom, border spacing 2 pt

The word "normal" in the specifications in Table 4.1 names the default style and links the style for chapter title to the default. If the default style changes, other styles related to it also change. Multiple versions of the "paragraph" style reflect different situations for using paragraphs. The style without an indent follows an example when no new paragraph is intended. The styles with "cell" as an extension have different indentation values for use in the cells of a table. The "hanging" indent means that the line after the first line is indented, as in a numbered list. The "border" is the rule beneath the table title and column heads. "Pt" is an abbreviation for "points," a measure of vertical space.

Figure 4.1 shows part of a draft page of this textbook with styles from the template applied.

Identifying Abbreviations

Abbreviations that are not familiar to readers must be identified the first time they are used. A parenthetical definition gives readers the information they need at the time they need it. After the initial parenthetical definition, the abbreviation alone may be used.

Researchers have given tetrahydroaminoacridine (THA) to patients with Alzheimer's disease. THA inhibits the action of an enzyme that breaks down acetylcholine, a neurotransmitter that is deficient in Alzheimer's patients. The cognitive functioning of patients given oral THA improved while they were on the drug.

If the abbreviation is used in different chapters of the same document, it should be identified parenthetically in each chapter, because readers may not have read or may not remember previous identifications. Documents that use many abbreviations may also include a list of abbreviations in the front or back matter.

Periods and Spaces with Abbreviations

Some abbreviations include or conclude with periods (U.S., B.A.); others, especially scientific and technological terms and groups of capital letters, do not (USSR, CBS, cm). The trend is to drop the periods. For example, NAACP is more common now than the older form N.A.A.C.P.

When an abbreviation contains internal periods, no space is set between the period and the following letter except before another word. At the end of a sentence, one period identifies both an abbreviation and the end of a sentence.

While working as a technical editor, she is also completing courses for a Ph.D.

FIGURE 4.1   Draft Section of *Technical Editing*, Formatted with the Template in Table 4.1.   The example shows level-one headings, paragraphs, and examples.

Templates and styles increase the efficiency of preparing documents for distribution because they eliminate the step of hard copy markup for design after the text is created. They also embed visual consistency. Many companies have standard templates for different types of documents—a template for a manual, another for an inhouse newsletter, and a third for reports. The writers use the templates in creating first drafts. Following a standard company structure may save time in organizing the material. Seeing the text as it will look in print—as the reader will see it—may help writers use visual cues effectively.

SGML

Styles and templates, like conventional copymarking, direct the appearance or format of the text. Most formatting, however, does not transfer in an optimal way to a second publishing medium, as from print to an online manual. The difference in size in a computer screen and a piece of paper requires some adjustments, as do other variations in screen and page design. If a company wants to publish all or part of a document in a second medium, some reformatting may be necessary. Furthermore, because software and hardware differ, the instructions may not work with different software or on different platforms. Over time even the same program will not recognize files created in its earlier versions. Thus, files, as well as programs and platforms, may become obsolete.

SGML, or Standard Generalized Markup Language, is a system of coding documents with identifying tags so that they can be distributed on any hardware and in any medium (print, CD-ROM, World Wide Web).

A tag is information, like copymarking marks, that is not part of the content. Angle brackets distinguish the tag from the content and tell the printer not to print those characters. Most tags occur in pairs—beginning and ending tags. A title would be marked in this way:

```
<title>Electronic Marking</title>
```

The slash in the end tag identifies it as the end of the title. Abbreviations within the brackets identify the part. The SGML tags, then, are comparable to the symbols of copymarking, but the tags are part of the electronic file and accompany the text in any form it takes whereas the marks on paper copy disappear once the text is set in its final form.

SGML is especially useful for companies that publish more than one version of a document or that expect documents to have a long lifespan. An SGML-coded print document does not have to be redesigned for publication as an online version. Instead the codes themselves can be adjusted to make the text suitable for the medium.

SGML is an international standard, accepted by the American National Standards Institute (ANSI) and the International Organization for Standardization (ISO). Because the coding is recognized internationally, SGML may facilitate international distribution of documents.

SGML has little to do with text editing; rather its uses are for production and information management. It works for these purposes because, unlike conventional copymarking, it is a descriptive markup system.

Descriptive versus Procedural Markup

Conventional markup, as described in Chapter 3, is procedural. Each mark directs a procedure (making the type bold, centering the type, inserting a letter). Procedural marks assume one type of output, probably a printed page. They mark the appearance and content of the document.

SGML is descriptive—it describes the structure of the document. SGML marks identify structural parts, called elements, including the title, headings at different levels, paragraphs, footnotes, and even details such as the copyright notice. It can also mark sections that identify kinds of content, such as abstracts, specifications, part numbers, and indexes. Instead of marking a title to be set in boldface (a procedure for changing the type style), the writer or editor would code the titles as titles (a description of a structural part).

Because the tags do not specify type and spacing, a document with SGML tags can be formatted for any output. A FOSI (Formatting Output Specification Instance) specifies formatting. Changing the format for another publication requires only a new FOSI. All the level-two headings might be 12-point bold italic type in print but become 14-point bold roman type on the computer screen, where italics are hard to read. The tags remain consistent; they are simply interpreted differently in the FOSI for different applications.

Because the tags are "generalized," not particular to one type of hardware or software, a tagged document can be distributed through any system, so long as the system has software known as an SGML parser or editor.

SGML Tags and Document Type Definition (DTD)

The SGML international standard specifies tags for about 200 elements of text and other identifiers, including details such as a contract/grant number, price, and acid-free paper indicator as well as author first name, author surname, foreword, list item, and paragraph. Users may define additional elements.

Tags can be typed in manually, but some software uses pull-down menus to let a person select the tags and keep them relatively invisible. Applying the tags from menus is somewhat like applying styles within a template. A writer, editor, or publication specialist can code the text. A typescript page would look similar to the source page for a hypertext page. (See Figure 4.2 in the next section.)

The document type definition, or DTD, is essentially a list of elements for a specific document. The DTD for a memo, for example, would include the sender, receiver, subject, and date lines as well as paragraphs for the body. Those paragraphs might be specified depending on the type of memo as problem statement, sales results, and conclusions. Each DTD is a kind of outline for each type of document. The DTD can specify details, such as that each level-one heading must be followed by a paragraph before a level-two heading or that there must be at least two level-two headings within a section.

The parser (software program) compares the actual document with the DTD to determine whether all the parts are present. It provides a good check of completeness at least from a structural standpoint.

SGML Documents as Databases

SGML tags do more than allow variations in format. They also identify document parts that can be searched and retrieved. The tags allow documents to function as searchable databases. Thus, SGML is a tool for information management.

Selected parts of a document can be retrieved for placement in another document. For example, the descriptions in various sources of the company's products could be compiled into a catalog. Or certain parts of a manual could be selected for a customized manual. The warnings could be updated in the manuals for all of the company's products if liability laws change. All the titles of the publications could be compiled for a bibliography. The value of SGML is proportional to the amount of material to manage and the variety of uses for the material.

SGML Coding and Editing

An SGML specialist or the editor or the writer may code the document. Even if the editor does not do the work directly, the fact of coding affects the editor's work. If each document is structured according to a DTD, and a FOSI specifies the format, the company's documents should achieve structural and visual consistency. If writers can easily import sections of existing documents into new documents, those already-edited sections represent less work for the editor. The editor should then be free to pay the greatest attention to style, paragraph structure, and other uses of language.

But SGML by itself will not help the editor with style, paragraph structure, and accuracy of the information. For those tasks, the editor relies on knowledge of language and communication, as discussed in Chapters 5 and following.

HTML

HTML, or Hypertext Markup Language, like SGML, includes codes with text in documents to give instructions. (HTML is a DTD of SGML.) HTML code permits display of text and graphics on the World Wide Web. HTML code looks like SGML code, but it functions more like styles and templates, as procedural rather than descriptive markup, to give instructions about the appearance of the document. It does not create a searchable database from the document.

Unlike styles and templates, though, HTML codes do not prescribe the exact look of the coded document. The appearance depends on the reader's browser, the software program that interprets the information. Different browsers create different output from the same input.

In addition to identifying elements to be formatted in certain ways, HTML identifies the links from one document to another or from one part of the document to another part. These links make hypertext functional; that is, they enable readers to select or ignore components of the text and conveniently use from it only what they need.

The HTML tags that link different documents are <A HREF> tags, where A refers to anchor and HREF refers to hypertext reference. The rest of the tag gives the file name (location) of the reference. The linking tags are underlined in the display. By clicking on the underlined text, a reader can jump to the document or section of the document that is linked.

Figure 4.2 shows the source code for the part of the Society for Technical Communication (STC) 1996 salary survey on the World Wide Web, and Figure 4.3 shows the page as displayed through a browser. (Different browsers would change the appearance of the page.) The source includes both the code and the text. The tags are enclosed in angle brackets. The following list identifies meanings of some of the tags:

H1=heading level one	TH=table heading	A NAME="top" identifies an anchor to the top of the document; it will be used later in a link
H2=heading level two	TD=table data	
P=paragraph	TR=table row	

Editors who work with hypertext documents should understand HTML well enough to recommend options for using the tags to increase the usability, correctness, and accuracy of documents. One editorial task may be to check links to be sure that they all lead to the intended places, just as an editor working with paper documents will check to be sure that all the parts are present. HTML documents also require editing for spelling, grammar, and consistency. Because the coding itself makes the page hard to read, editors most likely read the output as their browser displays it, and then return to the coded source to emend the text.

Online Editing: Policies and Methods

Styles, templates, SGML, and HTML all facilitate document development, visual and structural consistency, production, and distribution of printed and graphic information, but they do not do the other job of copymarking that Chapter 3 describes—marking the content to make it complete, accurate, correct, and consistent. This chapter has so far described electronic markup—identifying sections of a document to enable consistent formatting, publication in various media and on different platforms,

```
<HTML>
<HEAD>
<TITLE>STC Salary Survey</TITLE>
</HEAD>
<BODY BGCOLOR=#FFFFFF>
<IMG SRC="../gifs/stc_icon.gif" ALT="STC" BORDER=0 ALIGN="LEFT">
<CENTER>
<A NAME="top"> <H2> 1996<BR> Technical Communicator<BR> Salary Survey</H2></A>
<BR CLEAR=ALL>
<A HREF="../home.html" TARGET="_top"><IMG BORDER=0 ALIGN=TOP SRC="../gifs/home.gif"
ALT="Home"></IMG></A></CENTER><P>
The Society for Technical Communication recently surveyed a random sampling of its members
regarding their current salaries and benefits. Questionnaires were mailed to approximately 2,000
members residing in the United States or Canada who are employed in private industry as
writers/editors. More than 900 questionnaires were completed and returned, for a response rate of
45 percent.
Results are given separately for the United States and Canada.
<P><CENTER><IMG SRC="../gifs/red_rule.gif"></CENTER>
<H2> Salary Percentiles </H2>
The following tables show the salary data for technical writers/editors residing in the United States or
Canada. The salary data is presented by selected groupings and shown in percentiles. All figures are in
U.S. dollars unless otherwise stated.

The following definitions pertain:
<DL><DT><B>25%:</B> <DD> Twenty-five percent of the responses were below this value; seventy-
five percent were above this value.
<DT><B>Median:</B> <DD> Fifty percent of responses were below this value; fifty percent were
above this value.
<DT><B>Mean:</B> <DD> The value computed by averaging the tabulated responses.
<DT><B>75% Point: </B><DD> Seventy-five percent of the responses were below this value; twenty-
five percent were above this value. </DL>
<P>
<P><CENTER><IMG SRC="../gifs/red_rule.gif"></CENTER><P></PRE>
<H4> Median Salary, Mean Salary and Quartiles for Total U.S. and Total Canadian</H4>
<CENTER><TABLE BORDER=2>
<TR><TH></TH> <TH> 25% </TH> <TH> Median</TH> <TH>Mean</TH> <TH> 75% </TH></TR>
<TR><TH>U.S. Writer/Editors</TH> <TD ALIGN=center NOWRAP> $35,820</TD> <TD ALIGN=center
NOWRAP>$42,500</TD> <TD ALIGN=center NOWRAP>$43,782</TD> <TD ALIGN=center NOWRAP>
$50,000</TD></TR>
<TR><TH>Canadian Writers/Editors</TH><TD ALIGN=center NOWRAP> $40,000</TD> <TD
ALIGN=center NOWRAP>$45,000</TD> <TD ALIGN=center NOWRAP> $45,059</TD> <TD ALIGN=center
NOWRAP>$50,000</TD></TR>
</TABLE></CENTER>
<P><CENTER><IMG SRC="../gifs/red_rule.gif"></CENTER><P>
</PRE>
</BODY>
</HTML>
```

FIGURE 4.2   Source (HTML-coded) STC Salary Survey on the World Wide Web

Source: From the 1996 Technical Communication Salary Survey. http://www.stc-va.org/home.html
Reprinted by permission of the Society for Technical Communication

1996
Technical Communicator
Salary Survey

Home

The Society for Technical Communication recently surveyed a random sampling of its members regarding their current salaries and benefits. Questionnaires were mailed to approximately 2,000 members residing in the United States or Canada who are employed in private industry as writers/editors. More than 900 questionnaires were completed and returned, for a response rate of 45 percent. Results are given separately for the United States and Canada.

Salary Percentiles

The following tables show the salary data for technical writers/editors residing in the United States or Canada. The salary data is presented by selected groupings and shown in percentiles. All figures are in U.S. dollars unless otherwise stated. The following definitions pertain:

25%:
> Twenty-five percent of the responses were below this value; seventy-five percent were above this value.

Median:
> Fifty percent of responses were below this value; fifty percent were above this value.

Mean:
> The value computed by averaging the tabulated responses.

75% Point:
> Seventy-five percent of the responses were below this value; twenty-five percent were above this value.

Median Salary, Mean Salary and Quartiles for Total U.S. and Total Canadian
(All Canadian figures are in Canadian dollars)

	25%	Median	Mean	75%
U.S. Writer/Editors	$35,820	$42,500	$43,782	$50,000
Canadian Writers/Editors	$40,000	$45,000	$45,059	$50,000

FIGURE 4.3 STC Salary Survey on the World Wide Web

Source: From the 1996 Technical Communication Salary Survey. http://www.stc-va.org/home.html
Reprinted by permission of the Society for Technical Communication

and information retrieval. Online editing pursues the additional goal of improving the document contents, organization, and style.

To edit online for content, effectiveness, and correctness, an editor may manipulate the text itself by correcting spelling, rephrasing sentences, and cutting and pasting sections that are out of order. That is, instead of marking the paper copy for the writer to change, the editor implements the changes directly. Software, including spelling checkers, eliminates some of the tedious work of editing and increases its speed and accuracy. To query the writer electronically, editors may use annotation or hidden text functions of word processing and layout programs. Editors and writers send their work back and forth via networks or shared disks instead of by print and mail.

With such a powerful tool to support editorial tasks, it makes sense to use it in the most productive ways possible. Yet paper markup is not yet obsolete. Whether editors can and do edit online depends on several factors:

- preference: some editors prefer to work with paper, some with the computer
- length and complexity of the document: some editors can see the structure of long documents better on paper; on the other hand, some editors feel they can work with structure better if they can cut and paste electronically and see the effects of these changes immediately
- distribution medium: it makes sense to edit in the form that the readers will use—paper for printed documents, the screen for online documents—or at least to review the document in its intended form at some point during the editorial process
- company policies: companies may value the potential for increased productivity from online editing, or they may be wary about the risks of giving editors privileges to change the text

Policies govern the tracking of the document through its various stages of revision (version control) and records of changes.

Version Control

If the editor as well as the writer and perhaps a reviewer of the technical content can all change the files of a document electronically, some method will have to be established to ensure that anyone who works on the files uses the most recent version. Team members can waste a lot of time changing material that has already been modified by someone else or working on sections that have been eliminated from a revision.

Computers can provide version control automatically. According to David Farkas and Steven Poltrock, "the computer can keep track of who has (and has had) each section of the document, limit the distribution of certain sections, withhold all but 'read-only' access to parts of the document an individual is not authorized to change, and display the changes made by each individual."[1] Of course, the computer can only track the work that is done electronically, not what is marked on hard copy. Electronic version control works best in a setting where the writers and editors work and exchange documents electronically.

In small or print-based publication groups, the record keeping can be manual. One person might serve as the keeper of records. Each person working on the document would check with that person before beginning to work on it.

Alternatively, all people with access to the document can participate in record keeping. The simple log in Table 4.2 records who worked on what files and when. The filename changes incorporate a kind of record as the names add incremental information to suggest the version. When a publications team approves a version, old files (except for the original) can be purged (or stored in an out-of-the way folder) to minimize the chances of people working on old files.

However you manage version control, it's a good idea to preserve a copy of the original document with no changes. If the edited copy becomes corrupted in any way, the original remains intact so that the work on it doesn't have to begin from scratch. And, the original provides a source for checking the accuracy of editorial changes, particularly style changes. It may also be required for the company archives.

Tracking Changes

Because the editor is a language expert working on material written by a subject matter expert, editorial changes must usually be approved by the writer. On paper copy, the editor leaves a trail of emendations, and a writer can review them for accuracy. Computer changes may not be recognized. Some companies prohibit editors from changing the text "silently," fearing that they might introduce errors or change things that were not wrong. Unwarranted intervention is called "trespass." Version control, as described above, prevents inadvertent work on a version that has been superseded, but it does not provide a record of specific changes.

Editors can use the computer for records of changes. One option is to use the annotation and hidden text functions of word processing and page layout programs. The editor could write a nonprinting comment about a change right in the file, much as the editor attaches queries in conventional copymarking. In HTML the <!comment> tag offers the same option of writing nonprinting or nondisplaying information within the files. The writer could then review the comments to track and check the changes.

Editors can also keep records apart from the document file, particularly substantial changes. Those records become a development history and should be reviewed by the subject matter expert regularly so that any mistaken assumptions by the editor can be corrected before the text is changed further. Cover letters or email correspondence summarizing the changes and pointing to substantial or controversial ones can call attention to the parts that writers should review.

TABLE 4.2   Sample Log for Version Control

Document: tutorial styles		Location: Fdrive, Help folder	
date	reviewer	changes	new filename
8/17/97	DF	update specs in Table 4.3	stylesA.htm
8/18/97	SP	check links; add links to format and layout	stylesA1.htm
8/18/97	CR	spelling, grammar check	stylesA2.htm
8/19/97	team	review of individual changes	stylesB.htm

Finally, comparison programs or revision options will electronically compare two versions of a document to show exactly where there have been additions, deletions, or moves. So long as an editor maintains the original version intact, the electronic comparison will show exactly what has been inserted, deleted, or moved.

Enough options exist for tracking changes that the fear of trespass should not usually prohibit online editing, but editors must be willing to keep the records and respect the need to confer with subject matter experts whenever they are uncertain about meaning to reduce the chance of inadvertently introducing errors. Because it is so easy to change a text online, editors also need to be cautious about editing too much.

Electronic Copymarking

New tools are being developed for using computers to complete editorial tasks. In addition to the annotations and hidden text functions that word processing programs provide, some programs now allow editors to enter copymarking symbols into the computer file. Such programs could replicate the process of marking hard copy. Farkas and Poltrock (p. 115) also report on "digital ink," handwriting recognized by a computer. Editing symbols entered with digital ink can be executed by the computer (for example, a word marked for deletion could be deleted on command). The symbols themselves can be deleted once the directions have been completed and the editing has been approved.

Comparison of Hard Copy and Online Editing

Because electronic transmission of information is common and easy, it seems inevitable that editors will increasingly work on the computer for the full range of editing tasks. Besides collapsing two steps into one (marking, changing) and thereby increasing efficiency, editing online maintains clean copy for subsequent reviews. The writer, technical reviewers, and editor can read the current version without the distraction of copyeditor's marks and can also see the effect of the editorial changes directly. When the text will be published online, editors have to work with it at some point as it will be displayed. Furthermore, electronic markup with templates and markup codes should produce documents with structural and visual consistency.

All procedures have their limitations. Attention to templates and coding may distract from content development. Many companies use publication specialists for coding and template development in order to allow the writer and editor to focus on defining the need for information, choosing the best way to distribute that information, researching, and writing.

Relying on spelling and grammar checkers to do more than they are capable of doing may reduce the quality of the document. No checking program available at present is able to read for meaning nor to analyze whether information is coherent, readable, and usable. Automatic checking is a good first step in editing because it locates some surface errors. Editors can do a better job if the copy they work with is clean. But automated editing is better regarded as a first step or part of a final check than as the entire editing process.

Although the computer offers a useful tool to editors, it does not take the place of a language and information design expert reading carefully with understanding of the readers and uses for the document. When tools increase effectiveness and productivity, editors should embrace them. But if they divert editors or substitute incomplete automatic checks for the work of a specialist, then quality and productivity decline. As with any tool, understanding its uses enables one to use it effectively.

Summary

Desktop computers and hypertext have revolutionized the traditional process of writing, editing, and publishing documents. Instead of working primarily on paper, editors increasingly work on the computer. Computer markup for page design occurs at the beginning of document development rather than at the end and provides a template for document structure and appearance. The templates ensure some structural and visual consistency among a company's documents. SGML tags also facilitate information management. Templates do not address the quality of writing, however, and for that main goal of editing, editors continue to use conventional (procedural) markup on paper or revise and correct the text within the document file. Preserving a copy of the original version and keeping records of changes enable checks for the accuracy of online editing.

Note

1. David K. Farkas and Steven E. Poltrock. 1995. "Online Editing, Mark-Up Models, and the Workplace Lives of Editors and Writers." *IEEE Transactions on Professional Communication* 38.2: p. 112.

Further Reading

David K. Farkas and Steven E. Poltrock. 1995. "Online Editing, Mark-Up Models, and the Workplace Lives of Editors and Writers." *IEEE Transactions on Professional Communication* 38.2: 110–117.

National Information Standards Organization. 1991. *Electronic Manuscript Preparation and Markup.* Z39.59–1988. New Brunswick, NJ: Transaction Publishers. This book presents the international standard for SGML. It defines document structure and specifies the formal syntax. It is not for beginners.

National Center for Supercomputing Applications. "A Beginner's Guide to HTML." http://www.ncsa.uiuc.edu/General/Internet/WWW/HTMLPrimer.html

Discussion and Application

1. Definitions: Illustrate the difference in descriptive and procedural markup by marking the first two pages of this chapter (or other document your instructor provides) in both

ways. For example, in procedural markup, you would mark the title as 30-point italic Palatino with 5½ picas after. In descriptive markup, you would mark the title as a title.

2. Compare the documents in Figures 4.2 and 4.3 to relate HTML code to output. Identify codes for at least three elements of the output (Figure 4.3). Use the sample codes listed in this chapter to identify their functions. By looking at the output, what do you think B means?

   NOTE: You will need access to a computer for applications 3–6. For 3–5 you will need a sophisticated word processing program such as WordPerfect or Microsoft Word.

3. Use the document comparison or revision function of word processing or publications software to compare the original and revised versions of a document that your instructor provides.

4. Use the annotations function to query the writer about a puzzling point or need for more information or to seek approval for an editorial change.

5. Using the default template in a word processing program, write a memo to your instructor in which you discuss issues in electronic and paper-based editing with the topics of "efficiency" and "accuracy." Apply at least three styles in the way your instructor directs. If the default template does not include three styles, develop styles for these parts:

   - title, "Memo," centered, bold, in 12-point type, with 18 points of space after
   - level-one heading, left-justified, bold, 12-point type, with 12 points of space before, 4 points after
   - paragraph with indentation of 0.25", 10-point type, line length of 5 inches (left and right margins of 1.75" on an 8 ½" page), and tab stops of 1.2" and 3.5".

6. If you have access to the World Wide Web, open a site. Using the "view" menu, select "source." Then compare the HTML codes in the source document with the output on the screen.

7. Vocabulary: You should be able to define these terms: *HTML, SGML, procedural markup, descriptive markup, style, template*

P a r t 2

Basic Copyediting

Chapter 5
Basic Copyediting: An Introduction

Chapter 6
Copyediting for Consistency

Chapter 7
Spelling, Capitalization, and Abbreviations

Chapter 8
Grammar and Usage

Chapter 9
Punctuation

Chapter 10
Quantitative and Technical Material

Chapter 11
Proofreading

Chapter 5

Basic Copyediting: An Introduction

Basic copyediting is a necessary process in preparing a quality document for publication. It may be the only editing a document receives, or it may be a part of comprehensive editing. The copyeditor's task is to make the document correct, consistent, accurate, and complete. These standards enable readers to read it without being distracted by errors or inconsistencies.

- correct: spelling; grammar; punctuation
- consistent: spelling; terms; abbreviations; numbers (spelled out or figures); capitalization; labels on illustrations; matching of numbers and titles on the illustrations with the references in the text; visual design; documentation style; colors; icons
- accurate: dates; model numbers; bibliographic references; quotations (accurately restate the original); hypertext links
- complete: all the parts are present; all the illustrations that are referred to are in the document; sources of information are identified and acknowledged; all hypertext links lead somewhere

In addition, the copyeditor provides instructions about how to prepare the text for its final form. These instructions relate both to the words (whether to hyphenate, where to put punctuation) and to the form (whether to italicize, where to indent). For hardcopy transmission, the copyeditor marks the page using symbols for specific functions, such as insert, delete or italicize. (See Chapter 3.) The process of marking the choices is called copymarking or markup. If the document is transmitted electronically, the copyeditor may insert the choices electronically without marking. (See Chapter 4.)

The copyeditor focuses on details and production whereas a writer focuses on content. Good copyeditors know grammar and spelling. They read closely and can tell when a change in punctuation or wording might change meaning. In addition, they are alert to possible ambiguities in language and are willing to check with both the writer and with printed resources to verify their choices.

This chapter provides an overview of copyediting by describing its goals of correctness, consistency, accuracy, and completeness. This chapter also offers guidelines for copyediting books or manuals, illustrations, and hypertext documents to develop familiarity with the parts of these documents. Other chapters in Part II of this textbook review copyediting for consistency, spelling, grammar, and punctuation—attributes of all documents, whether verbal or visual, printed or online.

Document Correctness and Consistency

A correct document conforms to standard American English—that is, to the grammar, spelling, and punctuation that are accepted by language specialists and the usage panels of dictionary publishers. A consistent document avoids arbitrary and confusing shifts and variations in the use of terms, spelling, numbers, and abbreviations. It is also predictable in its visual features, such as placement of elements on the page or screen.

Copyediting for correctness and consistency is widely understood by writers and managers to be the most essential function of editing. Errors in language use are more apparent to readers than are errors of argument or organization. Errors distract readers from the content and can interfere with the effectiveness and usefulness of the document.

Chapters 6, 7, 8, and 9 review consistency, spelling, grammar, and punctuation in detail, but the following list identifies some basic standards of correctness and consistency.

- Groups of words punctuated as sentences are complete sentences.
- Punctuation is complete: sentences end with punctuation; quote marks and parentheses are closed.
- Subjects and verbs agree in number.
- Pronouns agree with their referents.
- Modifiers attach logically to the word or phrase they modify (they do not "dangle").
- Words are spelled and capitalized correctly.
- Numbers are spelled out or in figures according to a plan.
- Identifying information, such as running headers or navigation buttons in hypertext, is in the same place on every page or screen.

Technical documents are conservative in their use of language and require the highest standards of copyediting for correctness.

Document Accuracy

Accuracy refers to content, while *correctness* refers to the use of language. Errors in content, such as using the wrong term, date, or model number, can make a document partially or totally worthless to readers, legally incriminating, and even dangerous. Numbers and illustrations as well as words require editing for accuracy:

- quantitative data: model numbers, formulas, calculations (such as addition of figures in a column), measures, dates
- words: names, titles, terms, abbreviations, quotations
- illustrations: labels, cross-references, callouts
- organizational information: table of contents, index, menus and links in hypertext

Accuracy errors can occur even after a comprehensive edit or technical review because often the editor and reviewer have focused on the larger text structures—the argument, the meaning, the organization—and have only scanned details.

Paragraphs are easier to copyedit than quantitative data and names and titles, yet errors are as likely to occur in tables, reference lists, and display text such as headings. Editing requires a check of simple calculations, such as adding columns of numbers (just as you would look up the spelling of an unfamiliar word in a dictionary). If you discover an error, you should query the writer to determine whether the addition or one of the figures is wrong.

A check for accuracy requires familiarity with content and alertness to meanings. Even a copyeditor who is not a subject matter expert can spot probable errors. For example, in one long article on relationships between people, the writer referred to "dynamic" relationships. However, the quotations from other sources on the subject referred to "dyadic" relationships (relationships of dyads, or two people). The copyeditor wondered whether "dynamic" and "dyadic" relationships differed or whether the writer used one wrong term. Even without knowing advanced psychology, the copyeditor could spot an apparent discrepancy.

Not knowing for sure the meaning of the terms as used in the article, the copyeditor would have been premature in changing one term or in querying the writer immediately. A change could introduce an error. Frequent or unnecessary queries are irritating and raise questions about the copyeditor's competence. Instead, she looked for other text signals about meaning, such as parenthetical definitions of the terms or some other explanation of the difference. As she checked a quotation with the original source, she noted that the writer had substituted "dynamic" for "dyadic" in the quotation. With this evidence, her query to the writer could be specific. The copyeditor might also have consulted a psychology text index or glossary or a specialized dictionary of psychological terms to find definitions.

Inconsistencies, facts that seem to contradict logic, and omissions are clues to potential errors. A reference to a photographic process in 1796 raises a question about the date—you suspect (even if you do not know) that photography came later in history. The inconsistency in the following model numbers should send the editor to the product catalog or even to the products themselves if they are accessible:

JC238–9Y

JN174–2Y

R2472–3

An error in one type of text alerts you to pay close attention to all instances of similar text. For example, a reference list entry that cites a date and an author's initial(s) incorrectly is a clue to look carefully at the other entries. The error in one entry suggests the possibility that the writer was not careful in compiling the list. Likewise, an error in addition in one column of figures in a table requires a check of the other columns.

You rely on careful, alert reading more than on rules you can learn in a handbook when you copyedit for accuracy. Do not dismiss your quiet hunches that something may be wrong; do not take for granted that simply because a reputable writer and editor have previously reviewed the text that it is free from errors—nor that they are incompetent for having missed errors. They were concentrating on other text features. Be especially careful in reviewing numbers and dates, reference citations, and headings.

Completeness: Books, Manuals, Long Reports

A complete document contains all its verbal and visual parts as well as the necessary front matter and back matter. The copyeditor provides the quality control check to make sure that a document doesn't get published lacking its index or a table cited in the text. The copyeditor also checks that all the pages are included and in order, including blank pages, before the typescript or camera ready copy goes to the printer. (*Camera ready* means that the pages have been printed on a laser printer and will not be modified before they are reproduced. They will be photographed for reproduction.)

Parts of a Book, Manual, or Long Report

Books, manuals, and other long printed documents have three major sections: the preliminary pages or front matter, the body, and the back matter. Within these major divisions are subparts, some of which are optional. A part should be included only if it is functional.

Decisions about what parts a document will include are generally made before the copyeditor works on the copy. If the document is part of a set, such as a set of manuals for a company's products, other documents in the set help to establish the pattern. The editor who has worked on comprehensive editing may determine document specifications and list the parts for the copyeditor. The copyeditor, then, does not decide whether to include abstracts or a glossary but makes sure that the specified parts are included.

Only long documents will include all the parts identified in the discussion that follows here. Simple documents contain only some of these parts. A comprehensive style manual, such as *The Chicago Manual of Style,* offers guidance on how to prepare these parts and information on other, less common parts, such as dedications.

Preliminary pages

The preliminary pages appear before the text begins. The title and contents pages are included in all books, manuals, and long reports. Roman numerals identify their page numbers, with the half title (a short version of the title page) being page *i*. Blank backs of pages are counted in the numbering even if the number does not appear on the page. Parts other than the title and contents pages are optional.

Cover.   Includes both the type and the artwork, if any. It identifies the title and subtitle and may identify the author or editor and place of publication. On reports, it may also include the date and other publication information.

Half title page.   Includes only a short title or the main title without a subtitle, immediately inside the cover. The type is generally smaller than it is on the title page, and the back usually is blank.

Title page.   Includes the title and subtitle; may include the name of the author or editor. Title pages of reports generally contain more information; for reports or other short documents, the cover and title page may be the same page.

Copyright page.   Generally falls on the back of the title page; includes the copyright symbol, year of publication, and copyright holder. Here is an example:

© 1998, Allyn & Bacon
All rights reserved
Printed in the United States of America

No periods appear at the end of the lines. The copyright page also includes the address of the publisher and the ISBN (International Standard Book Number) for commercial publishers. This number identifies the country of publication and the publisher according to an international code.

Table of contents.   Lists main divisions and the page numbers where the divisions begin; may be labeled simply "contents."

List of tables and figures.   Lists the title, number, and page number of the illustrations; tables and figures may be listed separately and on separate pages as a "list of tables" and a "list of figures."

List of contributors.   In a multi-author book or periodical, lists the authors' names and other identifying material, such as position and place of employment. Alternatively, this information may appear for each contributor on the first or last page of the article or as a list in the back of the document.

Foreword.   Written and signed by someone other than the author or editor; introduces the document by indicating its purpose and significance. It is more common in academic and literary texts than in technical documents.

Preface.   Written by the document's author or editor; introduces the document by indicating its purpose, scope, significance, and possibly history. It often includes acknowledgments.

Acknowledgments.   Usually incorporated into the preface; names people who have helped with the publication, such as reviewers and typists, and their contributions.

Executive summary.   Summarizes contents including conclusions and recommendations for readers who are decision makers.

Body
The first page of the body of the document is numbered page 1 (arabic numeral).

Chapters; parts.   All should be present, in order; all pages for each subdivision should also be present and in order.

Titles, chapter numbers, author names.   The contents and pattern of identifying information should be standard for all the subdivisions.

Abstract.   Summarizes or describes the contents of the chapter or subdivision at the beginning; a standard part of academic publications, including research reports; usually paragraph length.

List of references.   Identifies the sources mentioned in the chapter or division. May be called "Works Cited" or "References," depending on the documentation style. May appear at the end of each chapter or in a reference section at the end of the book.

Half titles.   Sometimes placed between major parts of the document (groups of related chapters); identifies the part title.

Blank pages.   May be inserted in the typescript that goes to a printer to signal blank pages in the printed document. These blanks will probably be *verso* (back or lefthand) pages; used when all chapters or divisions begin on a *recto* (righthand) page. When editing electronically, insert page breaks for the blank pages, to keep the automatic page numbering accurate.

Running heads.   Headers in the top margin that name the chapter, author, or part title. They help people find their place in the book; they may differ on the recto and verso pages.

Illustrations.   Tables and figures. Illustrations are generally integrated into the chapters, though supplementary illustrations may appear in the appendix.

Back Matter
The back matter contains supplementary information. Page numbers continue sequentially from those in the body, in arabic numerals.

Appendix.   Provides supplementary text material, such as research instruments (for example, a questionnaire), tables of data, or troubleshooting guides. Each appendix is labeled with a letter or number and title (for example, Appendix A: Survey of Users).

Glossary.   Defines terms used in the document; a mini-dictionary that applies just to the document and its specific topic.

References.   Includes the complete publication data for the materials cited in the document, unless this information appears after each chapter.

Index.   Lists the key terms used in the text and the pages on which they are discussed or referred to; helps readers find specific information.

Copyediting Illustrations

Illustrations require copyediting for the same reasons that text does: to make the illustration correct, consistent, accurate, and complete. If the illustrations are camera ready, the copyeditor also reviews reproduction quality and readability. The marks used in copyediting illustrations are the same as the marks for text. The symbols identified in Chapter 3 for deletion, insertion, transposition, alignment, and type style work the same for visuals as for prose. The following list identifies the tasks in copyediting illustrations.

Correctness and Consistency
- Correct words for spelling, punctuation, capitalization, and hyphenation.
- Check the title of an illustration for a match with the contents. (If the title promises records from 1996 to 1999, the illustration cannot stop with 1998.)
- Compare references in the text with the illustration. The illustration must offer what the reference promises.
- Establish patterns for placement of callouts and titles. Align callouts if possible.

Accuracy

- Add columns of numbers to check the totals.
- Visually measure segments of graphs to determine whether they are proportional to their numeric values. If they look wrong, measure with a ruler or protractor.
- Compare line drawings with the actual product if you suspect distortion.

Completeness

- Check that all illustrations mentioned in the text are present and that the numbers go in sequence. (A gap in numbers suggests a missing illustration.)
- Make sure identifying information is present:
  parts on illustrations
  base measures (such as percent, liters) on tables and graphs
  meanings of shading, color, or line styles
  sources of information
  labels, numbers, titles
- In hypertext, if a graphic is "hot" (links to another part of the text), check that the link is complete.

Readability

- Eliminate excess ink—lines that aren't needed to separate rows and columns, measures after each item in a column rather than in a column head, superfluous callouts
- Use space to distinguish rows and columns.
- Align columns of numbers on decimals or imagined decimals.

Parts of Illustrations

The words attached to illustrations allow for cross-reference, identify and explain content, and provide information about sources. See Figure 5.1 for examples.

Labels, Numbers, and Titles

Illustrations are classified and labeled as either tables or figures. A table presents information in tabular form (in rows and columns). The information in the rows and columns may be quantitative or verbal. All other illustrations are considered figures, whether the illustration is a photograph, line drawing, bar graph, flowchart, line graph, or other graph. The identifying features of tables and figures, including labels, numbers, and titles, have specific functions.

Labels.   Labels ("table," "figure," or "step") and numbers enable easy cross-reference. Small and informal illustrations may not require labels and numbers if they appear right where they are mentioned and are not referred to again.

A consistent pattern of labeling illustrations should be established, and style guides may help. The American Psychological Association, for example, requires table titles and numbers at the top of the table and figure titles and numbers at the bottom. Not all documents in other disciplines will follow this particular pattern, but readers should be able to anticipate where to find specific kinds of information in illustrations. Record your choices about placement of labels and titles on your style sheet to encourage consistency.

In addition, labels on the illustrations themselves must be consistent with the references to them in the text. The reference to Figure 1 in the text must refer to the illustration actually labeled "Figure 1."

Numbers.   Illustrations are numbered sequentially through the document, but tables and figures are numbered independently. Thus, a document may include a Table 1 and a Figure 1. Double numeration is common in documents with multiple sections. "Table 12.2" indicates that the table is the second numbered table to appear in Chapter 12.

Illustrations of steps in a process should be numbered if the text refers to steps by numbers.

Titles.   Titles let readers identify the illustration without reference to the text. Because some readers will take shortcuts by reading only the illustrations, the illustrations should be as self-contained and self-explanatory as possible. The words in the title should refer to the content of the illustration in concrete terms. For example, "Sales of Chip #A422, First Quarter 1991" answers what and when questions and informs more specifically than "Sales" would. Titles of line drawings and photographs may require information on point of view, such as top view, side view, cross section, or cutaway view. It is never appropriate to name the type of illustration (photograph, line graph) in the title—the type is evident from the illustration itself.

Callouts, Legends, Captions, and Footnotes
Other identifying elements include callouts, legends, captions, and footnotes.

Callouts.   A verbal identifier, such as a part name, on an illustration is known as a *callout*. Callouts supplement the illustration information, and they link the illustration with the text. Callouts should be used selectively. If readers do not need to identify a part or procedure, delete the clutter of unnecessary words. Query the writer, however, before you make such decisions about what is necessary.

The callouts on the illustration must match the terms used in the text. If a part is a "casing" in the text, readers will search the illustration for a part called a casing, not a cover. Generally all parts that the text names should have callouts, and all the parts identified with callouts on the illustration should be referred to in the text.

The callout should be next to the part or procedure it identifies. If necessary, a straight line can connect the part and the callout. If the object has major and minor parts, that structural relationship can be shown by callout placement or type style. For example, all major parts might be identified on the left of the object while minor parts are identified on the right, or major parts could be named in capital letters while minor parts are set in lowercase letters. If the object has many parts crowded together, the parts may be numbered, with a numbered list of parts following. This type of callout is common in an owner's manual that identifies all parts of the mechanism.

For aesthetic purposes and also for the sake of readability, a standard typeface and size may be used for callouts on all illustrations in the document. The type must be large enough to be read but not so large as to overwhelm the illustration. Alignment of a vertical series of callouts, when practical, helps give the illustration a neat appearance.

Column and row headings.   Headings in a table identify contents in the same way that callouts do for figures. When the cells of a table display quantities, the unit of measure (for example, dollars or inches) in a heading eliminates the clutter of repeated measures in each cell.

Legends.   A *legend* explains shading, colors, line styles, or other visual ways to distinguish the elements on a graph from one another. The legend is a list of these variants with their meanings. The legend generally appears in a corner of the illustration. Because legends are separate from the elements they identify, they require an extra step in processing for readers. Label the components of the illustration, such as segments in a pie graph, when you can instead of depending on a legend.

Captions.   Brief explanations (a few lines of text) beneath (or above) the illustration are called *captions.* They let readers identify significant features and interpret the illustration without referring to the text. The caption efficiently presents essential points of information; long and complex interpretations belong in the text itself. The caption appears below the label, number, and title.

Footnotes.   Explanatory and source information notes may appear on illustrations for the same reasons they appear in text: to explain or to give credit. A superscript number or letter identifies a footnote. The note itself appears at the bottom of the illustration, above the label, number, and title if these are also placed at the bottom.

If the information in the illustration or the illustration itself comes from another document, cite the source, including author, title, publisher, date, and pages, at the bottom of the illustration. If the title is at the bottom, place the source note above the title. If an illustration is reprinted from copyrighted material, get written permission to reprint. Add to the source statement the words "Reprinted with permission" or other words that the copyright holder cites. This statement also appears above the title but below the explanatory footnotes. This requirement applies to illustrations you obtain or use online as well as to printed illustrations.

Figure 5.1 delineates the various parts of an illustration. There is no legend because the callouts identify the segments of the graph.

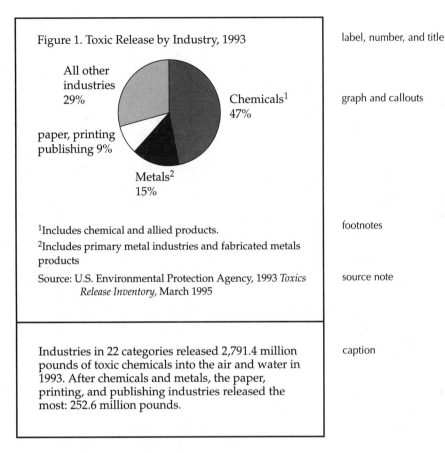

Figure 1. Toxic Release by Industry, 1993   label, number, and title

All other industries 29%

Chemicals[1] 47%   graph and callouts

paper, printing publishing 9%

Metals[2] 15%

[1]Includes chemical and allied products.   footnotes
[2]Includes primary metal industries and fabricated metals products

Source: U.S. Environmental Protection Agency, 1993 *Toxics*   source note
Release Inventory, March 1995

Industries in 22 categories released 2,791.4 million pounds of toxic chemicals into the air and water in 1993. After chemicals and metals, the paper, printing, and publishing industries released the most: 252.6 million pounds.   caption

FIGURE 5.1 Parts of an Illustration

Placement of Illustrations in the Text

The text should refer to the illustration before the illustration appears so the readers will know what they are looking at and why. Illustrations should appear as soon after the first mention as is practical, given the limitations of page size and page makeup. If readers have to flip pages to find the illustrations, they may not bother, and the illustrations will be wasted. The reference to the illustration may be incorporated into the sentences ("Table 3 shows that . . .") or placed in parentheses. If the parenthetical reference is incorporated into the sentence, the period for the sentence follows the closing parenthesis. If the parenthetical reference stands alone as a sentence, the period is inside, and "See" begins with a capital letter.

The paper, printing, and publishing industries account for 9% of the toxic waste released by all industries (see Figure 1). The paper, printing, and publishing industries account for 9% of the toxic waste released by all industries. (See Figure 1.)

Illustrations should face in the same direction the text faces; readers should not have to flip the document around to read the illustrations. If the illustration is too wide to fit on the page, it may be turned sideways (landscape orientation). Sideways illustrations on facing pages should face the same way so that readers have to turn the document only once. If an illustration appears on a verso (lefthand) page, place it with its bottom in the gutter (by the binding). On a recto (right hand) page, place the top of the illustration in the gutter. Readers will rotate the document clockwise a quarter turn to see any illustrations displayed at right angles to the text.

Instruction manuals often use a side-by-side arrangement of illustrations and text, with the illustration on the left and the explanation on the right.

At the typescript stage, illustrations may be appended to the end of a typescript to allow an appropriate placement when the pages are made up in camera ready form. You mark the text to show where the illustrations are to be inserted by writing the placement instructions in the margin. For example, write "Insert Figure 4.2 about here." If the text is camera ready, place the illustrations where they should appear.

Quality of Reproduction

If a typescript includes camera ready illustrations, the copyeditor determines whether the illustrations are suitable for printing. Photographs need a high contrast of darks and lights. Line drawings need to be drawn professionally or printed from a laser printer, because amateur drawings done with poor-quality pens produce lines with feathers and jags that printing exaggerates. Poor quality illustrations cheapen the document. For best reproduction, all illustrations other than photographs should be printed on a laser printer or drawn by a professional graphic artist.

Editors or writers may prepare the artwork, especially when the document and budget do not warrant full-scale printing. Amateurs can achieve professional results using computer drawing tools and prepared artwork, especially if they respect their own limits and keep the work simple.

The same principles of reproduction quality apply to online documents. Antialiasing, a procedure to minimize the jagged edges of some illustrations, may improve the quality. In addition, background colors or patterns should not distract from the illustration. Because graphics files are much bigger than text files, they will slow the speed with which a document appears on the user's screen. File size can be reduced by minimizing the number of colors and sometimes by changing the file type. Unless you are the computer specialist as well as editor, your job is not to make these changes but rather to notice and to suggest modifications when the visual quality seems poor or the document is slow to load.

Copyediting Hypertext

Hypertext documents, distributed online rather than in print, require copyediting for correctness, consistency, accuracy, and completeness for the same reasons that print documents do. Checking spelling may be easier before coding. Editing for grammar, punctuation, consistency of mechanics, and accuracy may also be easier before the document is ready for display.

Hypertext documents require more attention to navigation and access as well as to visual display for readability on screens that may not have high resolution. Visual consistency in hypertext documents helps readers maintain a sense of their place in a document. The same types of information, such as menus and returns to a home page, should appear in the same place on every screen and use the same cues of color and type. As with print documents, establishing styles and templates at the beginning helps with consistency. Also some standard screen layouts have become familiar, with functions and pull-down menus at the top of the screen and topics often in a column on the left or right. The bottom of the document includes some navigation aids—links to related documents or to points in the displayed document and an email address for the writer or web master. Variation from these familiar displays may confuse readers.

Because screens are usually harder to read than print, text must contrast sharply with the background. Italics—hard to read on a computer screen—should be minimized or eliminated.

The completeness check for hypertext can follow the same principles as the check for a print document and relies on the same types of information: document specifications and the page identifying the sections and structure of the document. Links should be tried. In addition, readers appreciate identifying information comparable to that on the copyright page of a printed book: the date of the last update and the name of the writer or web master. Some online documents include a copyright statement. This identifying information may appear at the bottom of the document or on the home page. As with print documents, sources of information should be cited.

How To Copyedit

Because copyediting requires attention to so many details, it may require more than one editorial pass or sweep. In one reading, you may not be able to keep your attention focused on grammar and spelling and all the elements of consistency, accuracy, and completeness, as well as attending to instructions for a printer, especially if the text is long and complex. So you may plan on going through the text two or three times, or making several passes, looking for different things each time: perhaps spelling, consistency, grammar, and punctuation the first time; illustrations and mathematical material the second time; and headings, type styles, fonts, and margins the third time. Check for completeness separately.

You can copyedit effectively either on hard copy or on a computer, or use a combination. Some editors like to see the entire document on hard copy before they begin editing, to get a sense of the whole.

1. **Gather information about the document.** Find out about its readers, purpose, conditions of use and any document specifications that are available. Specifications may include styles for the various levels of headings, parts to be included in the completed document, and information about printing and binding or other reproduction. These specifications will help you draw up your checklist for the completeness check. If a style sheet has been started, work with it in copyediting. (A style sheet lists choices of capitalization, spelling, use of numbers, abbreviations, hyphenation, and other mechanics.) Your supervisor may provide you with this information, or you may need to inquire about it. Clarify the production schedule and your specific duties. For example, by what date will the writer have the chance to review the editing? If you are editing on the computer, should you make corrections silently—that is, with no record of the changes—or should you insert comments that point to the changes? Are you expected to edit just for correctness and consistency, or do you also have responsibility for style?

2. **Survey the document.** A quick initial reading will identify features that need editorial attention, such as an extensive use of tables or an inconsistent documentation style. It will also reveal how extensively the document uses illustrations, lists, or other features. These preliminary impressions will help you make thoughtful line-by-line decisions.

3. **Run computer checks.** The spelling and proofreading checks will find errors and let you clean up the document before you begin line-by-line reading. Include a check for extra spaces between sentences. Using the replace feature of the word processing program, replace two spaces after a period with one. If the document has been prepared with a publishing template and styles (see Chapter 4), you may check at this point that the right styles have been applied to the different types of text elements.

4. **Edit the paragraphs and headings for correctness, consistency, and accuracy.** Read for meaning and make emendations only when you are sure that they will clarify, not distort, the meaning. Before you guess at a change, query the writer. If the text contains a large number of errors, you may need to make more than one pass to complete the correctness, consistency, and accuracy checks.

5. **Edit the illustrations and other nonprose text.** Edit for correctness, consistency, accuracy, completeness, and readability. Mark insertion points for illustrations. A graphic artist may provide instructions for treatment of illustrations—screening, doing color separations, cropping, or reducing illustrations. Attach these instructions to the hard copy.

Check equations themselves for accuracy, enumeration, and references in the text. (See Chapter 10 for more information on editing quantitative and technical material.)

6. Prepare the typescript for printing or other reproduction. Confirm that the document is complete—all the parts are present and in order. If the text is to be transmitted by paper, mark headings to define levels, indentations, alignment, and vertical space. (Alternatively, a graphic designer may mark the copy for the graphic design.) For electronic transmission, confirm that styles for various types of text have been applied correctly. For hypertext documents, check that links lead to where they promise.

At the point when the typescript is delivered to the printer or other vendor for reproduction, the copyediting is complete. The editing from this point forward is production editing—coordinating the rest of the production process to keep it on schedule. The copyeditor may be responsible for production editing, but the emphasis in editing changes from preparing the text to managing reproduction for distribution.

Summary

Copyediting aims for a correct, consistent, accurate, and complete document. Copyediting requires attention to the details of the text, which may be neglected in writing and comprehensive editing. Copyediting improves the quality of the document.

The copyeditor's best resources are knowledge of language, alertness to possible errors, and willingness to check when questions arise. Even the most experienced copyeditors have not memorized all points of grammar and the spelling of all terms; therefore, copyeditors have on their desks or on their computers the most essential reference materials: a good general dictionary; perhaps a specialized dictionary, such as a dictionary of scientific and technical terms or a medical dictionary; a handbook of grammar and usage; and style manuals. Easy access to resources makes it easy to use them. When these resources cannot provide answers to questions, the editor queries the writer or subject matter expert.

Further Reading

The Chicago Manual of Style. 1993. 14th ed. Chicago: University of Chicago Press.

Karen Judd. 1991. *Copyediting: A Practical Guide,* 2nd ed.Los Altos, CA: Crisp. See especially Chapters 1 and 3.

Robert Van Buren and Mary Fran Buehler. 1991. *The Levels of Edit.* Arlington, VA: Society for Technical Communication. Provides good lists of things to check in copyediting.

Discussion and Application

1. Find a book from a major commercial or academic press and identify the parts of it. Bring it to class and be prepared to point out its parts and their functions for that book. What parts have been omitted, and why? Are the parts labeled and ordered correctly?

2. Group work in class: In groups of about four, share the books you have collected for application #1. Determine:

   a. How many books and manuals have odd-numbered pages on the right side when the book is open and even-numbered pages on the left? How do you explain the patterns?

   b. What books use running heads? Do you see any patterns to them—that is, are left and right running heads the same? If the examples have different information in the running heads, is there any reason for the variations?

   c. What books have half-title pages? Are these pages at the front, or do they divide sections within the book? What is their function? How does a half-title differ from a title page?

   d. Does anything besides the half-title and title page precede the contents page?

   e. What other parts besides a title page and contents page precede the body of the book?

   f. What books have indexes? appendixes? glossaries? What are the different functions of these parts?

   Summarize: Make an observation to share with the whole class about the construction of books from their various parts. For example, one group might compare different uses of running heads, another group might consider different types of appendix material, a third group might observe different uses of half-title pages, and a fourth might notice a part that isn't mentioned in this chapter.

3. Compare a book and a technical manual or report. Find the book, as directed above. Also find a technical document, such as a computer manual or technical report. How do the two types of documents compare in the variety and arrangement of parts?

4. Vocabulary

   a. Distinguish between a preface and a foreword in the front matter by identifying their functions and writers. Note that both differ from an introduction, which is a substantive part of the book's body.

   b. Distinguish between a glossary and an index in the back matter.

   c. Define these terms. Look up unfamiliar terms in the glossary: *running head, half-title, style sheet, template*

5. The illustrations in Figure 5.2 appeared in instructions for fastening a preprinted sign to a car door. Use the guidelines in this chapter and the symbols in Chapter 3 to copyedit the illustrations. Look for correctness (especially spelling) and consistency of placement of different types of information. For example, distinguish callouts from instructions. Determine whether the callouts and instructions should be inside or outside the edges of the sign that is represented.

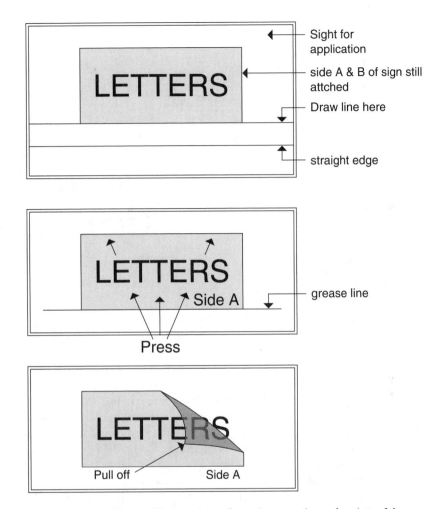

Sight for application

side A & B of sign still attched

Draw line here

straight edge

grease line

Press

Pull off   Side A

FIGURE 5.2   Three Illustrations from Instructions for Attaching an Adhesive Sign to a Vehicle.   These illustrations require a copyeditor's attention for consistency and correctness.

C h a p t e r **6**

Copyediting for Consistency

Consistency means that the same text or the same type of text or illustration is treated uniformly throughout the document. For example, an adjective that is hyphenated once in the text is hyphenated throughout the text, some pattern for spelling out numbers is followed, and one term, not a synonym, refers to the same process throughout the text.

Consistency gives useful information to readers. For example, the consistent type style and placement of headings helps readers perceive the organizational pattern. Inconsistencies may suggest changes in meaning where no change was intended. Arbitrary shifts can be disorienting. Copyediting for consistency also saves time and expense in production. For these reasons, one task of copyediting is to make consistent choices throughout a document.

Copyediting for consistency often means choosing among acceptable alternatives rather than applying rules. Copyeditors depend on dictionaries and style manuals to help them make their choices. These guides help copyeditors make style choices objectively and avoid tampering with a writer's style to suit their own preferences. Word processing programs help with establishing and editing for consistency.

This chapter discusses four types of consistency in documents—verbal, visual, mechanical, and content—and the use of style guides to achieve consistency.

Document Consistency

Table 6.1 lists and defines the four types of document consistency and subcategories. In addition to internal consistency (meaning that a document is consistent in itself), a document should be consistent with the document set to which it belongs, such as other manuals for the same type of equipment.

TABLE 6.1   Types of Internal Document Consistency

Type	Meaning
Verbal	
semantics	meanings of words: one term represents one meaning
syntax	structures: related terms, phrases, and clauses use parallel structure
style	diction, sentence patterns, and writer's voice: the document is consistent in its level of formality and sophistication and relationship to readers
Visual	
typography	typeface and type style are the same for parallel items, such as level-two headings
layout	repeated or sequential items, such as running heads and page numbers, appear in the same location on a series of pages
screen design	elements appear in the same position on related screens; color, banners, navigation buttons, icons, and type style have specific meanings and are used consistently
tables, figures	labels, titles, callouts, captions, and cross-references are treated the same for all similar illustrations; illustrations match content and purpose in quality, taste, and seriousness
Mechanical	
spelling	one spelling is used for the same term
capitalization	terms are capitalized the same in each related use
hyphenation	hyphenation is consistent for the same term and for like terms
abbreviation	use and identification of abbreviations follow a pattern
numbers	related numbers are consistent in punctuation, use of words or numerals, abbreviation, and arrangement
punctuation	items in a series and lists follow a pattern of punctuation; possessives and accents are the same on repeated uses of the same term
cross-references	all cross-references follow the same patterns of capitalization, abbreviation, and punctuation
documentation	references conform in arrangement, abbreviation, capitalization, underlining, and spacing to one acceptable pattern
lists	related lists match in capitalization, punctuation, spacing, signals such as numbering and bullets, and phrases or complete sentences
type style	variations such as italics, capitalization, and boldface signal specific meanings such as emphasis or terms used as terms
Content	references to the same object or event do not contradict one another

Verbal Consistency

Verbal consistency refers to words and their arrangement into sentences and paragraphs, as well as to meanings of words, structures, and style. Verbal consistency, whether semantic, syntactic, or stylistic, clarifies meaning.

Semantics

Semantics refers to meaning. Semantic consistency means that one term is used to represent a meaning. Referring to one switch as a "toggle switch" in one paragraph and an "on-off switch" in the next paragraph is confusing because the two terms indicate two switches. Semantic consistency refers to abbreviations and actions as well as to terms denoting things. Novice writers may insert semantic inconsistency because teachers have told them to avoid repetition. In technical writing, the desire for variety conflicts with the need for clarity. Technical writers and editors prefer consistency over variety for its own sake.

Syntax

Syntax refers to the structure of phrases and sentences. Syntactic consistency refers to parallel structure of related terms, phrases, and sentences. In the series "terms, phrases, and sentences," the choice of nouns for the items reinforces their relationship as three types of text that may have parallel structure. A nonparallel alternative might read: "terms, the phrasing, and whether sentences are parallel." The alternative seems sloppy, however, and the inconsistency of noun, gerund, and clause makes it more difficult to form a mental list of the levels at which structure matters. Likewise, series of steps in a procedure could be expressed either as commands or as statements but not as a mixture.

Style

Style refers to diction, sentence patterns, and the writer's voice. Stylistic consistency means that the document does not mix formal language with casual language nor a sophisticated approach to a subject and reader with a simplistic one. A variation in style surprises a reader and calls attention to itself rather than to the content. It forces readers to adjust their sense of who they are in relation to the subject. For example, a shift from third-person description to second-person instruction moves the reader from observation (outside the action) to participation (in the action). A shift in person is disorienting unless it reflects an intentional shift from learning to practicing.

Visual Consistency

Visual consistency refers to anything a reader can see on the page or screen, including typography, spacing, and illustrations. The copyeditor's function in establishing visual consistency may overlap with the graphic designer's function, but even if the copyeditor does not make the decisions about typography and layout, he or she marks the typescript for visual consistency.

Typography

Typographical consistency means that parallel parts of a document use the same typeface and type style. Levels of headings are distinguished by their spacing, typography, and capitalization. Typefaces, type styles, and type sizes are not altered except for a specific purpose.

Layout

Layout refers to the location of related items on the page and to the use of space. Page numbers appear in the same location on all pages, and the amount of space following a level-one heading is the same for each use. Figures are labeled consistently at the bottom or at the top, not sometimes at the top and sometimes at the bottom. If the document uses numerous illustrations paired with text, the spacing between text and illustrations is consistent. The top, bottom, left, and right margins as well as the space between columns, if any, are uniform from page to page. Indentation is the same not just for paragraphs but for other text elements, such as lists and block quotations. If the location of related items varies, it does so according to a pattern (for example, page numbers may appear in the upper left corner on a lefthand page but in the upper right corner on a righthand page).

Screen design

Screens in electronic publication benefit from consistent layout for the same reasons that print documents do: readers learn where to find kinds of information and thus use the screen functions more efficiently. Just as with print documents, some conventions for placement of different types of information have developed. For example, online help information is usually in the upper right corner of the screen. Effective screen design also requires consistent use of type style, color, and icons.

Figures and Tables

Patterns of using numbers and titles and of placing them on the visuals should be established and followed. Callouts (names of parts), captions, and cross-references will be like others for similar illustrations in typeface and type style, capitalization, and placement. (See Chapter 5.)

The content and reproduction of figures and tables should also be consistent with the overall purpose and quality of the document. If a document is typeset and printed on quality paper, a poor reproduction or an amateurish drawing is inconsistent. The insertion of a cartoon is inconsistent with the goals of a serious financial report.

Consistency of Mechanics

Mechanics include spelling, capitalization, hyphenation, abbreviations, numbers, and type styles. The following examples illustrate consistency issues that arise in these categories of mechanics.

Spelling

Dictionaries often provide alternative spellings for words. The first choice may indicate the preferred spelling, or the choices may be listed alphabetically. Spelling of terms should be constant throughout the document. Editors choose one from alternatives such as these:

acknowledgment	acknowledgement
insure	ensure
copy editing	copyediting

Capitalization

Capitalization depends on rules and conventions. Organizations may capitalize in different ways. For example, government writing capitalizes the words *federal* and *state* when used as modifiers, whereas most other writing would not. An editor might question capitalization in the following examples:

Crohn's Disease	Crohn's disease
Federal	federal
French fries	french fries
Helvetica	helvetica
World Wide Web	world wide web

Hyphenation

Hyphenation helps readers identify whether phrases function as adjectives or nouns, and some terms use a hyphen in their spelling.

long-term project	in the long term
two-meter length	length of two meters
computer-assisted	cross-references
e-mail	email

Abbreviation

Abbreviation choices include whether to abbreviate (some terms are better known by their abbreviations than by the spelled-out version), what abbreviation to use if there are alternatives, and how to identify the abbreviations for readers (parenthetically or in a list of abbreviations). Abbreviations such as the following will require consistent editorial choices.

a.m.	A.M.	A.M.
inches	in.	"
Pennsylvania	PA	Pa.

Numbers

The various possibilities for expressing numbers require editorial choices.

Commas	1000	1,000	
Numerals or words	five	5	
Dates	12 April 1998	April 12, 1998	
Inclusive numbers	411–414	411–14	411–4
Time	eight o'clock	8 a.m.	8 A.M.
Equation numbers	eq. 2.2	equation 2	
Illustration numbers	figure 1.1	figure 1	Figure 1

Punctuation

All of the following text features require decisions about punctuation.

Items in a series.   Should there be a comma before the final item?

> Four types of consistency are verbal, visual, mechanical, and content.
>
> OR
>
> Four types of consistency are verbal, visual, mechanical and content.

Lists.   Should list items be punctuated? (Also, should they be numbered or bulleted? Capitalized?)

1. Spelling;	• spelling
2. Capitalization;	• capitalization
3. Hyphenation; and	• hyphenation
4. Numbers.	• numbers

Possessives.   Should an *s* be added following the apostrophe?

> Paris' museums   Paris's museums

Accents.   Should accents be used?

> resume   résumé

Cross-references.   Cross-references within the text should be consistent in use of abbreviations, capitalization, and parentheses.

> See Figure 1   See Fig. 1
>
> (see Fig. 1).   (See Fig. 1.)

Documentation

The styles for citing sources differ from discipline to discipline. All citations should conform to the appropriate style for the discipline or publication. The style establishes the arrangement of items in an entry, capitalization, abbreviation, punctuation, and the use of italics. Styles are usually variations of the author-date and reference note styles. The following two entries conform to the author-date styles of the American Psychological Association and the Modern Language Association, respectively.

> Redish, J. (1995). Adding value as a professional technical communicator. *Technical Communication, 42*, 26–39.
>
> Redish, Janice. "Adding Value as a Professional Technical Communicator." *Technical Communication* 42 (1995): 26–39.

Lists

Lists should be consistent in these respects:

> punctuation
> numbering
> capitalization of the initial word in each list item

Type style

Type style is often a matter of visual consistency, but where it has to do with terms, it relates to mechanics. For example, will terms used as terms be italicized or placed in quote marks?

> The word *proofreading* is no longer hyphenated.
>
> The word "proofreading" is no longer hyphenated.

Consistency in all these types of mechanics eliminates some document noise and contributes to the readers' impression of a quality publication.

Content Consistency

The check for content consistency eliminates contradictory information. For example, if a part of a mechanism is said to be made of steel at one point, it should not be described subsequently as made of aluminum. If the initial date of a five-year project is given as 1993, readers will be confused if the ending date is identified as 1999. If readers are told to find a reference on page 18 of a document when that reference actually appears on page 19, they will be frustrated or may not find the reference at all. If a forecasting statement in an introduction establishes four major divisions in a document with five divisions, readers will have to revise their sense of the document, an unnecessary and confusing step in comprehension.

Inconsistencies in a document often result from revision or from product changes. Inconsistencies can also result from typographical errors or carelessness.

The check for content consistency somewhat overlaps comprehensive editing, but copyediting is particularly concerned with the details of content whereas comprehensive editing focuses more on the larger issues of organization, completeness of information, and style.

A Foolish Consistency...

Consistency aids interpretation, reduces noise, and communicates quality. However, copyeditors can go too far in their pursuit of consistency and create errors and clumsiness. One editor, reading the following sentence, saw that two of the three items in the series ended in *-al*.

> The reasons for the change are philosophical, technological, and economic.

Aiming for parallel structure, the editor added *-al* to "economic."

The reasons for the change are philosophical, technological, and economical.

The structure is parallel—but the use of the word is wrong. Changes might be economical, but reasons (which have no price tag) are economic.

Ralph Waldo Emerson observed: "A foolish consistency is the hobgoblin of little minds." A foolish consistency would be to insist that all the illustrations in a document be of the same type—for example, to refuse to use both photographs and line drawings. Such a consistency would be foolish because the two types of illustrations have different purposes. The goal of providing readers with the information they need in the most effective way supersedes any goal of making related parts of the document uniform.

Style Manuals

Editors make choices about mechanics based on conventions established for the document, the organization, and the discipline, as well as on conventions of standard American English (or the standards of language in the country where the document will be published).

Copyeditors use style manuals accepted in their discipline or organization in order to help make choices for a document. The style manuals establish conventions. Furthermore, they save a copyeditor from merely guessing or from wasting

---

BOX 6.1   A Word About "Style"

Style is a term with multiple meanings for editors. You will use it differently in different contexts.

Writing style: Writing style refers to the choices of words and sentence structures and the writer's attitude toward the reader and subject. These choices reveal the writer's personality and the formality of the document; they also affect whether the text is comprehensible.

Mechanical style: Style as it is used in "manual of style" or "style sheet" refers to choices about mechanics, such as capitalization, rather than writing style.

Type style: Type style is a variation of mechanical style. It refers to whether the type is bold or light, roman or italic, condensed or expanded.

Computer styles: In word processing, styles are format choices for different text elements, such as headings and paragraphs. A style includes instructions for typeface (font), size, style, indentation, spacing before and after, borders, and anything else from the format menu. This collection of choices is saved and given a style name. A user can apply all the formatting choices with the single act of selecting the named style for a portion of the text.

time and effort thinking through each situation in order to make a decision. Once an editor makes a choice based on a style manual, he or she records it on a style sheet for the document. Subsequent choices can then be consistent with the first choice.

Style manuals can be classified into three categories, ranging from very complete to very selective: comprehensive, discipline, and organization or house. The individual document style sheet is the most selective of all.

Comprehensive Style Manuals

Comprehensive style manuals cover style choices that might apply in any publishing situation, no matter what the discipline or document. Two style manuals widely used in technical communication are these:

The Chicago Manual of Style, 14th ed. 1993. Chicago: University of Chicago Press.

Government Printing Office Style Manual, 28th ed. 1984. Washington, DC: U.S. Government Printing Office.

The contents page from *The Chicago Manual of Style* (see Figure 6.1) shows how comprehensive and thorough is the manual's coverage. The guidelines for using numbers, for example, take eighteen pages. An editor willing to hunt through a comprehensive style manual should find the answer to almost any style question.

Discipline Style Manuals

Some disciplines publish their own style manuals, mostly to guide researchers who publish in academic journals. Other publications that are linked in subject matter to these disciplines or are read by readers familiar with the discipline should conform to the style manual for the discipline. For example, an inhouse research report in a psychological research center should conform to the *Publication Manual of the American Psychological Association.* Editors consult a discipline style manual not only for documentation style but also for conventions for labeling illustrations, hyphenating terms, using numbers, and other mechanics. Some of the widely used manuals are these:

biology, agriculture	*Scientific Style and Format: The CBE Manual for Authors, Editors, and Publishers.* 6th ed. 1994. Cambridge, MA: Cambridge UP. Used also in agricultural sciences. By the Council of Biology Editors.
mathematics	*Mathematics into Type,* rev. ed. 1979. Providence, RI: American Mathematical Society. By Ellen Swanson.
journalism	*The New York Times Manual of Style and Usage.* 1975. New York: New York Times Books. Edited by Lewis Jordan.

Contents

Preface   vii

Part 1   *Bookmaking*

1   The Parts of a Book   3
2   Manuscript Preparation and Copyediting   47
3   Proofs   105
4   Rights and Permissions   125

Part 2   *Style*

5   Punctuation   157
6   Spelling and Distinctive Treatment of Words   193
7   Names and Terms   233
8   Numbers   293
9   Foreign Languages in Type   317
10   Quotations   355
11   Illustrations, Captions, and Legends   385
12   Tables   405
13   Mathematics in Type   433
14   Abbreviations   459
15   Documentation 1: Notes and Bibliographies   487
16   Documentation 2: Author-Date Citations and Reference Lists   637
17   Indexes   701

Part 3   *Production and Printing*

18   Design and Typography   763
19   Composition, Printing, Binding, and Papermaking   789

Glossary of Technical Terms   831
Bibliography   861
Index   871

FIGURE 6.1   Contents Page from *The Chicago Manual of Style,* 14th ed.

Source: Reprinted by permission of the University of Chicago Press.

psychology	*Publication Manual of the American Psychological Association*, 4th ed. 1994. Washington, DC: American Psychological Association. Used widely by other disciplines, including business.
physics	*AIP Style Manual for Guidance in Writing, Editing, and Preparing Physics Manuscripts for Publication*, 4th ed. 1990. New York: American Institute of Physics.
government	The *GPO Style Manual*, a comprehensive manual, is also a specialized style manual, "intended to facilitate Government printing."

Five specialized style manuals are useful for editors who work with particular kinds of text:

scientific and technical reports	*Scientific and Technical Reports—Elements, Organization and Design* (revision and redesignation of ANSI Z39.18-1987). ANSI Z39.18–1995. Oxon Hill, MD: NISO Press. By the National Information Standards Organization.
metric system	*Guidance for Using the Metric System—SI Version.* 1975. Washington, DC: Society for Technical Communication. By Valerie Antoine.
computer documentation	*The Microsoft Manual of Style for Technical Publications.* 1995. Redmond, WA: Microsoft Press.
illustrations	*Science and Technical Writing: A Manual of Style.* 1992. New York: Holt. Philip Rubens, General Editor.
World Wide Web sites	*Web Style Manual.* By Patrick J. Lynch, Yale Center for Advanced Instructional Media. http://info.med.yale.edu/caim/Manual

Other specialized style manuals are identified by John Bruce Howell in *Style Manuals of the English-Speaking World: A Guide* (Phoenix: Oryx Press, 1983). The instructions for authors in a discipline's journals may name the preferred style manual. Reference librarians can also assist in locating either a professional association that may publish a style manual or the manual itself.

House or Organization Style Manuals

An organization that publishes frequently, whether or not it is a commercial publishing house, will make similar choices for most of its documents. Choices that will be appropriate for all publications from that house, such as technical terms and use of numbers, can be listed in a house style manual. The manual probably incorporates the preferred choices in a comprehensive or discipline style manual, but because the house manual is more selective, it answers specific questions without the clutter of extra information. Instead of eighteen pages on the use of numbers, for example, there may be only two pages that list those instances in which the organization will use numbers. Digital Equipment Corporation, Bell Laboratories, and the publishers of the *Oil and Gas Journal* are three organizations that have prepared their own style manuals, but such manuals are common wherever publication is frequent. They are prepared for inhouse distribution only. Figure 6.2 shows a page

---

Lists

Types of Lists   There are two types of lists:

- ordered lists
- unordered lists

In an ordered list, each item is numbered. Use an ordered list only when the order of the items is important. In general, procedural steps are the only type of list in which the order of the items is important.

At *all* other times, use an unordered list. If you list three features of a product, do not enumerate the features. Use bullets instead.

Use of Colon to   Use a colon to introduce a list if the introductory statement is a
Introduce a List   grammatically complete sentence and it is not followed by another introductory sentence.

Example: 3270 OPTIMIZER/CICS offers three major advantages:

- It operates transparently to CICS.
- It uses standard 3270 features.
- It uses minimal CPU resources.

Use a colon to introduce a list if the introductory statement contains *as follows* or *the following*, and the illustrating items (rather than another sentence) follow it directly.

Examples: The features of 3270 OPTIMIZER/CICS are as follows:

- transparent CICS operation
- standard 3270 compatibility
- minimal use of CPU resources

The following results occur after you press F3:

- The selected record is added to the file.
- The program writes a backup copy to disk.
- The primary menu appears.

FIGURE 6.2   Page from the *BMC Software Technical Publications Style Guide*

Source: Reprinted by permission of BMC Software, Inc.

from the BMC *Software Technical Publications Style Guide* prepared by BMC Software, Incorporated.

In establishing a style for a particular document, you will probably consult the house style manual first. It is most likely to address the choices that have to be made for that document. This manual will also help you achieve consistency with the document set. Once you make the choices, record them on a document style sheet, specific to one document.

Document Style Sheet

Editors record the choices they make with the help of style manuals on a document style sheet. This list of choices helps them make subsequent choices in the pages that follow. The style sheet does not list every editorial choice but only those that may recur in the document. It includes choices for which the editor had to consult a dictionary, handbook, style manual, or other source, as well as terms or situations for which several options are available. For example, because you know that state names are always capitalized, you would not list your marking of a lowercase *c* in California. But if you were unsure of the spelling and capitalization in Alzheimer's disease, you would list the term on the style sheet. The style sheet is not a record of changes but rather a guide to editing the pages that follow in the document.

The style sheet includes choices about punctuation, abbreviations, spelling, capitalization, hyphenation, and other matters of consistency. The style sheet may be only one page long if the document is short. One convenient way to prepare a document style sheet is to begin with a grid into which you can pencil the choices as you make them (see Figure 6.3). The words are roughly alphabetized so that you can find the choice quickly. You can include the page number of the first example so that you will always have an example to refer to. If you think you may change your mind later, include page numbers for all occurrences.

If your treatment of a term will differ depending on its part of speech, as from noun to adjective, include the part of speech on the style sheet. For example, the phrase *top-down* as an adjective is hyphenated, but as an adverb it is not. One might write, "Top-down editing helps an editor make decisions in light of the whole," and then be consistent in saying "That person edits top down." The style sheet would look like this:

top-down (adj.)

top down (adv.)

In addition to listing terms in the alphabetical sections, you can include sections on the sheet for punctuation, numbers, and formatting.

A second way to prepare a style sheet is on computer. You can keep your style sheet file open while you work on the document you are editing so that you can easily add or refer to terms. You can alphabetize as you add entries using the sort function. Page numbers of occurrences are not so important with this method because of the computer's search function.

A-D	E-H	I-L
backup copies	*email* *half title*	*inhouse (adj.)* *in house (pred. adj.)*
M-P	Q-T	U-Z
proofreading *markup* *online*	*typeface* *type size* *type style*	*World Wide Web; the Web*
numbers	style, spacing	miscellaneous
numerals with measures *Spell one to nine* *dates: 11 April 1998*	*Cross-refs: lc* *figure 3.1 ; see figure 2.6* *roman type (no caps)*	*terms:* *document = the entire* *publication, incl. hypertext* *text = prose part of the doc.* *labels: double numeration* *figure 3.1*

FIGURE 6.3   Grid-Type Style Sheet for a Document

Part of the style sheet for this textbook appears in Figure 6.4.

Either way you prepare your document style sheet, using a grid or the computer, you should be able to determine choices you have made previously and be consistent throughout the document. The style sheet accompanies the document through subsequent stages of production, including proofreading. It gives other people working on the document a basis for answering questions and making decisions.

Using the Computer to Establish Consistency

The computer is a good tool for achieving consistency. You can use the search and replace functions to locate variations in spelling, capitalization, abbreviations, and numbers and then apply one choice in all instances. You can also use the styles function of word processing programs to create the same typographical and spacing choices for related types of text. (See Chapter 4.) Your document template includes

spelling, hyphenation, capitalization	type style and punctuation
backup copies	clause following a colon: no initial cap
camera ready (adj)	introductory phrase or term: comma
cross-references: Chapter 9, figure 4.1	follows (Thus, editing...)
copyediting	terms used as terms: italicize
double-spaced (adj)	
double-spacing (pred adj) 16.7	**illustrations**
email	cross-references in text: init cap (Figure
half title	3.1, see Figure 3.1)
hard copy	labels: double numeration (Figure 3.1)
inhouse (adj)	
in house (pred adj)	**terms**
Internet	document = the entire publication, includ-
lefthand (adj)	ing hypertext
markup	text = prose part of the document
metric system	
online	
proofreading	
righthand	
roman type (not capitalized)	
typeface	
type size	
type style	
World Wide Web, the Web	

FIGURE 6.4   Partial Style Sheet for *Technical Editing*

styles for each type of text that is repeated more than once, such as headings of the various levels, paragraphs, and bulleted lists. If you keep your document style sheet in a word processing file, you can always keep it alphabetized, complete, neat, and handy for ready access as you edit.

Summary

One of the characteristics of a quality publication is that readers can count on consistency in word choice, visual elements, mechanics, and content. Consistency helps readers to locate and comprehend the information they need. Within all the categories of consistency—verbal, structural, visual, and content—editors can make some choices from among acceptable alternatives. These choices are not simply personal preference, however, but reflect conventions established and recorded in dictionaries and style manuals.

Further Reading

Rudy Domitrovic. 1977. *How To Prepare a Style Guide.* Washington, DC: Society for Technical Communication.

David K. Farkas. 1985. "The Concept of Consistency in Writing and Editing." *Journal of Technical Writing and Communication* 14(4): 353–364.

Laura J. Gurak. 1992. "Toward Consistency in Visual Information: Standardized Icons Based on Task." *Technical Communication* 39 (February 1992): 33–37.

Patrick J. Lynch. 1995. *Web Style Manual.* Yale Center for Advanced Instructional Media. http://info.med.yale.edu/caim/stylemanual-1.html

Discussion and Application

1. Locate two discipline style manuals from the list in this chapter or by asking a reference librarian. Compare the directions for citing references and for spelling out numbers in the two manuals. What similarities and differences do you note?

2. Explain why a document style sheet should not list every editorial change made in a document.

3. What features in the following two paragraphs will require a decision about consistency of mechanics? Why would the typo "continous" not appear on the style sheet?

   Assuming that the paragraphs are the introduction to a ten-page report that must be edited in its entirety, make a one-page grid-type style sheet for these paragraphs so that subsequent decisions may be consistent. If you have questions about some of the choices, determine what printed resources you could use to find answers.

   > Ignitron Tubes were first used in dc-arc welding power supplies. An ignitron tube is a vacuum tube that can function as a closing switch in pulsed power (using a pulse of current rather than a continous current) applications. A closing switch is a switch that is not *on* or will not pass current until it is triggered to be on. When the tube is triggered to be "on," it provides a path for the current. The ignitron tube is turned on by a device called an "ignitor." The ignitor sits in liquid mercury inside the vacuum tube. The ignitron tube creates a path for the current by vaporizing mercury. More mercury will be vaporized as the current crosses the tube. When there is no current the vaporized mercury goes to the bottom of the tube and turns back into liquid.
   >
   > When an Ignitron Tube turns on and passes current, this process is referred to as a shot. The greatest amount of current that ignitrons can presently handle is nearly 1,000,000 amps. The tube can handle this amount of current for five shots before the tube fails. Industry wants a tube that will handle 1000 shots before it fails.

Chapter *7*

Spelling, Capitalization, and Abbreviations

If you have ever found a spelling error in a printed document, you know the feeling most readers have when they discover errors. Making (or missing) errors in spelling and capitalization is like having a piece of spinach stuck between your front teeth when you ask your boss for a raise—the person responds by laughing, inwardly or outwardly. Such errors mean something is amiss; someone was not careful. Credibility suffers. Readers' comprehension may suffer, too, if only because they are distracted from the content.

Finding that lone error in a document that is printed or displayed on a screen is easier than finding all of the errors in copyediting. The error sticks out in the published document because most of the rest of the document is correct and probably attractive. In a typescript, however, you are distracted from one error by others and even by the mess of a heavily copyedited page. Or you may become so familiar with a typescript that your eye reads the words as correct even when they are not. Finding typos on a computer screen may be even harder than finding them on a printed page.

Even if you are a good speller, use a spelling checker, and know the rules of capitalization and abbreviation, perfecting a typescript can be a challenge. Knowing the issues and principles of spelling will increase your efficiency. Checking dictionaries and style manuals is a necessary strategy of a careful editor. Recording the correct version (or your choice when there are alternatives) on the document style sheet will help you make the same choice consistently throughout the document and will save the time of rechecking when the word occurs again.

This chapter identifies ways to recognize misspelled words and alerts you to times when you should check capitalization and abbreviations.

Spelling

Spelling is notoriously difficult in English because of the many irregularities and inconsistencies. Imagine trying to teach a non-native speaker how to spell and pronounce these words: *tough, though, through.* Or try to explain to someone why the *sh* sound in English can be spelled in at least ten different ways, as illustrated by the words *shower, pshaw, sugar, machine, special, tissue, vacation, suspicion, mansion,* and *ocean.* Editors depend on guidelines and tools to achieve correct spelling.

Guidelines and Tools

The following list outlines resources and practices to help you edit for spelling.

1. Use a spelling checker on your computer. One of the most valuable tools the computer offers editors is the spelling checker. By matching words in the document with its database of words, a computer can recognize words that do not match and that may be misspelled. For each mismatched word, you must choose whether to change the spelling or to let it stand. Spelling checkers may suggest alternative spellings, but the user must key in changes that are not in the database.

A computerized spelling check is a good first step in editing any document in electronic form. If spelling errors can be corrected at the beginning, they will not distract the editor from correcting other errors. Running the check as the last step in editing can also help to catch errors that have been introduced in editing.

A computerized spelling checker will not relieve you of responsibility for checking spelling. Because it matches only characters and does not read for meaning, it will not know when words are correctly spelled but misused. It will approve *type script* even though you meant *typescript* because the two words match words in its database, and it will not catch *in* substituted for *on.* A computer would approve the spelling in this sentence:

Their our know miss steaks in this sent ants.

The memory may also not contain all the words in the document. A typical desk dictionary contains 200,000 words, whereas a typical computerized dictionary stores only 80,000 words, including variations (such as different tenses) of the same word. Thus, the computer may question words that are spelled properly, especially technical terms and names.

Be cautious about accepting the spelling checker's recommendations for change. The newsletter *Simply Stated,* from the Document Design Center, reported in March 1989 the amusing results of a spelling change implemented by a computer. The Barclay Banks in England used a spelling checker on a document prepared for depositors. When the checker did not recognize "Barclay," it substituted "Broccoli." Depositors were confused—and amused—when they received correspondence from the Broccoli Bank!

Programs can recognize signs of possible grammar problems, but they cannot identify certain errors, such as lack of subject-verb agreement or dangling modifiers.

For example, a *to be* verb followed shortly by a past participle might signal a dangling modifier; on the other hand, the construction might be grammatically correct. Possible ways to arrange words in sentences are so numerous that huge amounts of computer memory would be required to store all the possible versions of typical grammar errors to be available for matching with the document in the active file. Programs that promise to find grammar errors need informed users to decide whether the errors are real and how they should be corrected.

Don't depend entirely on the computer for your spelling check, but do use it to save time and to catch misspelled words that you might otherwise miss.

2. Use a dictionary. Look up the spelling of a term that is unfamiliar to you. Use a subject dictionary (such as a dictionary of medical terms) for specialized terms. When you look up a word, mark it in your dictionary to encourage review when you are using the dictionary another time. You can also add the word to your own list of commonly misspelled words and to the document style sheet to enable a quick check another time you are uncertain about the spelling.

3. Keep a list of frequently misspelled words. Even good spellers have difficulty learning the spelling of certain words. Table 7.1 lists words that are frequently misspelled; you may have other words to add to your individual list. (In addition to these frequently *misspelled* words, see the list of frequently *misused* words, later in the chapter.) The list itself reveals problem areas:

endings (-*ar*, -*or*, or -*er*; -*ment* or -*mant*; -*able* or -*ible*; -*ance* or -*ence*; -*ceed* or -*cede*; -*ary* or -*ery*)

consonants—single or double

the schwa ("uh") sound—created by *a, e, i, u*

ie and *ei*

TABLE 7.1   Frequently Misspelled Words

accommodate (double *m*)	changeable	irrelevant	precede (compare *proceed*)
achievement	complement	irresistible	questionnaire (double *n*)
acknowledgment (no *e* before the suffix)	convenient	judgment	receive ("*i* before *e* except after *c*...")
	defendant	liaison	
acquire	embarrass (double *r* and double *s*)	license	resistance
a lot (two words, like *a little*)	feasibility, feasible	lightning	separate (*a* in the second syllable)
		maneuver	
all right	forty (not *fourty*)	miniature	severely (keep the final *e* before the suffix)
analysis	gauge	mischievous	
approximately	grammar (the ending has an *a* in it)	misspell	sincerely
argument		occurred	subpoena
assistance	grievous	parallel	subtly
basically	harass	personnel	surveillance
category (*e* in the second syllable)	inasmuch as	plausible	undoubtedly
	indispensable	pastime	

When you check spelling, look especially for the endings and consonants and the schwa sound. Spelling rules and mnemonic devices, like those in guidelines 4 and 6, will help you spell these words.

4. Develop mnemonics. A mnemonic is a memory device, usually an association. By remembering the association, you can remember the spelling. Here are seven examples:

> *acco**mm**odate:* When you accommodate someone, you do more than is expected; there are more *m*s than you expect.
>
> *all right:* It's two words, like *all wrong.* Likewise, *a lot:* two words, like *a little.*
>
> *li**ai**son:* The "eyes" on either side of the a look both ways, the way a liaison looks to both sides.
>
> *princi**pal*** (a school official): The principal is (or is not) your pal.
>
> *station**er**y* (writing paper): Remember, er as in paper.
>
> *de**ss**ert* (sweet food): More of the letter *s* than in *desert* (a dry region); a dessert tastes so good that you want more.
>
> *affect, effect: Effect* is usually a noun, while *affect* is usually a verb. An auditory mnemonic: th**e e**ffect (long *e* for both words); *the* precedes a noun, not a verb. A visual mnemonic: an *e*ffect is the result or the *e*nd.

5. Learn root words, prefixes, and suffixes. Words are frequently formed from basic words and from the addition of suffixes and prefixes. Recognizing the structure of the words can help you spell them. Here are some examples:

> *sub**poena:*** Although you don't hear the *b* when you pronounce this word, you can recognize the prefix *sub-,* meaning "under." A *subpoena* requires a person to give testimony in court, *under penalty* if he or she refuses.
>
> *mi**ss**pell:* Instead of wondering whether the word has one *s* or two, consider that the prefix *mis-* is added to the root word *spell.* The structure of the word explains the double consonant. The words *misfortune, mistake,* and *misconduct* have only one *s* because their root words do not begin with s. The same principle applies to words with the *dis-* prefix: dissatisfied and dissimilar, but disappear and disease.
>
> *i**n**a**n**imate:* One *n* or two? The root of this word is a Latin term, *animalis,* meaning "living." The prefix in- means "not." An inanimate object is not living. There is only one *n* in the prefix, and the root word does not begin with an *n,* thus the single *n* in the first syllable. There is only one *n* in the root word, as in *animal.*
>
> *mi**ll**imeter:* The root *mille* means "thousand." A millimeter is one thousandth of a meter.

Common roots in English are sometimes similar enough to be confused:

> *ante* (before); *anti* (against): *ante*bellum (before the war); *anti*dote
>
> *hyper* (above, over); *hypo* (less, under): *hyper*active; *hypo*glycemia (low blood sugar)
>
> *macro* (large); *micro* (small): *macro*cosm; *micro*scope
>
> *poly* (multiple); *poli* (city): *poly*gamy (multiple marriage); metro*poli*tan

6. Learn spelling rules. Spelling rules treat especially the addition of suffixes to words. You can find a more complete discussion of rules in a spelling textbook, such as *Spelling Fifteen Hundred.*

final *e:* A word ending in a silent e keeps the e before a suffix beginning with a consonant: *completely, excitement, sincerely.* Exceptions: *acknowledgment, judgment, argument, truly.* If the suffix begins with a vowel, the silent e disappears: *loving, admirable. Changeable,* however, keeps the *e* to establish the soft *g* sound.

final consonant: When a word ends with a consonant, the consonant is doubled before a suffix beginning with a vowel: *planned, transferred, occurrence, omitted.* If the final syllable is not stressed, the final consonant is not doubled: *inhabited, signaled. Busing/(bussing), focusing/(focussing),* and *canceled/ (cancelled)* are correctly spelled with one consonant or with two.

able/ible: The suffix *-able* generally follows words that are complete in themselves, such as remarkable, adaptable, moveable, changeable, readable, respectable, and comfortable. The suffix *-ible* follows groups of letters that do not form words. There is no *feas, plaus,* nor *terr;* thus, we write *feasible, feasibility, plausible,* and *terrible.* Words that end in a soft s sound also take *-ible: responsible, irresistible, forcible.* A mnemonic: words that are *able* to stand alone take the suffix *-able.* Exceptions: *inhospitable, indefensable, tenable.*

ie/ei: The familiar rhyme works most of the time: *i* before *e* except after *c* and when sounded like *a* as in *neighbor* and *sleigh.* Exceptions should be memorized: *Neither foreigner seized the weird financier at leisure on the height.*

7. Respect the limits of your knowledge. Editors have made mistakes by substituting familiar words for similar but unfamiliar words. You can imagine the impatience of a writer in psychology whose editor changed *discrete behavior* to *discreet behavior,* not knowing the term *discrete* was defined to mean "separate." Or the frustration of the writer in a social service organization whose editor changed *not-for-profit* to *nonprofit* when *not-for-profit* was the legal term used in the application for tax-exempt status. Or the writer in biochemistry whose editor changed *beta agonists* to *beta antagonists.* In all cases, the first term was correct in its context; the editor substituted a term that was more familiar but incorrect.

Wise editors respect the limits of their knowledge and check a reliable source, such as a dictionary or the writer, before changing the spelling of an unfamiliar term.

Frequently Misused Words

The following groups of words are so similar in appearance or meaning that they are frequently confused.

affect, effect: Both words can be nouns, and both can be verbs. Usually, however, *effect* is a noun, meaning "result," and *affect* is a verb, meaning "to influence."

Because a noun can take an article but a verb does not, an auditory and visual mnemonic may help: th*e e*ffect. Because a verb is an action, the beginning letters of *a*ct and *a*ffect can serve as a visual mnemonic.

Your training in writing will have an effect on your performance in editing.

The Doppler effect explains the apparent change in the frequency of sound waves when a train approaches.

Class participation will affect your grade.

assure, insure, ensure: All three words mean "to make secure or certain." Only *assure* is used with reference to persons in the sense of setting the mind at rest. *Insure* is used in the business sense of guaranteeing against risk. *Ensure* is used in other senses.

The supervisor assured the editor that her salary would be raised.

The company has insured the staff for health and life.

The new management policies ensure greater participation in decision making by the staff.

complement, compliment: A complement completes a whole; a compliment expresses praise.

In geometry, a complement is an angle related to another so that their sum totals 90 degrees.

The subject complement follows a linking verb.

Your compliment encouraged me.

continually, continuously: Continually suggests interrupted action over a period of time; continuously indicates uninterrupted action.

We make backup copies continually.

The printer in the lab stays on continuously during working hours.

discreet, discrete: Discreet means "prudent"; *discrete* means "separate."

A good manager is discreet in reprimanding an employee.

We counted seven discrete examples of faulty reporting.

farther, further: Farther refers only to physical distance; *further* refers to degree, quantity, or time.

The satellite plant is farther from town than I expected.

The company cannot risk going further into debt.

fiscal, physical: Fiscal refers to finances; *physical* refers to bodily or material things.

The tax return is due five months after the fiscal year ends.

imply, infer: Imply means "to suggest"; *infer* means "to take a suggestion" or "to draw a conclusion."

These figures imply that bankruptcy is imminent.

We inferred from the figures that bankruptcy is imminent.

its, it's: Its is a possessive pronoun; *it's* is a contraction for *it is.*

The company is proud of its new health insurance policy.

It's a good idea to accept the major medical option on the health insurance policy.

lay, lie: Lay is a transitive verb that takes an object: you lay something in its place. *Lay* can also be the past participle of *lie. Lie* is always an intransitive verb. It may be followed by an adverb.

A minute ago, I lay the flowers on the table, and they lie there still. I will lay the silverware now. Then I will lie down.

personal, personnel: Personal is an adjective meaning "private" or "one's own"; *personnel* is a collective noun referring to the persons employed or active in an organization.

The voltage meter is my personal property.

Take the job description to the personnel manager.

principle, principal: A principle is a basic truth or law; a *principal* can be a school official (noun), or it can mean "first" or "primary" (adjective).

Our company is managed according to democratic principles.

Writing ability is the principal qualification for employment.

stationery, stationary: Stationery refers to writing paper; something *stationary* is fixed in place or not moveable.

Our business stationery is on classic laid paper.

The projection equipment in the seminar room is stationary.

their, there, they're: Their is a possessive pronoun; *there* is an adverb designating a place (note the root *here*) or a pronoun used to introduce a sentence; *they're* is a contraction for *they are.*

Their new book won a prize for design.

There are six award-winning books displayed over there on the table. They're all impressive.

whose, who's: Whose is a possessive pronoun; *who's* is a contraction for *who is.*

Whose house is this?

Who's coming to dinner?

your, you're: Your is a possessive pronoun; *you're* is a contraction for *you are.*

I read your report last night. You're a good writer.

International Variations

Spellings differ in countries even where the language is the same. For example, British and American English spell words differently. A British person would write "whilst" and "amongst" whereas an American would write "while" and "among." Choose the spelling for the countries where most readers reside.

Capitalization

The basic rules of capitalization are familiar: the first word of a sentence, titles of publications and people, days, months, holidays, the pronoun *I*, and proper nouns are capitalized. Proper nouns include names of people, organizations, historical periods and events, places, rivers, lakes, and mountains, names of nationalities, and most words derived from proper nouns.

Other capitalization choices require more specialized knowledge. Capitalization conventions vary from discipline to discipline. For example, if you work for the government, conventional usage dictates that you capitalize *Federal, State,* and *Local* as modifiers, but in other contexts, such capitalization would be incorrect.

Generally, capitalize nouns that refer to specific people and places but not general references:

- Capitalize names of specific regions (the *Midwest*) but not a general point on the compass (the plant is *west* of the city).
- Capitalize adjectives derived from geographic names when the term refers to a specific place (*Mexican* food) but not when the term has a specialized meaning not directly related to the place (*french* fries, *roman* numerals, *roman* type).
- Usually capitalize the specific name of an organization (Smith *Corporation*) but not a reference to it later in the document by type of organization rather than by specific name (the *corporation*).
- Capitalize a title when it precedes a name (*President Morales*) but not when it is used in general to refer to an officer (the *president*). And, capitalize names of disciplines that refer to nationalities (*English, Spanish*) but not to other disciplines (*history, chemistry*).

The trend over time is toward less capitalization. In the eighteenth century, many nouns, not just proper nouns, were capitalized. When words are introduced into the language, they are often capitalized to distinguish them as new words. For example, in 1997, World Wide Web and Web were usually capitalized to distinguish them from general meanings of web, not because they are proper nouns. Gradually "web" without a capital letter will be understood from the context to refer to networked computers that serve and receive graphical files, not to something a spider makes.

The use of capitalization has declined in part because excessive capitalization distracts readers and interferes with readability. Capitalizing for emphasis encourages word-by-word reading rather than reading for overall sense. The repeated capitals also look like hiccups on the page. Typing words in all capital letters destroys the shape of the word (all words become rectangles) and therefore obscures information that readers use to identify words. Other techniques of typographic emphasis, including boldface type, type size, white space, and boxing, are more effective.

Distracting Capitalization

Visitors Must Register All Cameras with the Attendant at the Entry Station.

VISITORS MUST REGISTER ALL CAMERAS WITH THE ATTENDANT AT THE ENTRY STATION

Preferred Styles

Visitors must register all cameras with the attendant at the entry station.

Visitors must register all cameras with the attendant at the entry station.

Abbreviations

An editor considers whether to use abbreviations or to spell out the terms and inserts identifying information for unfamiliar abbreviations. A related editorial task is to make consistent decisions about capitalization and periods within abbreviations. The best reason for using an abbreviation is that readers recognize it; perhaps it is more familiar than what it represents. Abbreviations may also be easier to learn than the full name, and they save space on the page. *NASA* is now more familiar to readers, and also was easier to learn when first coined, than is the spelled-out title, National Aeronautics and Space Administration. When an abbreviation becomes a word in its own right, it is an acronym. *NASA* has reached that status.

Abbreviations in the spelling of names should be respected. If a company is known as Jones & Jones (with the ampersand appearing on company signs and on the company letterhead), do not spell out the "and" to make it more formal.

Identifying Abbreviations

Abbreviations that are not familiar to readers must be identified the first time they are used. A parenthetical definition gives readers the information they need at the time they need it. After the initial parenthetical definition, the abbreviation alone may be used.

> Researchers have given tetrahydroaminoacridine (THA) to patients with Alzheimer's disease. THA inhibits the action of an enzyme that breaks down acetylcholine, a neurotransmitter that is deficient in Alzheimer's patients. The cognitive functioning of patients given oral THA improved while they were on the drug.

If the abbreviation is used in different chapters of the same document, it should be identified parenthetically in each chapter, because readers may not have read or may not remember previous identifications. Documents that use many abbreviations may also include a list of abbreviations in the front or back matter.

Periods and Spaces with Abbreviations

Some abbreviations include or conclude with periods (U.S., B.A.); others, especially scientific and technological terms and groups of capital letters, do not (USSR, CBS, cm). The trend is to drop the periods. For example, NAACP is more common now than the older form N.A.A.C.P.

When an abbreviation contains internal periods, no space is set between the period and the following letter except before another word. At end of sentence, one period identifies both an abbreviation and the end of a sentence.

> While working as a technical editor, she is also completing courses for a Ph.D.

Because they represent complete words, initials substituting for names and abbreviations of single terms require spaces after the periods (for example, Mr. T. J. Corona). But single abbreviations consisting of multiple abbreviated words do not have spaces within them (see the next section, on Latin terms). When in doubt, check a dictionary for the conventions of capitalization and punctuation for an abbreviation. Most standard desk dictionaries include a list of abbreviations as part of the back matter.

Latin Terms

Abbreviations that represent Latin terms are used primarily in scholarly documents and may be unfamiliar to readers or seem pretentious outside a scholarly context. The Latin abbreviations in the list that follows are common in academic work. Editors who work in an academic environment should know them. Abbreviations of Latin terms use periods following the abbreviated term. Editors who work with user manuals should convert Latin abbreviations, except for the well-known A.M. and P.M., to their English equivalents.

> A.M. (*ante meridiem,* before midday); P.M. (*post meridiem,* after midday): these abbreviations may be set in capital letters, lowercase letters, or small caps. The periods distinguish A.M. from the abbreviation for amplitude modulation, but small caps or context may do as well, and the periods may be omitted.
>
> *ca.* (*circa,* about): used in giving approximate dates
>
> *cf.* (*confer,* compare): used for cross-references
>
> *cwt.* (*centum-weight,* hundred-weight): a hybrid of Latin and English; used in measuring commodities
>
> *et al.* (*et alii,* and others): used primarily in bibliographic entries in humanities texts. Note that *et,* being a complete word and not an abbreviation, takes no period.
>
> *etc.* (*et cetera,* and other things, and so forth)
>
> *ff.* (*folio,* on the following page or pages)
>
> *i.e.* (*id est,* that is): used in adding an explanation
>
> *e.g.* (*exempli gratia,* for the sake of an example; for example)
>
> *N.B.* (*note bene,* note well)
>
> *Ph.D.* (*Philosophia Doctor,* Doctor of Philosophy): the Latin terms explain the capitalization and use of periods in this abbreviation.
>
> *v.* or *vs.* (*versus,* against)

The overused *etc.* should be restricted to informal writing—or omitted altogether. It is often a meaningless appendage to a sentence, added when a writer is unsure whether the examples suffice. It may signal incomplete thinking rather than useful information for readers. In addition, it is redundant if a writer has also used *such as,* which indicates that the list following is representative rather than complete.

Measurement and Scientific Symbols

Three systems of measurement are widely used: the U.S. Customary System, the British Imperial System, and the International (metric) System. Specific abbreviations are recognized for the units of measure in all three systems. An editor checks that abbreviations used for measurement conform to the standards.

The metric system, which provides a system of units for all physical measurements, is used for most scientific and technical work. It is also the common system for international use. Its units are called SI units (for *Système International,* in French). The abbreviations for the metric system do not use periods. In the U.S. and British systems, the conventions about periods with abbreviations vary. For example, the abbreviation for inch may or may not include a period (*in.* or *in*). The abbreviation for pound (*lb.*) uses a period because it stands for the Latin word *libra.* The abbreviation is always singular even if the quantity it stands for is multiple. Thus, *3 lb.* is correct; *3 lbs.* is not.

Check a list of abbreviations in a dictionary or style guide to determine usage, and record your choice on the style sheet. Because the abbreviations for measurements are standard, they do not require parenthetical identification. Within sentences of technical and scientific texts, measures used with specific quantities may be abbreviated or spelled out. When measures are used alone, without specific quantities, they are spelled out. A sentence would read "The weight is measured in kilograms." In tables and formulas, however, abbreviations are preferred. As always, the document should be consistent.

Like the abbreviations for measurements, scientific symbols for terms in chemistry, medicine, astronomy, engineering, and statistics are standard. Consult a specialized style guide to determine the correct symbols. (Chapter 10 discusses quantitative and technical material in detail.)

States

The United States Postal Service recognizes abbreviations consisting of two capital letters and no periods for the states and territories (for example, CA, NC). The old system of abbreviations for the states—usually consisting of capital and lowercase letters with a period—is obsolete. Use the postal service abbreviations on envelopes because both human and machine scanners recognize them quickly. In the headings of letters, either the abbreviation or the spelled-out state name is appropriate, but be consistent. Within the text of a document, spell out names of states. In references and bibliographies, the use of postal abbreviations, non-postal abbreviations, or spelled out state names is dependent on the particular style in use.

Summary

Correct spelling, capitalization, and abbreviations increase the accuracy and clarity of writing. Experience with language and understanding of its rules will help editors work quickly and accurately on these aspects of a document, but good editors will also depend on dictionaries and style manuals. These resources establish conventions and aid in consistency.

Further Reading

Sylvia A. Gardner. 1992. "Spelling Errors in Online Databases: What the Technical Communicator Should Know." *Technical Communication* 39: 50–53.

J. N. Hook. 1986. *Spelling Fifteen Hundred,* 3rd ed. San Diego: Harcourt Brace Jovanovich.

Discussion and Application

1. Use copymarking symbols to mark spelling corrections in these paragraphs.

    A feasability study is a way to help a person make a decision. It provides facts that help a researcher decide objectively weather a project is practical and desireable. Alot of research preceeds the decision.

    An invester deciding whether to purchase a convience store, for example, must investigate issues such as loans, lisences, taxes, and consummor demand. One principle question for a store in a residentail nieghborhood concerns the sale of liquer. People who come into the store might be asked to compleat a brief questionaire that inquires about there wishes.

    Undoubtably, the issue of personal is signifigant as well. Trustworthy employes are indispensible. The investigator can check employmant patterns at the store and in the area overall.

    After the study is complete, the investigator will guage the results and make a judgement. Some times the facts are ambigous. Intuition and willingness to take risk will influence the decision. If the project looks feasible, the recomendation will be to procede with the purchase.

2. Which of the spelling errors in the paragraphs in exercise 1 would a computerized spelling checker *not* catch? Why?

3. From the lists of frequently misspelled and misused words in this chapter, or from your own list of difficult words, select three, and develop mnemonics for remembering their spelling or use. Alternatively, analyze the spelling of these words by analyzing their roots.

4. Get acquainted with your desk dictionary as an editorial resource. What information beyond correct spelling does it provide an editor? Check the front matter and appendixes. Also look for help within the word lists on measurements and on symbols and signs. Name your dictionary, and cite at least three types of information other than spelling and usage. Compare your dictionary with a classmate's to get a sense of how different dictionaries may help an editor.

5. Correct the following sentences for spelling, misused words, capitalization, and abbreviation. Consult a dictionary and style manual when you are uncertain. Use copymarking symbols.

   a. Development will continue in the Northern part of the city.

   b. Do not spill the Hydrochloric Acid on your clothes.

   c. Do swedish meatballs go well with French fries?

   d. A research project at the sight of a major city landfill has shown how slowly plastic decomposes; i.e., a plastic bottle takes 100–400 years to decompose.

   e. The Society for Technical Communication has planned feild trips to three Hi Tech company's. Later in the year some students will attend the Annual Conference of the STC. All of these plans for travelling requie some extensive fund raising this Fall. The sponsers have proposed the establishment of an Editing Service. Student edtors would aquire jobs thru the Service and return 15 per cent of there earnings to the group.

   f. A minor in computor science in combination with a major in english can make a student an attractive canditate for a job in technical communication.

   g. Capitol investmant in the company will raise in the next physical year.

   h. A prevous employe was the defendent in an embarasing and costly law suit charging sexual harrasment.

   i. Only 10 m. seperates her house from mine. I wish it were further.

Chapter 8

Grammar and Usage

Grammar refers to relationships of words in sentences: subjects to verbs, modifiers to items modified, pronouns to antecedents, and so forth. Grammar describes conventional and expected ways for the words to relate. For example, subject and verb agree in number. Grammar changes over time, but the conventions are used widely enough to be described as "rules."

Usage refers to acceptable ways of using words and phrases. Matters of usage include, for example, whether *he* can be used as a generic pronoun, whether to use a singular or plural verb with *data,* and when to use *fewer* and *less.* To establish some consensus about usage and to monitor the trends in language, in the mid-1960s the editors of *The American Heritage Dictionary* formed a usage panel consisting of language experts. Such experts determine whether new words or words used in new contexts are acceptable. Some dictionaries contain usage notes. Professional associations may also state their own conventions of usage.

Grammar is ultimately about meaning, not about rules. Grammar rules describe words in relationship to each other. Facts and ideas become meaningful in relationship to each other, not in isolation, and so do words. Errors in grammar may result in misreading or misunderstanding because relationships between parts of sentences seem unclear, making the sentences hard to interpret. Because errors are relatively easy for readers to identify, they can also distract readers and diminish their trust in the content of the document. Errors in usage are less serious than errors in grammar, but they raise questions about the care with which the document was created.

Handbooks for writers provide detailed information on grammar and usage. Most major publishers offer a handbook. If you do not have one, find out if one is preferred in your organization or in your discipline before you select one. Otherwise, use a current edition of a handbook from a major publishing house. For a comprehensive review of grammar, study *Rhetorical Grammar* by Martha Kolln. Dictionaries identify parts of speech and summarize grammar rules.

This chapter reviews the basic concepts and vocabulary of grammar. Its purpose is to help grammar make sense as a whole by defining it as descriptions of relationships of words in sentences. It also reviews vocabulary so that you will be able to discuss options for sentence construction and punctuation in a professional way. The chapter begins with brief definitions and then describes sentence structure. The chapter then discusses the common rules for the relationships of words in sentences and the errors that result from breaking those rules. Chapter 9 reviews other grammar errors that relate to punctuation. This chapter reviews material you have probably learned previously. If you encounter unfamiliar terms, please consult the glossary or a handbook.

Definitions

The definitions below are for quick review and reference. If you need more detailed definitions, please consult a handbook of grammar. Table 8.1 defines grammatical divisions of sentences. Table 8.2 defines the parts of speech (words).

Sentence Structure

Sentence structure helps readers interpret the meaning of sentences. Sentences have two parts: subject and predicate. The subject states the topic of the sentence. The predicate comments about the subject. Each part contributes necessary meaning.

> The new computer has increased productivity by 40 percent.
>
>   subject   predicate

The subject of the sentence identifies the agent or recipient of the action in a sentence. The simple subject is the subject without its modifiers. In the example sentence, "computer" is the simple subject. Modifiers are the article "The" and the adjective "new."

The predicate tells what the subject does or what happens to the subject. It makes a statement or asks a question about the subject. The predicate always includes a verb. The simple predicate is the verb or verb phrase. In the example sentence, "has increased" is the simple predicate.

Verbs and Sentence Patterns

The predicate in the sample sentence includes more words than the verb phrase. These words form the *complement*, words that "complete" the comment on the subject. The verb determines the relationship between the subject and the rest of the predicate, or the type of complement. Verbs may be categorized as *transitive, intransitive, linking,* and *to be.* All of those words suggest functions: a transitive verb

TABLE 8.1   Functional Parts of Sentences (Grammatical Divisions)

Term	Definition, function, and examples
Subject	The subject is the noun (or noun form) in the sentence about which the predicate says something.
Predicate	The predicate comments on the subject. It always contains a verb but it may also include a complement and modifiers.
Complement	The complement, a part of the predicate, completes the sense of the verb. Its type depends on the nature of the verb.
direct object	A direct object completes a transitive verb. It answers the implied "What?" or "Who?" question of the transitive verb. Our supervisor created <u>our Web page.</u> ["Our Web page" answers, Created what?]
subject complement	A subject complement completes a linking or *to be* verb. It is either a noun or adjective. The Web page is <u>colorful.</u> ["Colorful" is an adjective that modifies "page."]
object complement	An object complement describes or identifies a direct object. The new chip makes our computer <u>obsolete.</u> ["Computer" is the direct object of "makes"; "obsolete" describes the computer.]
Modifier	Modifiers are words or phrases that describe or limit a noun, verb, or other modifier. Modifiers may be words (adjectives, adverbs) or modifying phrases (infinitive, prepositional, or appositive phrases).
Appositive	An appositive phrase renames a noun just named in the sentence.
Clause	A clause is a group of words that contains both a subject and a verb. Sentences in formal English require at least one clause.
independent	An independent clause can form a sentence. It is also called a main clause.
dependent (or subordinate)	A dependent clause contains a subject and verb *plus* a subordinating conjunction or relative pronoun. A dependent clause by itself is a sentence fragment. It must be joined to an independent clause.
restrictive	A restrictive clause begins with a relative pronoun (*who, which, whomever, that*). It is essential to the meaning of the sentence because it "restricts" the meaning of the noun to which it refers.
nonrestrictive	A nonrestrictive clause also begins with a relative pronoun. Although it provides meaning, it is not essential to defining the subject of the sentence.
Phrase	A phrase is a group of words that function together as one grammatical unit. A phrase does not contain both a subject and verb. Phrases can serve as subjects, predicates, and modifiers. Phrases are classified according to their headword.
noun phrase	The cute puppy
prepositional phrase	on the sidewalk
verb phrase	barked excitedly
infinitive phrase	to get attention.

TABLE 8.2   Parts of Speech

Term	Definition, function, and examples
Noun	A noun names a person, place, thing, or idea. *Roberto, Canada, machine, life,* and *Marxism* are all nouns. A noun functions in a sentence as subject, object of a verb or preposition, complement, appositive, or modifier.
Verb	A verb denotes action or state of being. A verb tells what the subject of the sentence does or what is done to it.
Adjective	An adjective modifies (describes or limits) a noun or pronoun. *Contaminated, careful, flexible, thorough, clean,* and *strong* are adjectives. Adjectives show comparisons with the suffixes *-er* and *-est* (*strong, stronger, strongest*). The articles *a, an,* and *the* are adjectives.
Adverb	An adverb modifies a verb, adjective, other adverb, or the entire sentence. Adverbs are frequently formed by adding the suffix *-ly* to an adjective. The surgeon stitched <u>carefully.</u> ["Carefully" modifies (describes) "stitched."] <u>Surprisingly,</u> the rates declined. ["Surprisingly" modifies the entire sentence.] The team played <u>very well.</u> [Both "very" and "well" are adverbs; "well" modifies "played," and "very" modifies "well."]
Pronoun	A pronoun substitutes for a noun. A relative pronoun makes a clause dependent.
personal	*I, you, he, she, they*
relative	*who, whoever, which, that* (relative pronouns make clauses dependent)
indefinite	*each, one, neither, either, someone*
Preposition	A preposition links its object (noun) to another word in the sentence. Common prepositions are *about, after, among, at, below, between, from, in, of, on, since,* and *with.* The word following the preposition is its object. The preposition plus its object and modifiers form a prepositional phrase. A prepositional phrase is a modifier, either adjectival or adverbial.
Conjunction	A conjunction joins words, phrases, or clauses.
coordinating conjunctions	Coordinating conjunctions join elements of equal value. *and, but, or, for, yet, nor, so*
subordinating conjunctions	Subordinating conjunctions make words that follow "subordinate" to the main clause. A subordinating conjunction makes a clause dependent. *after, although, as, because, if, once, since, that, though, till, unless, until, when, whenever, where, wherever, while* (etc.)
Conjunctive adverb	Conjunctive adverbs modify an entire clause; unlike subordinating conjunctions, they do not make the clause dependent. *consequently, however, moreover, besides, nevertheless, on the other hand, in fact, therefore, thus* (etc.)
Verbal	Verbals convert verbs into words that function as nouns or adjectives, but these words still retain some properties of verbs. Like verbs, they have implied subjects and may take objects or complements.
gerund	A gerund functions as a noun. It is formed by the addition of *-ing* to the verb. <u>Editing</u> on the computer saves time.
participle	A participle functions as an adjective. It is formed by the addition of *-ing* or *-ed* to the verb. In the example, "Planning" describes "she." <u>Planning</u> to edit on the computer, she established document templates before the writers began their work.

carries over to a direct object, but an intransitive verb does not. Linking verbs "link" the subject and predicate, and *to be* verbs identify the predicate of the sentence with the subject. The following paragraphs will explain these functions further, but for now you should understand that part of the function of the verb is to establish the relationship between the subject and the comment in the predicate. These relationships form the basis of some important grammar rules.

Transitive Verbs

Transitive verbs "carry over" to a direct object, one type of complement. The object answers the implicit question of the verb: "What?" or "Who?" The sentence would be incomplete without the answer. Most verbs are transitive.

> The <u>computer</u> <u>has increased</u> productivity.
>
> subject   transitive verb   direct object

The sentence would not make sense unless it answered the question, What has the computer increased?

Intransitive Verbs

An intransitive verb forms a complete predicate and does not require a complement. It does not carry over to a direct object.

> The hard drive <u>crashed</u>.
>
> The car <u>crashed</u> into the fence.

In neither case does the sentence raise the question "What?" or "Who?" Nothing more is necessary in the sentence to complete the meaning of the verb. The prepositional phrase, "into the fence," in the second sentence is not necessary to complete the meaning but rather is an adverbial phrase that answers the question "Where?"

Linking Verbs and To Be Verbs

Linking verbs (such as *look, seem, appear,* and *become*) and *to be* verbs (*is, are, was, were, am, have been*) require a complement that reflects back on the subject. Thus, the complement that follows a linking verb is a subject complement. This complement is either an adjective that describes the subject or a noun (or noun phrase) in the same class of items as the subject. The complement can substitute for the subject if it is in the same class of things. (All items in a "class" share some common feature, such as biological life. A class of living things might include people, animals, and plants.)

> The <u>equipment</u> <u>looks</u> imposing. ["Imposing" is an adjective that describes "equipment."]

The new <u>hard drive</u> <u>is</u> <u>reliable</u>. ["Reliable" is an adjective that describes "hard drive."]

The supervisor is an <u>expert</u> on computer graphics. ["Expert" is a noun that can stand in the place of "supervisor" because both words are in the category of "people at work."]

Dictionaries identify verbs in two classes, transitive and intransitive, based on whether the verb takes a direct object. Thus, dictionaries classify linking verbs as intransitive. The distinction between linking and intransitive verbs is that the linking verb is followed by a complement that says something about the subject (either an adjective or a noun) whereas the intransitive verb is followed, if at all, by an adverb (not something that reflects on the subject).

To summarize this discussion of verbs and sentence patterns, a subject and predicate form the two essential parts of the sentence. Unless the predicate includes an intransitive verb, the predicate also requires a complement to complete the meaning. The complement for a transitive verb is a direct object. Linking and *to be* verbs require a subject complement, either an adjective that describes the subject or a noun that can substitute for or be identified with the subject.

Verb Types and Their Complements

verb type	complement
transitive verb	direct object (answers What? or Who?)
intransitive verb	(none; the predicate may include an adverb)
linking/to be verb	subject complement (reflects back on the subject—noun or adjective)

Adjectives, Adverbs, and Modifying Phrases

The simple subject and simple predicate form the core of the sentence, but most sentences include modifiers as well. Modifiers affect meaning more than structure (their value is semantic more than syntactic). Modifiers "modify" the meaning of the subject, verb, or complement by describing it and thereby limiting the number of features that it can have. For example, in the phrase "feasibility report," *feasibility* is an adjective that establishes the subject as one particular type of report.

Modifiers may be adjectives and adverbs, but they may also be phrases. A phrase is a group of words that functions as a grammatical unit. The phrase may function as a noun (subject, object, or subject complement), verb, adjective, or adverb. For example, in the previous sentence, the prepositional phrase "as a noun" functions as an adverb modifying "functions." Phrases that function as adjectives or adverbs are modifying phrases. These phrases may be prepositional phrases, participial phrases, infinitive phrases, and appositive phrases.

Phrases are identified by their headword. For example, a prepositional phrase is introduced by a preposition, such as *by, from, to, for,* or *behind.* A noun phrase includes a noun and modifiers. An infinitive phrase includes the infinitive form of a verb, identified by *to*. A participial phrase includes a participle (a verb with an *-ing* or *-ed* ending that functions as a modifier). An appositive phrase renames the term just mentioned. Some common types of phrases and their functions in sentences are illustrated here:

Prepositional phrase	The package arrived *from New York City.*	adverbial: modifies "arrived"
Noun phrase	We obtained *the required papers.*	direct object of "obtained"
Infinitive phrase	We will try *to complete* the proposal tomorrow.	direct object of "will try"
Participial phrase	The technician *taking readings* discovered a faulty procedure yesterday.	adjectival: modifies "technician"
Appositive phrase	OKT3, *a monoclonal antibody,* helps to prevent a rejection crisis in kidney transplants.	adjectival: modifies "OKT3"

Prepositional phrases may be adjectival as well as adverbial, and noun phrases may be subjects as well as objects.

A phrase never contains both a subject and a verb. A group of words with both a subject and verb is a clause. You will read more about clauses in the next chapter.

Relationships of Words in Sentences

Grammar describes the accepted ways of using words in relationship to one another to form sentences. The aim of grammar is clarity. Although structuring sentences according to the rules will not ensure that the resulting sentences can be understood, breaking these rules will almost certainly hinder understanding.

In this section, the principles of relationship are defined. Common errors that violate these grammatical principles are summarized in Table 8.3 at the end of the section.

Subjects and Predicates

The most basic rule for forming sentences is that each sentence must contain a subject and a verb. A group of words that lacks both a subject and a verb is only a phrase. A phrase used to substitute for a sentence is a sentence fragment. Fragments are generally more difficult to understand than complete sentences because some essential information is missing. Subjects and verbs must agree with one another in number, and they must make sense together.

Subjects and Verbs Agree in Number

A sentence will be easier to understand if the subject of the sentence agrees in number with the verb. A singular subject takes a singular verb; a plural subject takes a plural verb.

- Singular subject, singular verb:

    The Twenty-First Amendment gave states the right to set minimum age requirements for purchasing alcohol.

    Adolescent drinking is sometimes an expression of rebellion against the parental authority figure.

- Plural subject, plural verb:

    Age limits on drinking encourage clandestine drinking by the young.

    Correlations between stress and alcoholism are high for all age groups.

An agreement error mixes a singular subject with a plural verb or vice versa.

The identity of words as singular and plural is not always obvious. Editors need to read closely, using the signals of sentence structure as well as knowledge of meanings of words.

- Collective nouns, such as *staff* and *committee*, may be regarded as either singular or plural, depending on whether you think of the group as a single entity or as a collection of individual members.

    The committee is prepared to give its report.

    All staff are responsible for completing their sick leave forms.

- The subject of the sentence, not the modifier closer to the verb, determines whether the verb is singular or plural. In the following sentence, it would be easy to read "fees" as the subject because it is next to the verb.

    An increase in parking fees was recommended. ["Increase" is the subject of the sentence.]

    NOT: An increase in parking fees were recommended.

- When the demonstrative pronoun *there* begins a sentence, the verb agrees in number with the noun that follows the verb.

    There have been thefts in this location.

    There has been talk of hiring an all-night guard.

Subjects and Predicates Must Make Sense Together

Subjects and predicates must work together logically as well as grammatically. When your ear and intuition tell you that the sentence doesn't make sense, consider whether the subject can do what the verb says or if the subject complement can be identified with the subject. If not, the error is faulty predication.

> These observations have concluded that 70 percent of new employees will be promoted within three years.

The sentence doesn't make sense because an observation cannot conclude; only a person can. The writer meant that the observations have shown or that, on the basis of the observations, researchers have concluded. In the following sentence, the same problem is evident.

> In some cases, a master's degree can fill the position.

A degree cannot fill a position; only a person can. The writer meant that in some cases a person with a master's degree can qualify for the position.

Faulty predication occurs when a human agent is required but missing from the subject. This statement does not mean that all subjects or agents of action must be human. Inanimate subjects can be agents of some actions.

> These studies have revealed a relationship between diet and disease.
>
> The report explains the marketing problems.
>
> The sentence makes sense.

Rules cannot explain when inanimate subjects can perform specific actions. One learns the conventions of a language by experience with it.

A Subject Complement Must Reflect on the Subject

Linking verbs and *to be* verbs link the subject of the sentence to a subject complement in the predicate that is identified with the subject. The subject complement must either modify the subject or be able to substitute for it.

Faulty predication occurs when the modifier or noun in the predicate cannot modify or substitute for the noun in the subject.

> The <u>application</u> of DDT <u>is</u> one of the best chemicals to eradicate the rootworms.

"Application" is the subject, and "one of the best chemicals" is the complement. However, "application" is not a chemical, so the complement is faulty. A simple revision eliminates the redundant "application."

> <u>DDT</u> <u>is</u> one of the best chemicals to eradicate the rootworms.

Now the complement can be identified with the subject because DDT is a chemical.

Another revision eliminates "chemicals." It also changes the structure of the sentence by using a transitive verb ("eradicate") and an object rather than a *to be* verb and a subject complement.

The application of DDT could eradicate the rootworms.

Faulty predication is about semantic relationships between the subject and predicate: between subject and transitive verb or between subject and complement with a linking verb.

Verb Tense and Sequence

Verbs communicate relationships of time. The *tense* of a verb refers to the time when the action takes place—present, past, or future. This information helps readers create a sense of chronology. The six tenses in English are present, past, future, present perfect, past perfect, and future perfect.

Tense	Examples	Use
Present	I edit	Indicates action that takes place now or usually
	I am editing	Indicates continuous action in the present
Past	I edited	Indicates an action that has been completed
	I was editing	Indicates continuous action in the past
Future	I shall edit	Places the action in the future.
Present perfect	I have edited	Indicates a recently completed action
Past perfect	I had edited	Places the action before another action in the past
Future perfect	I shall have edited	Indicates an action that will be complete before some other event in the future

The Verb Tense Must Accurately Indicate the Time of the Action

Use the present tense for action that occurs in the present or for timeless actions. Most descriptive writing is in present tense.

The page layout program includes functions for HTML markup.

The past tense is identified by the past participle of the verb (formed by adding *-ed* to the verb unless the verb is irregular). It can also be identified by the helping verb *was*. Use the past tense only for action that occurred at a specific time in the past.

The accident occurred last night.

BUT: Newton illustrates the concept of gravity with the example of an apple falling from a tree.

The future tense is identified by the helping verbs *will* or *shall*. Use the future tense for action that will take place in the future

When the new equipment arrives, we will convert to desktop publishing.

The "perfect" tenses (past, present, future) place action at a point of time in relationship to other action. In the sample sentence, both the editing and the revision occurred in the past, but the editing occurred before the revision.

I <u>had already edited</u> the chapter when the writer unexpectedly <u>sent</u> me a revision.

    past perfect tense   past tense

Shift Tenses Only for Good Reasons

Unnecessary shifts in tense disorient readers by requiring them to move mentally from the present to the past or future.

On the other hand, tenses may be mixed in a document when it describes actions that occur in different points in time. For example, a trip report may include statements in the past tense that recount specific events on the trip and also statements in the present tense that describe processes, equipment, or ongoing events.

In our meeting, President Schober told us that STC plans a new recruiting drive this fall. The organization presently attracts technical communicators who are well established in their jobs. The recruiting drive will target new technical communicators.

The paragraph includes verbs in the present, past, and future tenses, but the tenses all identify accurately when the action occurs. A confusing shift is an arbitrary shift among tenses.

Modifiers

Modifiers (adjectives, adverbs, and modifying phrases) are meaningless except in combination with the words they modify. Readers must be able to perceive the relationship of the modifier to what it modifies. Some uses of modifiers obscure their relationship to the words being modified. Some common misuses are called *misplaced* modifiers and *dangling* modifiers.

Misplaced Modifiers

A modifier, in English, generally precedes the noun it modifies. Readers would expect to read "tensile strength," for example, but would be startled and confused by "strength tensile." Adverbs generally follow the verbs or adverbs they modify.

If modifiers are placed too far from the term they modify, the sense of their meaning may be lost, and they may seem to modify something other than what is intended.

> Based on annual usage, 458 copies would have to be purchased by walk-in customers to reach the break-even point each day.

Because of its placement at the end of the sentence, the modifying phrase, "each day," seems to modify "to reach the break even point." However, logic suggests that the writer is not concerned about breaking even each day but about the number of copies to be purchased each day by walk-in customers. Readers can figure out the meaning, but they shouldn't have to reread to do so.

> Based on annual usage, 458 copies would have to be purchased each day by walk-in customers to reach the break-even point.

Only and *both* are frequently misplaced. The restaurant owner who posted the sign, "Only three tacos for $1.59," really meant to say "Three tacos for only $1.59." In the following two sentences, "both" serves first as an adjective modifying "cancer cells" and as an adverb modifying "to kill...and to carry." Placement of the modifier determines its function. The first sentence inadvertently claims that there are only two cancer cells.

> The monoclonal antibodies serve to kill both cancer cells and to carry information.

> The monoclonal antibodies serve both to kill cancer cells and to carry information.

Dangling Modifiers

A modifier, by definition, attaches to something. If the thing it is intended to modify does not appear in the sentence, the modifier is said to "dangle." Instead of attaching to what it should modify, it attaches to another noun, sometimes with ridiculous results.

> By taking a course and paying court fees, the charge for DWI is dismissed.

Only a person can take a course and pay a fee, but no person is in the sentence. The modifier, "By taking a course and paying court fees," dangles. The phrase modifies "charge," the subject of the sentence. The sentence says that the charge will take a course and pay a fee.

A modifier that dangles may be a verb form: a participle, gerund, or infinitive. (Participles, which function as adjectives, are formed by the *-ing* and *-ed* endings of verbs: for example, *studying, studied.* Gerunds, which function as nouns, are formed by the *-ing* ending. Infinitives are verbs plus *to:* for example, *to study.*) These verb forms, called verbals, function as adjectives, nouns, and adverbs, but they maintain some properties of verbs. They have implied subjects and may take objects. The implied subject of a verbal in the modifying phrase should be the same as the subject of the sentence that the phrase modifies. That is, the subject of the sentence should be able to perform the action implied in the verbal of the modifier.

Readers of the following examples, expecting that the subject of the verbal is the subject of the sentence, will be surprised to learn that a hose can change oil or a tax return can complete an application.

> <u>Changing</u> the oil, a worn radiator hose was discovered. [dangling modifier with a participle]
>
> <u>To complete</u> the application, the tax return must be attached. [dangling modifier with an infinitive]

Dangling modifiers often appear at the beginning of sentences, but shifting positions will not eliminate the problem because the implied subject of the verbal will still not be in the subject position of the sentence. Rearrangement can only disguise the problem.

> The charge for DWI is dismissed by taking a course and paying court fees. [The modifier still dangles.]

One way to correct a dangling modifier is to insert the implied subject of the verbal into the sentence.

> By taking a course and paying court fees, a DWI offender may have the charge dismissed.

An alternative is to rephrase the sentence altogether to eliminate the modifying phrase (that is, to replace the phrase with a clause that contains its own subject and verb).

> If a DWI offender takes a course and pays court fees, the court will dismiss a DWI charge.

One frequent cause of dangling modifiers is a modifying phrase in combination with the passive voice. *Passive voice* means that the subject of the sentence receives the action of the verb rather than performing the action. "Charge" in "The charge may be dismissed" receives the action of dismissal; the sentence is in passive voice. If the subject is receiving rather than performing action, it cannot perform the action implied by the verbal in the modifier. One way to reduce the possibility of dangling modifiers is to use the active voice rather than passive voice.

Pronouns

A pronoun substitutes for a noun that has already been named or is clear from the context. Pronouns relate to other words in the sentence by number, case, and reference.

Pronouns Agree in Number With Their Antecedents

The noun that precedes the pronoun is the antecedent, meaning that it "comes before" the pronoun. Pronouns may be singular (*he, she, it, him, her, one*) or plural (*they, their*). If the antecedent is *people,* the pronoun will be *they;* but if the antecedent is *person,* the pronoun will be *he, she,* or *he or she* (depending on whether you know the gender of the person). The pronouns with "one" in them, including *everyone, anyone, none,* and *no one,* are singular. So are pronouns with "one" implied, such as *each, either,* and *neither.* In formal usage, they should be used with other singular antecedents and verbs.

> Every<u>one</u> should complete <u>his or her</u> report by Monday. [singular possessive pronoun with singular antecedent]
>
> NOT: Everyone should complete <u>their</u> report by Monday. [plural possessive pronoun with singular antecedent]
>
> Neither of the options is satisfactory. [singular verb with singular subject, "neither"]
>
> None of the students was prepared.

Informally, however, the singular pronouns are interpreted as plural. *Their* is increasingly accepted when following a pronoun with "one" in it, especially as a way of avoiding pronouns of gender. Pronouns with "one" in them are interpreted as referring to groups of people rather than individuals. Unless you know your readers and co-workers approve of this usage, however, it is safer to be conservative—or to avoid the issue altogether by using plural subjects.

> All applicants should submit their transcripts.
>
> The options are not satisfactory.

Relative pronouns (*who, which, that*) used as subjects take the number of their antecedent.

> Applicants who send samples of their writing will have the best chance to be called for interviews.
>
> The editor who is my supervisor has helped me learn how to estimate time for various tasks.

*Choose Pronoun Case According to the Pronoun's Role
in the Sentence*

Pronouns may be subjects of clauses (*I, we, you, he, she, they, anyone, each*); objects of prepositions (*us, me, him, her, them*); or possessives (*my, mine, your, yours, their, theirs*). Some pronouns can serve as either subjects or objects (*it, every one, anyone, one*). Their *case* describes their role in the sentence, whether subject, object, or possessive. Objects and subjects cannot substitute for one another. You may have been chastised as a child for saying "Her and me are going out," using pronouns in the objective case for the subject function. The rule governing case requires "She and I are going out." Likewise, saying "The coach picked he and I" uses pronouns in the subjective case as objects. The transitive verb requires objects: "The coach picked him and me." If the pronoun is the subject of a sentence, use a pronoun that serves as a subject; if the pronoun is the object of a verb or preposition, use a pronoun that serves as an object. People seem especially self-conscious about using *I* and *me*. "Carlos and myself will conduct the study" errs because the personal pronoun in the subjective case is "I." The pronouns containing *-self* only appear when the noun or pronoun they refer to has already been named in the sentence. "Carlos made photocopies for me and himself." In the subjective case, *myself* is used only for emphasis: "I myself will clean up the mess."

Who and *whom* also create confusion. *Who* is a subject; *whom* is an object. If a person does what the verb says, use *who*. If the person receives the action of the verb, use *whom*.

> The person who trained the new computer operator was sent by the computer company.

> The person whom we trained can now perform all the necessary word processing functions.

The Pronoun Must Refer Clearly to the Noun It Represents

Pronouns with more than one possible antecedent may confuse readers.

> Cytomegalovirus (CMV) is a danger to those whose natural immunity has been impaired, such as persons with transplanted organs. Their immunity is deliberately weakened to prevent it from attacking the donor tissue.

Is "it" the virus or the immunity? Rereading will aid interpretation, but readers, more conditioned to thinking of viruses attacking tissue than immunity attacking tissue, may read "it" as "virus." The sentence demonstrates an error in pronoun reference.

Table 8.3 summarizes some of the more common grammatical errors that arise when the principles discussed in the preceding sections are violated.

TABLE 8.3   Common Grammatical Errors

Error	Definition
subject-verb agreement	The subject and verb do not agree in number. A singular subject is used with a plural verb, or a plural subject is used with a singular verb.
faulty predication	The predicate does not comment logically on the subject. Either the subject cannot complete the action that the verb defines, or the subject complement cannot be identified with the subject.
dangling modifier	The modifier (participle, gerund, or infinitive phrase) defines an action that the subject of the sentence cannot perform.
misplaced modifier	Misreading is likely because the modifier is separated from the item it is intended to modify. The solution is to move the modifier.
pronoun-antecedent agreement error	The pronoun does not agree in number with the antecedent.
ambiguous pronoun referent	It is unclear which of two previously named items in the sentence or a previous sentence the pronoun might refer to.
pronoun case error	A pronoun in the subject case is used in the object position or vice versa.
tense error	The tense does not accurately represent the time of the action.
tense sequence error	The times of actions as implied by the verb tenses contradict logic or the time shifts arbitrarily.

Conventions of Usage

Expressions in common use may technically violate grammar rules but still be acceptable. Likewise, some expressions in common use are not acceptable in professional writing. A few conventions that technical communicators use frequently are reviewed here: expression of quantities and amounts, use of relative pronouns, and idiomatic expressions.

Quantities and Amounts

The adjectives *fewer* and *less* and the nouns *number* and *amount* have specific uses and are not interchangeable. The use is determined by whether they refer to measurable quantities, such as pounds, dollars, or liters, or to indefinite amounts, such as weight, money, and milk. *Fewer* and *number* are standard usage when the reference is to measurable quantities.

> Fewer people attended the trade show this year than last, but the number of products sold was greater.

When the reference is to indefinite amounts, *less* and *amount* are standard.

The amount of gasoline consumed rose 50 percent.

This shipment of apples shows less contamination from pesticides than the shipment in April.

Relative Pronouns

Who refers to persons; *which* and *that* refer to things. It would be preferable to say "The people who came to the meeting brought tape recorders" rather than "The people that came…"

In formal usage, *which* and *that* are distinguished according to whether the clause they begin is restrictive or nonrestrictive. A restrictive clause limits the noun's meaning and is necessary to the sentence. A nonrestrictive clause provides additional information about the noun, but it does not restrict the meaning of the noun. Some usage experts specify the use of *that* with a restrictive clause and *which* with a nonrestrictive clause. (See also Chapter 9.)

restrictive clause use that

The supplies <u>that</u> we ordered yesterday are out of stock. [Only those supplies ordered yesterday, not all supplies, are out of stock.]

nonrestrictive clause use which

Murray Lake, <u>which</u> was dry only two months ago, has flooded. [Even without the information about the dryness of the lake, readers would know exactly which lake the sentence names.]

Idiomatic Expressions

Some expressions in English cannot be explained by rules alone. In fact, their usage may contradict logic or other conventions. These expressions are idiomatic. It would be difficult, for example, to explain why we say *by* accident but *on* purpose, or why we can be excited *about* a project, bored *with* it, or sick *of* it. The use of prepositions, in particular, is idiomatic in English. One simply learns the idioms of the language through reading and conversation. Idioms are especially difficult for writers for whom English is a second language. These writers may be perfectly competent in English grammar but have too little experience with the language to know all its idioms. A copyeditor quietly helps.

Guidelines for Editing for Grammar

No one memorizes all the rules of grammar that handbooks define. Neither do editors have time to consult a handbook each time they are uncertain about grammar. A systematic approach to editing for grammar will improve your efficiency and effectiveness.

1. **Edit top down,** from the largest structures in the sentence to the smallest—from clauses to phrases to words.

   a. Begin with the subject and predicate (the sentence core). Assess whether subjects and verbs agree in number and whether they work together logically. If the verb is a linking verb, assess whether the subject complement can be identified with the subject.

   b. Check phrases for accurate modification and parallelism. If the sentence contains participles or infinitive phrases, make sure that the subject of the sentence can perform the action the modifier implies.

   c. Check individual words—pronouns, modifiers, numbers.

2. **Increase your intuition about grammar by studying sentence structures.** Grammar rules describe the use of words and phrases within sentence structures. When the structures are familiar, grammar will not seem like hundreds of discrete rules, mostly focused on error, but rather like a coherent description of ways in which the parts of the sentence relate. Knowing the patterns will help you learn and remember the details.

3. **Edit for sentence structures, not just to apply rules.** Aim to clarify relationships among words in the sentence, not just to be "correct."

4. **Use a handbook for reference.** To use it efficiently, learn enough of the vocabulary of grammar to know what to look up.

Summary

Editing sentences according to the patterns that grammar describes will make the document easier to understand because the patterns will clarify relationships between words. Editing for grammar is a basic job for a copyeditor. Knowing how different types of words function in sentences will speed up editing, but editors also need the help of dictionaries and handbooks.

Further Reading

Charles T. Brusaw, Gerald J. Alred, and Walter E. Oliu. 1997. *Handbook for Technical Writing.* 5th ed. New York: St. Martin's.

Martha Kolln. 1996. *Rhetorical Grammar: Grammatical Choices, Rhetorical Effects.* 2nd ed. New York: Allyn & Bacon.

Also see handbooks of grammar and usage published by major publishing companies.

Discussion and Application

1. *Sentence patterns:*

   Describe the predicate in each of the following sentences. Identify whether the verb is transitive, intransitive, or linking. If necessary, consult a dictionary.

   Identify whether the remainder of the predicate consists of a direct object, a subject complement, or an adverb (or adverbial phrase). When you find subject complements, determine whether they are nouns or adjectives. Note that some verbs can be either transitive or intransitive.

   a. The Dow Jones Industrial Average fell by 53 points.
   b. I will lie down.
   c. She laid the newspaper down.
   d. A virus is a pathogen.
   e. The accident occurred yesterday.
   f. The program works well.
   g. The students worked the algebra problems.
   h. The marketing strategy seems successful.
   i. All members are present.

2. *Transitive and intransitive verbs*

   The sentences in a and b are both common, but which is "correct"?

   a. She graduated from college.
   b. She graduated college.

   Is the verb "graduated" transitive or intransitive? (Check your dictionary.) How does your answer define what should follow the verb? How does "from college" function in sentence a? How does "college" function in sentence b? How would an editor determine the better choice of sentences?

3. *Tense*

   a. For the sentences in exercise #1, identify whether the verb tense is present, past, or future.
   b. Edit this paragraph with particular attention to verb tense.

   We take great pleasure in welcoming you to our staff. We hope our relationship is one of mutual understanding and support. The owners had many years of experience in the operation of successful and profitable businesses. We were fortunate in the past with our choices for our staff, and we sincerely hope that you will follow in this path.

4. *Parts of speech*

   Identify the parts of speech of all the underlined words in the following sentences. If necessary, consult a dictionary or a handbook of grammar and usage. Note with regard to sentence 3 that a gerund (noun formed from a verb) can take an object just as a verb can.

   [1]Genomes are complete sets of genetic instructions for assembling and sustaining the life of an organism. [2]Not surprisingly, the human genome is over-

whelmingly complex. ³Biotechnology will soon be capable of mapping and sequencing the human genome. ⁴A genome map will show scientists precisely where genes are located on a chromosome. ⁵This map will provide a basis for predicting genetic diseases. ⁶This project will be expensive. ⁷It will cost millions of dollars each year.

5. *Functional sentence parts*

Identify the following sentence parts in each of the sentences in exercise 4: simple subject, complete subject, complete predicate, and modifiers.

6. *Subject and verb*

The following sentences contain errors in the verb or complement. For each sentence, first identify the simple subject and then the simple verb. (Remember that the grammatical subject may differ from the topic of the sentence.) If the verb is a linking or *to be* verb, identify the complement. Then identify and correct errors in subject-verb agreement or faulty predication.

a. The overall condition of the facilities are good to very good.
b. The resources dedicated to repair is minimal.
c. The record of all courses attempted and completed appear on the transcript.
d. A wide range of extracurricular activities are available to students.
e. Shipment of factory sealed cartons are made from our warehouse via the cheapest and fastest way.
f. The benefit of the annuity to the investor will be a source of additional retirement income.
g. The income from the annuity does not indicate that it will offset living expenses.
h. The agriculture industry is susceptible to pest problems.
i. Overloading on the library floor has moved approximately 20,000 volumes off campus to avoid structural damage to the building.
j. The interaction between the chemical mechanism and the dynamic mechanism appears to be the two important factors behind the depletion of the ozone layer.
k. The purpose of this section of the report is to increase the fatigue strength of an already welded joint.

7. *Dangling modifiers*

The following sentences contain dangling modifiers. Edit the sentences by inserting missing subjects into sentences or by converting modifiers to clauses. You may create two sentences from one if necessary for clarity. Which of the original sentences are written with passive voice verbs?

a. When preparing copy for the typesetter or when correcting errors on the screen, the cursor can be easily moved with the mouse.
b. Rather than make marks on the copy, the change can be placed in the computer for a faster and neater job.
c. The cost of production can be reduced by purchasing software and hardware for desktop publishing.

    d. A mosquito bit Lord Carnarvon on his left cheek five months after entering King Tut's tomb.

    e. Growing up to five feet long and weighing over 600 pounds, natives on the Moluceas Archipelago use the shells of the giant man-eating clam as children's bathtubs.

    f. By using lead-free gasoline, harmful lead oxides and lead chlorides and bromides are not released into the atmosphere as is the case with leaded (regular) gasoline.

8. *Misplaced modifiers*

The following sentences contain misplaced modifiers. Edit to show where the modifier should go. In your own words, explain the difference between a misplaced modifier and a dangling modifier.

    a. Only smoke in the break room.

    b. Racquets with safety thongs and bumpers are only allowed.

    c. Journalists must be able to operate equipment used to produce the stories such as computers.

9. *Pronouns: case, number, antecedent*

The following sentences include pronoun errors of various types. Identify the error and edit to correct the sentence.

    a. A positive attitude allows the waitperson to laugh at oneself and learn from their mistakes.

    b. If an investor wants to sell their shares of stock, they are sold at the market price at the time of sale.

    c. He and myself will conduct a workshop on investments.

10. *Usage*

    a. What questions of usage do the following sentences raise?

        (1) Use the express lane if you have less than ten items.

        (2) The project to edit the employee handbook could be divided between several students.

    b. Consult your dictionary and handbook to determine what advice they may offer about usage on the following two issues:

        (1) *Hopefully:* Is the adverb *hopefully* misused as a substitute for "it is hoped" as in the sentence "Hopefully, we will finish before Friday"? Define the grounds on which the use of *hopefully* could be considered a matter of usage rather than of grammar.

        (2) "The reason is because…": Check *The American Heritage Dictionary* under "because" for a usage note. Remembering sentence patterns with linking verbs and guidelines for subject complements, explain why the structure might be considered grammatically incorrect. Discuss how usage rather than grammar may determine whether a structure is acceptable.

$$Chapter\ 9$$

Punctuation

In the previous chapter, you learned that grammar describes relationships of words in sentences and that sentence patterns follow from the type of verb. This chapter continues the review of sentence patterns and explains how punctuation identifies the patterns. By signaling the structures in sentences, punctuation provides readers with clues to meaning.

A specific goal of the chapter is that you will understand these terms: *clause* (independent or main and dependent or subordinate), *phrase, conjunction* (coordinating and subordinating), *relative pronoun, restrictive and nonrestrictive modifier,* and *parallel structure.* As an editor, you will use these terms in making decisions about punctuation.

The chapter begins by explaining the value of punctuation in signaling meaning. It then illustrates sentence patterns and their punctuation, as well as the punctuation of sentence elements other than clauses. Finally, it reviews the mechanics of punctuation. If you learn the basic principles and terms in this chapter, read closely for meaning when you edit, and consult a handbook when you have questions, you should be a good editor for punctuation.

The Value of Punctuation

Editors and writers sometimes feel that rules of punctuation are annoying and even arbitrary. It is tempting to rely on the ear, on habit, and on intuition. Many writers have not used the language of grammar since elementary or junior high school days, and yet their own writing is reasonably correct from the standpoint of punctuation. Of what value are the rules?

Punctuation helps readers to read a sentence accurately. Internal marks—commas, semicolons, dashes—as well as end punctuation reveal sentence patterns and relationships. Incorrect punctuation may create document noise. Although readers may overlook the occasional error or patiently reread a sentence with an unclear structure, frequent errors will interfere with comprehension.

To editors, knowing where to punctuate and why contributes to accurate editing. Our ears are good but limited tools for editing. A novice may be satisfied to place commas where one might "breathe" or "pause" (recalling the advice of beginning instruction about grammar); yet such punctuation may send faulty signals about structure. A professional editor knows that punctuation is logical—not biological! Knowing to insert a comma before a coordinating conjunction because the sentence consists of two independent clauses is more precise than choosing to insert a comma at a good place to breathe or pause. You are a technical expert on punctuation just as the person whose work you edit may be a technical expert on computer hardware. Editing and speaking like the expert you are encourages others to respect you as an expert and encourages you to edit expertly.

Clauses, Conjunctions, and Relative Pronouns

Punctuation corresponds to sentence structure. In the previous chapter you reviewed the sentence patterns that follow from the use of transitive, intransitive, and linking verbs. You learned that each sentence contains a subject and predicate. These components create the core of a sentence, the *clause,* which always includes both a subject and a verb. Sentences may contain more than one clause. The correct joining of the clauses in sentences depends on whether the clauses are independent or dependent and on the joining words, called *conjunctions.*

This chapter continues the analysis of sentence patterns by considering ways in which clauses may be joined in single sentences. Specific punctuation marks show the different ways to join clauses. The terms that describe structural parts of the sentence, including *clause, conjunction,* and *relative pronoun,* are defined and illustrated in this section.

Independent and Dependent Clauses

A *clause* is a group of words containing a subject and a predicate. The subject is a noun or noun substitute; the predicate tells what is said about the subject and must contain a verb. The clause is the fundamental unit of a sentence—all sentences must contain a clause. (Exceptions to this pattern are acceptable in advertising copy and informal writing, but most technical writing follows conventional rules.) Sentences may contain two or more clauses.

Clauses may be independent or dependent. An *independent clause* (sometimes called the main clause) includes the subject and predicate and can stand alone as a sentence.

A *dependent clause* (sometimes called a subordinate clause) contains a subject and predicate, but by itself it is a sentence fragment. (It is dependent on or subordinate to an independent clause.) Like an independent clause, it contains a subject and predicate, but it also includes a subordinating conjunction or relative pronoun. The inclusion of a subordinating conjunction or relative pronoun makes the clause dependent on an independent clause. The following list summarizes the definitions and gives examples.

	structure	descriptor	example
clause	subject + predicate	an essential part of a sentence	
independent (main) clause	subject + predicate	may stand alone as a sentence	The <u>editor</u> <u>proofread</u> the text.
dependent clause	subordinating conjunction or relative pronoun + subject + predicate	by itself is a sentence fragment	Although the <u>editor</u> <u>proofread</u> the text,...

Subordinating conjunctions and relative pronouns are defined and discussed on the next few pages. For now, note in the example ("Although the editor proofread the text,...") that the dependent clause, because of the subordinating conjunction "although," makes readers expect that the thought will be completed elsewhere. This clause is dependent on the independent clause that presumably will follow.

When you are determining whether a group of words is a clause, make sure the verb is a verb and not another part of speech formed from a verb. Gerunds and participles, for example, are formed from verbs, but they function as nouns and modifiers. Thus, "The experiment being finished" is not a clause because it contains no verb. To turn this group of words into a clause, an editor could substitute a verb, *is* or *was*, for the participle "being."

The subject of a sentence in the imperative mood—phrased as a command—is understood to be "you." "Turn on the computer" is an independent clause because it means "[You] turn on the computer."

In order to punctuate sentences correctly, you must be able to distinguish a clause from a phrase. A phrase is a group of related words, but it does not contain a subject and predicate. Phrases function in a sentence as the subject or predicate, as modifiers, or as objects or complements. Because they do not contain both a subject and verb, however, they cannot be punctuated as independent clauses (or as sentences).

Conjunctions

When a sentence contains more than one clause, the clauses are often joined by a conjunction. As the root word *junction* suggests, the purpose of the conjunction is "to join." The type of conjunction, coordinating or subordinating, determines whether a clause is dependent or independent.

Conjunctions may be *coordinating*, joining independent clauses or other sentence elements (such as the terms of a compound subject) of equal grammatical value. Only seven words can function as coordinating conjunctions, all with three or fewer letters. Three of the words rhyme. You can memorize them.

| **coordinating**
conjunctions | *and, but, or, for, nor, yet, so* |

Conjunctions may also be *subordinating,* joining a dependent clause to an independent clause. They establish that one clause is less important than another, or subordinate to it. If a clause contains a subordinating conjunction, it is a dependent or subordinate clause. You probably won't memorize this list, but being familiar with it will help you make punctuation decisions accurately and quickly.

| **subordinating**
conjunctions | *after, although, as, because, if, once, since, that,*
though, till, unless, until, when, whenever, where,
wherever, while (etc.) |

Recognizing conjunctions and their type will help you determine whether the clause you are punctuating is independent or dependent.

Relative Pronouns

Relative pronouns relate to a noun already named in the sentence. They introduce dependent clauses (dependent because readers must know the noun already named in the sentence to understand the meaning of the clause). Like subordinating conjunctions, relative pronouns make clauses dependent.

| **relative**
pronouns | *that, what, which, who, whoever, whom,*
whomever, whose |

Clauses beginning with relative pronouns may be the subject of the sentence, or they may be modifiers.

Whoever comes first will get the best seat. [The clause is the subject of the sentence.]

The library book that Joe checked out is overdue. [The clause modifies "book."]

These groups of words introduced by relative pronouns are called clauses, not phrases, because they contain subjects and predicates (the pronoun is the subject). They are dependent clauses because the relative pronoun must relate to something in another clause. In the first example, "Whoever" is the subject of the main clause as well as the dependent clause.

To summarize, a clause contains a subject and verb, and each sentence must contain at least one. Subordinating conjunctions and relative pronouns make clauses dependent. A dependent clause must be attached to an independent clause to form a complete sentence.

Sentence Types and Punctuation

The number and type of clauses in a sentence determines its type and its internal punctuation. The four types of sentences are simple, compound, complex, and compound-complex. Each type of sentence uses particular punctuation marks to signal the clauses. The punctuation patterns are summarized in Table 9.1.

Sentence type	Structure	Example
simple	one independent clause (subject + predicate)	Rain fell.
compound	two independent clauses	Acid rain fell, and the lake was polluted.
complex	one independent clause plus one dependent clause	Because the factory violated emission standards, its owners were fined.
compound-complex	two independent clauses plus a dependent clause	The factory that violated emission standards was fined; therefore, the owners authorized changes in procedures and equipment.

TABLE 9.1   Punctuation Options for Sentence Types.   The shaded boxes with solid borders represent independent clauses. The clear boxes with dotted borders represent dependent clauses. Compound-complex sentences, not represented here, combine the patterns for compound and complex sentences. Note that the semicolon appears only in the compound sentence that lacks a coordinating conjunction.

sentence type	structural pattern	descriptor
simple sentence 1 independent clause	▭ .	one clause
compound sentence 2 independent clauses	▭ , and ▭ .	coordinating conjunction
	▭ ; however, ▭ .	no coordinating conjunction
complex sentence 1 independent, and 1 dependent clause	Although…, ▭ .	introductory subordinate clause
	▭ although… .	concluding subordinate clause
	▭ , which… .	nonrestrictive clause
	▭ that… .	restrictive clause
	Subject , which…, predicate .	internal nonrestrictive clause
	Subject that… predicate	

Punctuating Simple Sentences: Don't Separate the Subject and Verb with a Single Comma

Simple sentences contain one independent clause. The punctuation of simple sentences is end punctuation only unless there are introductory phrases or interrupting elements within the sentence. No single commas should separate the subject from the predicate because separation interferes with perception of the predicate as a comment on the subject.

Paradoxically, while a single comma separates, a pair of commas does not—the comma pair functions like parentheses. A pair of commas may enclose an interrupting modifier or an appositive phrase.

> WRONG: Some student editors with good grades in English, assume they know grammar well enough to edit without consulting a handbook. [the comma separates subject and verb.]

> CORRECT: Leigh Gambrell, a good student in English, edits effectively for grammar. [The pair of commas sets off the modifying phrase.]

Simple sentences may contain a compound verb or object. Nevertheless, they are still simple sentences (because they contain only one subject) and generally should be punctuated as such. If a comma is inserted before a second verb appears, it separates this verb from its subject.

Both of the following sentences use an unnecessary comma that separates the second verb from the subject:

> This draft prunes away some wordiness, and attempts to present the remaining information more efficiently.

> The writer prepared the draft of the text on the computer, but drew in the visuals by hand.

Perhaps the commas in the sentences appeared because the writer expected to breathe after the long beginning. Perhaps "and" or "but" seemed to begin a clause rather than the second part of a compound verb or object. However, it is illogical to separate parts of the sentence that belong together. The sentence type is simple; the comma should be deleted except when it adjusts rhythm or emphasis.

Punctuating Compound Sentences: Determine Whether There Is a Coordinating Conjunction

Compound sentences contain two independent clauses. Internal sentence punctuation (comma, semicolon, colon, or dash) reveals the structure of the sentence, signaling the end of one clause and the beginning of another. These punctuation marks cause readers to anticipate a new subject and a new verb.

There are two ways to join independent clauses in a compound sentence depending on whether they are joined by a coordinating conjunction:

1. Use a comma with a coordinating conjunction.
2. Use a semicolon (or sometimes a colon or dash) without a coordinating conjunction.

If the independent clauses are joined by a coordinating conjunction, separate them with a comma.

> Joe catches typos when he proofreads galley proofs, but he is less careful about word-division errors.

If the sentence contains no coordinating conjunction, however, insert a semicolon or colon between the two clauses. The semicolon separates statements relatively equal in importance.

> Good proofreaders check for omissions as well as for typos; they also note details of typeface, type style, and spacing.

> The dynamic mechanism theory explains sudden and drastic concentration changes of ozone; the chemical mechanism theory cannot account for these huge depletions.

The colon is appropriate if the first clause introduces the second, if the second explains the first, or if you could insert "namely" after the first clause. A colon creates a sense of expectation that some additional, qualifying information will follow.

> One fact remains clear: the field of editing lacks a consistent vocabulary.

> Comprehensive editing is rhetorical: it begins with an editor's understanding of the document's audience, purpose, and context of use.

The Conjunctive Adverb in the Compound Sentence

Clauses in compound sentences are often linked by the conjunctive adverbs *consequently, however, moreover, besides, nevertheless, on the other hand, in fact, therefore,* and *thus.* Conjunctive adverbs both join and modify. They are often mistaken for subordinating conjunctions and thus invite punctuation errors. *However* and *although* are easily confused because they both signal contrasting information, but they are different parts of speech and affect clauses differently. If you use only a comma between clauses in a compound sentence whose second clause begins with a conjunctive adverb, you will have an error called a comma fault. A semicolon must be used with a conjunctive adverb in a compound sentence, following the pattern of separating independent clauses with a semicolon unless they are joined by a coordinating conjunction.

> COMMA FAULT: Our organization transmits text to the printer electronically, therefore, we save the costs of rekeyboarding.

CORRECT PUNCTUATION: Our organization transmits text to the printer electronically; therefore, we save...

The first sentence in the following pair consists of two independent clauses (a compound sentence). Because the sentence has no coordinating conjunction, the clauses are separated with a semicolon. The second sentence in the pair consists of an independent plus a dependent clause introduced by the subordinating conjunction *although*. Because the sentence pattern is complex rather than compound, it requires no internal punctuation.

We have no leather recliners in stock; however, we do have vinyl ones.

We have no leather recliners in stock although we do have vinyl ones.

If *however* does not introduce a clause but is only an adverb, it is set off in the sentence with commas.

We have no leather recliners in stock.

We do, however, have vinyl ones.

The second sentence is a simple sentence (with one independent clause) and therefore requires no semicolon.

You will make the right decision about punctuating clauses within a sentence if you can identify which clauses are independent and which are dependent. Although you do not need to memorize the entire list of subordinating conjunctions, you may need to memorize the common conjunctive adverbs and recognize that they are *not* subordinating conjunctions.

Punctuating Complex Sentences

Complex sentences contain both an independent and a dependent clause. There may be no punctuation between the clauses, or they may be separated by a comma. Semicolons and colons should not be used to separate independent from dependent clauses.

An introductory clause with a subordinating conjunction is followed by a comma. A comma is not necessary if the dependent clause follows the independent clause.

Although we use a computerized proofreading program, we also depend on a human proofreader to correct errors that the computer cannot identify.

We depend on a human proofreader although we also use a computerized spelling checker.

We depend on a human proofreader because the computerized spelling checker does not comprehend meaning.

Dependent clauses formed by relative pronouns are modifiers. Whether a modifying clause should be separated from the main clause with a comma or not depends on whether the modifying clause is restrictive or nonrestrictive.

Restrictive Clause (Essential to the Definition)

A *restrictive clause* restricts the meaning of the term it modifies. The term may suggest a class of objects, but the restrictive modifier clarifies that the term in this sentence refers only to a subclass.

> Concrete that has been reinforced by polypropylene fibers is less brittle than unreinforced concrete.

In this sentence, the term "concrete" refers to all concrete. The clause "that has been reinforced by polypropylene fibers" clarifies that the sentence is not about all concrete but just the concrete that has been reinforced. The clause restricts the meaning of "concrete" as it is used in the sentence.

To illustrate visually, the circle below represents all concrete. The segment represents the subject of the sentence—the concrete that has been reinforced. The information in the clause has restricted the meaning of "concrete."

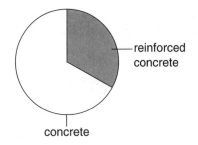

Because the restrictive modifier must be understood in conjunction with the term it modifies, it would be illogical to separate it from the term with a comma. If the clause is restrictive, do not use a comma.

Nonrestrictive Clause (Meaningful but Not Essential)

A nonrestrictive clause adds additional information, but it does not restrict the meaning of the term it modifies. Although removing the clause from the sentence would cost the sentence some meaning, a reader would still understand the meaning of the term. Because the term can be identified on its own, it does not need to be joined to its modifier. Use a comma to separate a nonrestrictive clause from the term it modifies. Use a pair of commas if the dependent clause falls within the sentence rather than at the beginning or end.

> Research in the reinforcement of concrete with polypropylene fibers led Shell Chemicals to develop their "caracrete" material, which essentially consists of polypropylene fibers and concrete.

> Caracrete, which consists of polypropylene fibers and concrete, can be used in...

Usually a proper name or specific term will be followed only by a nonrestrictive clause. "Caracrete" defines a material that is not further subclassified. The nonrestrictive clause that follows tells readers something about the material, but it does not define just one type of caracrete. On the basis of the information in the restrictive clause, we could not draw a circle representing "caracrete" and identify a pie slice that represented just one type of caracrete.

Formal usage requires that a restrictive clause begin with *that*, and a nonrestrictive clause with *which*. Because most writers do not know or do not apply that distinction, it is an unreliable clue about whether the clause is restrictive or nonrestrictive. Usually you can determine whether the clause is restrictive or nonrestrictive by close reading of the sentence on its own and in context. If you cannot, query the writer to clarify meaning.

When the relative pronoun in a restrictive or nonrestrictive clause refers to people rather than to objects, use the pronoun *who* rather than *that*. Never use *which* to refer to people.

> The proposed computer center would benefit walk-in customers, who would receive inexpensive and prompt service. [nonrestrictive clause, relative pronoun referring to people]

The table below summarizes the distinction between restrictive and nonrestrictive clauses.

	function	punctuation	preferred usage
restrictive clause	essential to the definition	no comma	that
nonrestrictive clause	meaningful but not essential	comma	which

Punctuating Compound-Complex Sentences

To punctuate sentences that contain two independent clauses and a dependent clause, you apply the rules both for punctuating compound sentences and for punctuating complex sentences. Begin with the biggest structures, the two independent clauses. If the clauses are joined by a coordinating conjunction, separate them with a comma preceding the conjunction; otherwise, use a semicolon (or possibly a colon or dash). Next look at the dependent clause. If the dependent clause introduces the independent clause, place a comma after it. If the dependent clause modifies the subject, determine whether it is restrictive or nonrestrictive, and punctuate accordingly.

Punctuating Phrases

Phrases often require punctuation within the sentence, usually commas. The commas mark the beginnings and endings of phrases and signal whether they are restrictive or nonrestrictive modifiers.

Series Comma and Semicolon

A series is a list within a sentence of three or more items such as terms, phrases, or clauses. Parallel structure means that items with related meanings share a common grammatical form. The items in the list may be a series of prepositional phrases ("by planning, by scheduling tasks, and by evaluating"), or nouns, adjectives, infinitive phrases, clauses, or other structural units that may appear in a series. The structural similarity helps to identify the items in the list and to establish that they are related.

The items in the series are generally separated by commas, with the conjunction *and* preceding the final item. The commas signal to readers where one item ends and another begins. In fact, conventions about whether to place a comma before the final item in the series have changed over time and differ, to some extent, among disciplines. However, handbooks usually recommend the comma. The comma is especially important for clarification when one item in the series includes a conjunction. A coordinating conjunction may signal that one of the items in the series has two components as well as signal the end of the series. The comma establishes with certainty that the series is ending.

Some readers may stumble over the following sentence.

> Good hygiene, such as sneezing into a tissue, washing hands before handling food and cleaning utensils and cutting boards thoroughly, eliminates most of the risk of transmitting bacteria in food handling.

On first reading, a reader may regard "food and cleaning utensils" as objects of "handling." Rereading and reasoning will confirm that the writer is not advising washing hands before handling cleaning utensils, but reading time has been wasted. One problem is that the series is built on *-ing* words ("sneezing," "washing," "cleaning"), but other *-ing* words ("handling," "transmitting") interfere with the recognition of their relationships. A comma before the final item could eliminate the need to reread.

> Good hygiene, such as sneezing into a tissue, washing hands before handling food, and cleaning utensils and cutting boards thoroughly, eliminates most of the risk of transmitting bacteria in food handling.

Journalism, unlike technical communication and educational publishing, advises writers not to use the comma before the final item in the series. If you are editing according to journalistic style, you will not use the comma. Appreciation of the possible confusion resulting from its absence, however, should encourage you to structure sentences to minimize the possibility of misreading.

When items in a series contain subdivisions separated by commas, semicolons may be used to separate the main items.

> The editing course covers sentence-level issues such as grammar and spelling; whole-document issues such as style, organization, and format; and design and production issues such as typography and printing.

When you check the punctuation of a structurally complex series, look at the main divisions first and then consider the smaller divisions.

Commas with a Series of Adjectives (Coordinate Adjectives)

If adjectives in a series both or all modify the noun, use a comma between them but not after the final one.

> The older, slower models will be replaced. [The models are both older and slower.]

Sometimes one adjective must be understood as part of the term modified. The phrase *theoretical research* identifies a type of research that differs from *empirical research*. The phrase *recent theoretical research* does not use a comma between the two adjectives, "recent" and "theoretical," because "recent" modifies the whole phrase "theoretical research" rather than "research" alone.

Parallelism

Parallelism means that related items share a grammatical structure. Faulty parallelism results when the structure of items in a series shifts. Faulty parallelism creates confusion for readers because the structural signals contradict the meaning of the words. Because of faulty parallelism, readers may not interpret the sentence correctly. In the following sentence, the shift from the noun form, "setting," to the verb form, "remove" and "install," reveals a shift from description to instruction, inappropriate in a list of steps.

> FAULTY PARALLELISM: The major steps in changing spark plugs are setting the gap on the new plugs, remove the spark plug wire, and install the new plugs.

The noun form is the correct one to follow the *to be* verb, because the information that follows such a verb either modifies or substitutes for the subject of the sentence.

> The major steps in changing spark plugs are setting the gap on the new plugs, removing the spark plug wire, and installing the new plugs.

What appears as faulty parallelism is often incorrect punctuation. You may recognize faulty parallelism first but edit the sentence by restructuring. For example, the punctuation in the following sentence signals three research tasks, but analysis reveals that the punctuation is misleading.

MISLEADING PUNCTUATION: In researching the career of an agricultural scientist, I consulted the *Occupational Outlook Handbook,* Standard and Poor's *Index of Corporations,* and interviewed a research associate at the Agricultural Experiment Station.

As you analyze the structure of sentences in order to punctuate, it may help to list key terms in outline form. The outline, which preserves the three-part series, reveals the faulty parallelism in the sentence.

consulted

Standard and Poor's

and interviewed

The first and third items are verbs, but the second is the name of a book. Thus, you might try a second outline.

consulted a and b

and interviewed...

The second outline reveals how the predicate includes a compound verb, not a series of three, with the first verb taking two objects. Two items are joined by a coordinating conjunction, without punctuation.

In researching the career of an agricultural scientist, I consulted the *Occupational Outlook Handbook* and Standard and Poor's *Index of Corporations* and interviewed a research associate at the Agricultural Experiment Station.

The punctuation identifies the structure of the sentence as a simple sentence with a compound verb. The verbs are joined by the coordinating conjunction "and," without punctuation. However, because of the complexity (compound object and compound verb), it will help readers if you create two independent clauses here by repeating the "I."

In researching the career of an agricultural scientist, I consulted the *Occupational Outlook Handbook* and Standard and Poor's *Index of Corporations,* and I interviewed a research associate at the Agricultural Experiment Station.

A series of three is often easier to understand than a double compound. If you can preserve the meaning accurately by converting a double compound to a series of three, choose the series of three.

original (faulty parallelism)	Thousands of openings will occur each year resulting from growth, experienced workers transferring, or retiring.
double compound	Thousands of openings will occur each year resulting from growth and from experienced workers transferring or retiring.
series of three	Thousands of openings will occur each year resulting from growth, transfers, and retirements.

When the items in a series are prepositional or infinitive or noun phrases, the word that introduces the phrase (for example, the preposition *to* or the article *the*) may or may not be repeated for each item. The choice depends on the complexity of the sentence. It may be sufficient to use the introductory word just once, in the first item.

> The accountant is responsible for preparing the balance sheet, analyzing the data, and reporting findings to the manager.

The series is short, and the three responsibilities are evident. In a more complex series, however, the repetition of the introductory word with each item will help readers recognize when they have arrived at subsequent items on the list.

> Damaged O-rings have been tested in order *to determine* the extent of failure in the material and the temperature at which failure occurred and *to evaluate* the effect of joint rotation.

Be consistent in your choice.

Introductory and Interrupting Phrases

Introductory phrases are usually followed by a comma unless they are quite short. The comma signals that one main part of the sentence, the subject, is about to begin. Phrases at the ends of sentences usually are not preceded by commas. Those phrases are part of the predicate rather than a separate division.

> In the appendix to our style guide, we have listed the correct spellings for specialized terms that are frequently used in our publications. [Introductory phrase followed by a comma]

Interrupting phrases (and words) are set off by a pair of commas or none, but not by a single comma. A single comma would interrupt another structural part. A pair of commas acts like parentheses. You "close" the commas just as you close parentheses.

Use a pair of commas for an interrupting adverb, for an appositive phrase, and for another modifying phrase that is nonrestrictive.

Natural fibers, such as cotton, silk, and wool, "breathe" better than fibers man-ufactured from oil byproducts. [A pair of commas sets off the prepositional phrase beginning with "such as."]

On May 6, 1996, we introduced our accounting software at the convention. [A pair of commas sets off the year.]

If the internal phrase, such as an appositive, is restrictive, don't use any commas.

Plants such as poison ivy may cause allergic dermatitis.

"Such as poison ivy" identifies a small category of plants. It restricts the meaning of "plants" to those that are like poison ivy. Other plants do not cause allergic dermatitis.

Punctuation Within Words

The punctuation discussed thus far reveals sentence structure. Punctuation within words also gives readers useful information. This category of punctuation includes the apostrophe and the hyphen.

The Apostrophe

Apostrophes show either possession or contraction. The apostrophe before a final *s* distinguishes a possessive noun or pronoun from a plural. For example, *engineer's report* and *bachelor's degree* show that the second term in some way belongs to the first. Whenever you could insert the conjunction *of* in a phrase, the noun or pro-noun is likely to be possessive and should have an apostrophe. Although we prob-ably wouldn't say "report of the engineer" or "degree of a bachelor," it would not be wrong to do so, and the "of" points to the need for an apostrophe.

Its and other pronouns (*his, hers, theirs, whose*) are exceptions to the use of an apostrophe to show possession. The word *it's* is the contraction for "it is"; *its* is the possessive, as in "its texture" or "its effect." *Who's* means "who is."

Many writers ignore the apostrophe or confuse them with plurals. You may have seen signs like the one in the grocery store announcing "banana's for sale" or the building directory identifying an office for "Veterans Adviser's." A visual mne-monic device to remember whether plurals or possessives require the apostrophe is this: for a *pos*sessive, use an *ap*ostrophe (except for pronouns).

Words can be both plural and possessive. In these cases, you form the plural be-fore you show the possessive. For nouns whose plural ends in *s*, the *s* will precede the apostrophe, as in "students' papers" (more than one student) or "pipes' corro-sion" (more than one pipe). Irregular plurals are an exception: "women's studies."

The plurals of years or abbreviations do not require an apostrophe (the *s* alone forms a plural):

the 1920s, the late 1800s

IQs, YMCAs

Finally, with a compound subject, use the possessive only on the second noun.

Mark and Steve's report

The Hyphen

The hyphen shows that two words function as a unit. Often new terms introduced into the language are formed from two familiar words. These new terms are compound nouns and compound verbs, such as *cross examination* and *cross-examine*. Compound terms may be hyphenated, "open" (that is, two words, unhyphenated, such as *type style*) or "solid" or "closed" (that is, a single word, such as *typesetting*). Hyphens are also frequent in modifiers when two terms join to modify another, but not all modifiers are hyphenated.

You can often determine which compounds are hyphenated by consulting a dictionary. Check the dictionary, for example, to determine whether to hyphenate "cross" compounds. You will find inconsistencies: *cross section* but *cross-reference* and *crosscurrent; crossbreed* but *cross-index; halftone* but *half note* and *half-truth*. Where the dictionary is incomplete, you will have to apply general principles, such as those discussed here and in more detail in a style guide.

Noun Forms

Noun forms are more likely to be open or solid than hyphenated. Noun phrases formed from two nouns are usually open or solid. The nouns *problem solving* and *decision making,* for example, are not hyphenated; however, the nouns *lovemaking* and *bookkeeping* are closed up as one word. Noun phrases formed from a noun and an adjective are more likely to be hyphenated or solid. Thus, the "self" compounds—for example, *self-service, self-study,* and *self-treatment*—are hyphenated; so are the "vice" compounds—*vice-president. Hydrochloric acid,* however, and many other compounds of noun plus adjective, are open. Fractional numbers, such as *one-half* and *two-thirds,* are hyphenated.

Adjective Forms

Compound adjectives are hyphenated if they precede the word modified and if they are formed in the ways listed below:

- Adjective or noun + past participle (if the compound term precedes the noun):

  country-smoked ham   BUT: The ham was country smoked.

  computer-assisted editing   BUT: The editing was computer assisted.

  dark-haired girl   BUT: The girl was dark haired.

- Noun + present participle:

  decision-making responsibility   BUT: decision making (noun)

  interest-bearing account

- Compounds with "all," "half," "high," or "low" (whether they precede or follow the noun):

  all-purpose facility, all-or-none reaction, all-out effort, all-around student

  half-raised house, half-blooded   BUT: halfhearted effort, halfway house

  high-energy particles, high-grade disks   BUT: high blood pressure

- Compounds with "well" (if they precede a noun; note that compounds with "well" are not hyphenated in a predicate):

  well-researched report   BUT: The report was well researched.

  Do not use a hyphen in compound adjectives in these situations:

- The suffix -*ly*:

  highly motivated person, recently developed program

- Two proper names:

  Latin American countries

- Two nouns:

  blood pressure level

Unit Modifiers with Numerals

When a numeral forms part of a compound adjective, it should be hyphenated when it precedes the noun it modifies. Thus, we would write "two-unit course" or "six-foot fence." The hyphen is particularly important if two numerals are involved.

  three 2-liter bottles

A predicate adjective, however, is not hyphenated.

  The fence is six feet tall.

Make sure that the phrase is a modifier and not simply a measure.

  a two-semester course   BUT: a course lasting two semesters

If you have a series of unit modifiers, repeat the hyphen after each numeral.

The foundation offers two- and three-year scholarships.

Finally, spelled-out fractions are hyphenated.

three-fourths full, one-half of the samples

Color Terms
If one color term modifies another, do not hyphenate.

bluish gray paper

If the two color terms are equal in importance, however, insert a hyphen.

blue-gray paper

Prefixes and Suffixes
Generally, prefixes and suffixes are treated as part of the word and spelled solid although some prefixes are hyphenated to prevent misreading. For example, the prefix *re-* is frequently hyphenated if the word that follows begins with an *e*—*re-enter, re-enact, re-examine, re-educate.* The dictionary will show both hyphenated and solid alternatives for some of these types of words. In that case, you should choose one form and place it on your style sheet so that you can be consistent in future choices.

When in Doubt, Leave It Out
If you judge that readers will read accurately without the hyphen, chances are you don't need one—except for unit modifiers with numerals. This principle also acknowledges a trend in language development away from hyphens. In the history of a word's development, the two familiar words that form a new compound may be hyphenated initially so that readers will see their relationship. When, over time, the compound becomes familiar, the hyphen isn't needed for this information and often disappears. This trend is one reason why *copyediting* is spelled solid in this book though you will see it spelled elsewhere both as *copy editing* and *copy-editing. Proofreading*, an older term, has become a permanent compound spelled solid though it began as *proof reading* and then became *proof-reading*. Increasingly *email* replaces *e-mail*. Likewise, the words *longterm* and *fulltime* (and even *shortterm* and *parttime*) are increasingly spelled solid. Within a document, you will always hyphenate consistently.

Marks of Punctuation

The discussion in this chapter so far has emphasized punctuation as a clue to meaning. As an editor, you also need to know the conventions of placing the marks of punctuation. This section reviews common uses of quotation marks, parentheses, dashes, colons, and ellipsis points.

Quotation Marks

In editing for a North American audience, place quotation marks outside a comma or period but inside a semicolon or colon.

> The prosecutor described the defendant as "defiant and uncooperative."

> The defendant was described by the prosecutor as "defiant and uncooperative"; however, the defendant's behavior changed once he took his place on the witness stand.

British usage is to place quotation marks inside the comma and period as well as inside the semicolon and colon.

Do not use quotation marks with a block quotation (one that is indented and set off from the rest of the text). In such a situation, they are redundant, as the block form identifies the material as a quotation.

If parentheses enclose an entire sentence, include the end-of-sentence punctuation within the parentheses. But if parentheses enclose only part of a sentence, place the period or other punctuation outside the parentheses.

> Twelve reports in the library deal with air pollution and acid rain in various ecosystems (e.g., the urban environment, streams, forests, and deserts). [The period is outside the parenthetical part of the sentence.]

> Twelve reports in the library deal with air pollution and acid rain in various ecosystems.

> (Examples of ecosystems are the urban environment, streams, forests, and deserts.) [The entire sentence is parenthetical; the period is inside the parentheses.]

Attach parenthetical cross-references and references to visuals to a sentence, or create a separate sentence for the cross-reference within parentheses. Do not let a parenthetical cross-reference "float" in a paragraph without sentence punctuation.

> The survey revealed that 40 percent of the employees favor a profit-sharing plan for retirement (see Table 1).

> OR: ...plan for retirement. (See Table 1.)

Dash

A dash—the length of two hyphens without space around them—can substitute for parentheses or can show a break in thought. In the sentence you just read, they function as parentheses. Commas could have been used instead, but the dashes are more emphatic. Dashes also signal additional information at the end of the sentence that helps a reader interpret the significance of the primary information in the sentence.

> Some state prison systems apply the policy of risk-group screening for AIDS only to pregnant women—a very small number of inmates.

Dashes have their proper and formal uses, but overuse will diminish their power to emphasize, and constant breaks in thought will make the writer seem immature or disorganized. There are no rules about how many are too many; you will have to trust your hunches and common sense.

Colon

The colon can introduce a list after a clause, but it is often used superfluously and therefore incorrectly. Do not use a colon between a preposition and its object nor between a verb and its complement or object. (See Chapter 8 and the glossary for definitions of these terms.)

> WRONG: Natural fibers, such as: cotton, silk, and wool, are comfortable next to the skin.

> WRONG: Three types of communication are: written, oral, and graphic.

> CORRECT: A complete technical communicator is competent at three types of communication: written, oral, and graphic.

In the sentences labeled "wrong," the colon separates the preposition from its object or the *to be* verb from its subject complement. The preposition requires the object, and the *to be* verb requires the complement. Punctuation should not separate necessary parts of sentences. In the sentence labeled "correct," the verb and complement are complete before the colon.

Ellipsis Points

Ellipsis points (three periods) indicate that some words have been omitted from a quotation. They are rarely used at the beginning or end of the quotation. If the ellipsis comes at the end of a sentence, insert a fourth period.

Typing Marks of Punctuation to Emulate Typesetting

Some typing habits linger from the days of typewriters, but word processing programs enable the creation of characters that emulate professional typesetting. For example, to create the em dash, you can press certain key combinations to create a solid line of em dash length rather than the broken line created by two hyphens. Likewise, you can use key combinations to create curly "smart quotes" rather than straight quotes, and ellipsis points that won't break at the ends of lines. Check the instructions for your software to find out how to create these characters if your company produces its own camera ready copy instead of using the services of a typesetter.

Although many people type two spaces after a period or colon (a remnant of typewriter days), professional typesetters use only one space.

Method of Editing for Punctuation

An effective method of editing for punctuation parallels the method of editing for grammar: a top-down method that begins with clauses, then phrases, and finally words. Edit for the largest structures first.

1. **Look for clauses and determine the sentence type:** simple, compound, complex, or compound-complex. Punctuate the clauses.
2. **Look for phrases:** introductory, internal, series. Identify restrictive and nonrestrictive phrases and check for parallel structure. Use punctuation to clarify the relationship of the phrases to the other words in the sentence.
3. **Look for punctuation in words:** hyphens, apostrophes.

Summary

Correct punctuation clarifies meaning because it reveals structures in sentences and relationships of ideas. To punctuate correctly, an editor must understand sentence patterns and the punctuation that identifies the different sentence types. Punctuation of phrases and words also clarifies meaning and supports efficient reading.

Further Reading

Charles T. Brusaw, Gerald J. Alred, and Walter E. Oliu. 1997. *Handbook for Technical Writing.* 5th ed. New York: St. Martin's.

Martha Kolln. 1996. *Rhetorical Grammar: Grammatical Choices, Rhetorical Effects.* 2nd ed. Boston: Allyn & Bacon.

Handbooks of grammar and usage from major publishing companies.

Discussion and Application

1. *Phrases and clauses*

   Determine which of the following groups of words are phrases, which are clauses, and which combine a phrase with a clause.

genetic instructions

have recently determined

researchers have determined the cause

on the bottom

in the event that the program crashes

2. *Sentence types*

Identify the sentence type—simple, compound, complex, or compound-complex—for each of the following sentences.

[1]Since 1974, thousands of papers have addressed the subject of ozone depletion. [2]The first hard evidence that proved a problem existed did not surface until 1985; in that year, Dr. Joe Farman of the British Antarctic Survey team reported finding a hole in the ozone directly over Antarctica. [3]In order to learn more about this Antarctic ozone phenomenon, scientists around the world have joined forces to create the National Ozone Expedition (NOZE). [4]Each year since 1986, these scientists have braved the cold of Antarctica in order to study the ozone depletion patterns that occur there every spring.

3. *Sentence types and punctuation*

Punctuate the sentences to clarify the structure of clauses and phrases within them. Articulate the reason for each mark of punctuation you insert or the reason for deleting a mark.

a. I had hoped to find a summer job in the city however two weeks of job hunting convinced me that it was impossible.

b. We came to work today in the rain; although we all preferred to stay in bed.

c. Most of our studies have focused on the response of cultured mammalian cells to toxic inorganics such as cadmium, these in vitro studies benefit from the ease and definition with which cultured cells may be manipulated and from the absence of complicating secondary interactions that occur in vivo. Perhaps the greatest advantage of working with cultured cells however is that it is often possible to derive populations that vary in their response to the agents in question. Because the mechanisms involved in cell damage or protection are often altered specifically in such cells; they provide tools invaluable to identification of these mechanisms and to a definition of their importance in the overall response of the cell

d. *This intentional fragment appears in a college catalog course description. Explain why the colon won't work, and suggest alternative punctuation.*
The study of contemporary techniques of music: modes, synthetic scales, serialism, vertical structures, with a term project.

e. Since computer skills, including page design have become essential to professional writers, college writing courses include a computer component.

f. Other substances inhibit digestion of insects, alter insect reproduction to make it less efficient or interfere with insect development.

4. *Restrictive and nonrestrictive modifiers*

In each of the following sentences, identify the dependent clause that begins with a relative pronoun. Determine whether the clause functions as a restrictive or nonrestrictive modifier, and punctuate it accordingly. Change "which" to "that" when the modifying clause is restrictive. When you cannot tell for sure, explain the difference in meaning if the modifier is restrictive rather than nonrestrictive.

   a. Adding polypropylene fibers to concrete increases its resistance to dynamic loading which is characterized by high strain rates that result from explosive impact or earthquake loading.

   b. The death of 3,000 white-tailed deer in 1962 resulted from heavy overpopulation and range abuse which led to malnutrition and its various side effects.

   c. Americans admire intelligence, which has practical aims, but intelligence which ponders, wonders, theorizes, and imagines is suspect.

   d. The photoconductive cells measure the amount of radiant energy and convert it to electrical energy which is then interpreted by the computer and displayed on the meter.

   e. Present lab equipment allows pulse energy experiments, which require 300,000 kW or less of electric power. New equipment would increase the potential of the lab.

   f. A high deer population which continuously feeds on the seedlings of a desired tree species can severely retard the propagation of that species.

   g. Pulse welding can join many metals which are impossible to join by conventional welding methods.

5. *Parallelism, series, and compounds*

Each of the following sentences contains faulty parallelism or a double compound punctuated as a series of three. The present punctuation and item structure do not accurately reveal the overall sentence structure. Punctuate and use parallel structure to clarify the relationship of the items. If you cannot determine the writer's meaning well enough to punctuate, write a query to the writer to elicit the necessary information.

   a. The rehabilitation center is raising funds to purchase Hydro-Therapy bathing equipment, wheelchairs, and to renovate the Hydro-Therapy swimming pool.

   b. The disease is typified by a delay in motor development, by self-destructive behavior, and it leads to death.

   c. The virus may cause AIDS, cancer, or even kill.

   d. The literature component of the professional writing major allows students to develop their sensitivity to language, texts, and their ability to read critically.

   e. This anthology is for psychologists, students of psychology, and of other related fields. The articles deal mostly with contemporary problems in psychology, minority, cultural, and other underrepresented groups.

   f. These instructions are written for automobile owners who do their own minor repairs, know the primary parts of an engine, and the use of tools.

6. *Parallel structure*

For each of the following, make sure all related items in a series have parallel structure.

   a. The old copiers produce copies of inconsistent quality, sometimes too light and at other times a black smudge of toner.

   b. Some qualities that are needed are the ability to communicate orally and written, and possess good judgment and tact.

   c. We have investigated ways of raising capital for building the Ronald McDonald House. We have spoken to banks about loans, to a foundation about getting grants, as well as fundraisers.

   d. Please send information on the following:
- the national office's program for low-interest loans
- the matching funds program
- any other helpful advice on funding

   e. The ten-year cost is $5,000 for renting and $3,864 if they were to buy.

   f. A teenager may reveal a drug habit by skipping classes, falling grades, and losing friends.

   Be careful, this one is tricky. Hints: A verb + "-ing" can form a participle (adjective) or gerund (noun substitute); "fall" is an intransitive verb, but "skip" and "lose" are transitive.

7. *Plurals and possessives*

Which of these phrases require an apostrophe to show possession?

   a. bears paw
   b. two years experience
   c. two years ago
   d. jobs requirements
   e. requirements for the job
   f. its colors
   g. joints rotation
   h. Steves Café
   i. Masters degree
   j. two months allowance
   k. policies cash value
   l. employees cafeteria

8. *Internal sentence punctuation*

Correct the punctuation in the following sentences, and explain why each change was necessary. Some sentences require more than one change. If the sentence is punctuated correctly, cite the principle that verifies its correctness.

   a. The copy center, like the check cashing service and the convenience store would be open 24 hours each day.

   b. Harris / 3M offers a 36 month leasing plan. For the proposed copiers for the center, the 6055 and the 6213. Lease costs would be $702 per month which includes maintenance.

   c. If the target goal of 36,000 copies per month billed at $.045 is met the equipment would pay for itself in 2.5 years sooner if the monthly allowance is exceeded.

   d. The fixed costs involved with this project; electricity, ventilation and floor space are not considered.

   e. There are two types of ultraviolet (UV) radiation; UV-A and UV-B. UV-A radiation which is frequently used for tanning beds, is lower in energy (longer in wavelength) than UV-B therefore it is considered safer than UV-B radiation. Some experimenters; however, believe that UV-A is just as damaging as UV-B; although higher doses of UV-A are required.

   f. Many theories about the depletion of ozone have been proposed but at present, two main hypotheses are widely accepted; the chemical mechanism and the dynamic mechanism.

   g. Chlorofluorocarbons (CFCs) such as the refrigerant freon, are very inert compounds, however, when CFCs drift into the upper atmosphere ultraviolet light will activate them.

9. *Compound terms and hyphenation*

Determine whether the compound terms and modifiers in the following sentences should be hyphenated, solid, or open.

   a. While studying day to day read outs from the Total Ozone Mapping Spectrometer, Dr. Wilson discovered that drastic depletions of ozone in a short period of time are not at all out of the ordinary.

   b. The manual includes step by step instructions.

   c. We took a multiple choice exam.

   d. The grant requires a semi-annual progress report.

   e. Type the sub-headings left justified.

   f. The left justified headings are not recognizable as level one headings.

   g. The pipe is two meters long.

   h. The valves in the two, three, and four meter pipes have corroded.

   i. Cross fertilization joins gametes from different individuals. The parents may be different varieties or species.

   j. This text book also serves as a reference book.

   k. The architect designed a multi purpose cafeteria for the school.

10. *Punctuation and meaning*

The following pairs of sentences differ only in punctuation. Explain the difference in meaning that results.

   a. A style sheet is a list of the general style—spelling, capitalization, abbreviation, hyphens—and unfamiliar or specialized terms.
   A style sheet is a list of the general style—spelling, capitalization, abbreviation, hyphens, and unfamiliar or specialized terms.

   b. When you make an entry on your style sheet, write the page number of the first occurrence (or every occurrence if you think you may change the style later). You will then be able to check your choice in its context.
   When you make an entry on your style sheet, write the page number of the first occurrence, or every occurrence, if you think you may change the style later. You will then be able to check your choice in its context.

11. Examine a sample of your own writing. Prepare some goals for yourself for structuring and punctuating sentences. Determine:

   a. sentence patterns: what types of sentences do you most use?

   b. punctuation: is your work punctuated correctly?

12. *Word play*

Tell why these terms are named as they are.

   a. Why is there a "junction" in a conjunction?

   b. Why are some pronouns "relative"?

   c. Why are some clauses "subordinate"?

   d. How is "punctuation" related to "punctuality"?

   e. How does a "period" relate to time?

   f. How is a pronoun "pro" the noun?

   g. What does a restrictive modifier "restrict"?

   h. How does a modifier "modify"?

   i. What is "semi" about a semicolon?

   j. How does a "dash" dash?

Chapter 10

Quantitative and Technical Material

When the material is mathematical, statistical, or technical, the copyeditor's job is the same as when the material consists solely of words. The object is to establish that the material is correct, consistent, accurate, and complete. In addition, the copyeditor prepares the text for production by marking for italics and capitalization, distinguishing letters from numbers, and indicating spacing and other unusual elements.

Because of the potential complexity of quantitative and technical material, which may contain many symbols and unusual type devices, the chances of errors occurring are greater than with paragraphs and sentences. As a copyeditor of quantitative and technical material, you do not have to understand fully the meaning of equations and statistics. However, you must be aware of the implications of changing symbols, capitalization, or punctuation, and of the need to mark clearly for typesetting. In addition to checking accuracy and form, copyeditors establish correct grammar and proper punctuation in mathematical and other technical documents.

This chapter introduces some principles for using numbers in math and statistics as well as basic standards of measurement and some scientific symbols. It also introduces guidelines for displaying and marking mathematical material in equations and in tables. The chapter will not make you an expert math and statistics editor, but if you learn the principles outlined here and check the applicable style manuals as you work, you can be competent in and confident about marking this type of text, and you will have a start on developing expertise. You can increase your value as an editor if you develop specialized ability to edit quantitative and technical material.

Using Numbers

The conventions for the treatment of numbers in technical texts differ from those in humanities texts, and to some extent, they differ among the technical disciplines.

The conventions concern questions of style such as whether to spell out a number or use a figure and whether to use the metric, British, or U.S. system of measurement. A style guide will provide specific guidelines within a discipline, but some generalizations about using numbers in technical documents should be noted.

1. **Use figures for all quantifiable units of measure,** no matter how small, rather than spelling out the numbers. Also, use a figure whenever you abbreviate the measure.

2 m	1 ½ in	0.3 cm
12 hours	$1 million	18 liters

   Figures aid readers in comprehending and in calculating. In documents that are not scientific or technical, however, spelling out both the number and the measure is common if the number is lower than 10 or possibly lower than 100, depending on the style guide.

2. **Do not begin a sentence with a figure.** Either rearrange the sentence to avoid the figure at the beginning or spell out the number.

3. **Given a choice, use the metric system of measurement** rather than the British or U.S. system. Don't mix systems. For example, do not describe something in both inches and centimeters. Readers may be confused or misled by the dual usage and may misjudge the measures.

4. **Set decimal fractions of less than 1.0 with an initial zero** (0.25, 0.4). An exception is a quantity that never equals 1.0, such as a probability (p) or a correlation coefficient (r).

   Check a comprehensive or discipline style manual for specific directions on the treatment of numbers in dates, money, and time.

5. **Convert treatment of numbers in a translation** according to usage in the country where the document will be used. There are cultural variations for dates, time, and money. See Chapter 15.

Measurement

Three systems of measurement are widely used: the U.S. Customary System, the British Imperial System, and the International (metric) System. In the U.S. and British systems, the standard measures are the yard and the pound, but these systems vary in some measures and in the expression of them. For example, a billion in the British system equals a million million, whereas in the U.S. system it equals a thousand million. The metric system is used for most scientific and technical work.

The International System of Units, an extension of the metric system, is a standard system of units for all physical measurements. Its units are called SI units (for *Système International,* in French). The International System has seven fundamental units, listed here.

Quantity	Unit	Abbreviation
length	meter	m
mass	kilogram	kg
time	second	s
electric current	ampere	A
temperature	kelvin	K
luminous intensity	candela	cd
amount of substance	mole	mol

Other measures, based on these fundamental units, are derived units. That is, they are multiples or parts of the fundamental units. Because the metric system is a decimal system, all the derived units are multiples of 10. The prefix indicates the multiple: a kilometer equals 1,000 meters, and a centimeter equals 0.01 meter. Use only numbers between 0.1 and 1,000 in expressing the quantity of any SI unit; use the derived units for smaller or larger numbers. For example, 10,000 m equals 10 km. SI units established for other physical quantities are used primarily in science and engineering. The following list presents a few of these units. The SI units and symbols pertinent to a given field are likely to be listed in that field's handbook.

Quantity	Unit	Symbol or Abbreviation
acceleration	meter per second squared	m/s^2
electric resistance	ohm	Ω
frequency	hertz	Hz
power	watt	W
pressure	newton per square meter	N/m^2
velocity	meter per second	m/s

As copyeditor, you ensure that the symbols used for these measurements are correct and that they are used consistently. Whereas the units are lowercase, even when they represent names, some abbreviations include capital letters, especially if they represent names. For example, *watt* and *newton* as are lowercased as units but capitalized as abbreviations. The units and abbreviations are set in roman type and without periods.

The abbreviations are never made into plurals. Thus, six watts would be written as "6 W," not "6 Ws." However, if the measures are written in prose, the words are formed into plurals according to the same rules that govern other words. Thus, "6 watts" is correct in a sentence. Fractions of units are always expressed in the singular whether they are expressed in symbols or in prose: 0.3 m and 0.3 meter.

A hyphen is used between the measure and the unit when they form a modifier but not when they simply define a measure. Thus, both "The trial lasted 10 seconds" and "the 10-second trial" are correct.

Marking Mathematical Material

Documents that include fractions, equations, or other mathematical expressions require close copyediting attention. To save space, stacked fractions have to be converted for inline presentation. Equations may have to be displayed, numbered, or broken. In addition, copymarking must clarify characters that could be confusing (for example, the number *1* and the letter *l*, the number *0* and the capital letter *O*, the unknown quantity *x* and the sign for multiplication) and indicate the position of subscripts and superscripts on the page, as well as noting italics and capitalization. Because mathematical material is expensive to copyedit, typeset, and print, good copyediting can save expense later in production. Fortunately, the computer has eased the copyeditor's job. Writers are likely to use programs that generate equations and other mathematical material.

Fractions

Fractions may be set "stacked" or "solid" (inline). The inline version substitutes a slanted line, or *solidus*, for the horizontal line of the stacked fraction.

Stacked	**Solid (Inline)**
$\dfrac{1 + (x - 3)}{y + 2}$	$[1 + (x - 3)]/(y + 2)$

The stacked version is preferable for comprehension because it conveys visually the relationship of the numbers. However, it takes more space in typesetting. If a fraction appears in a sentence, it may have to be converted to inline form in order to avoid awkward line breaks and extra space between the lines. Complexity in fractions, however, will justify the extra space for stacking.

When fractions are converted to inline form, parentheses may be necessary to clarify not only the numerator and denominator but also the order of operations, which will affect the result obtained. (Operations within parentheses are completed before other operations.) In the previous example, the denominator is the quantity (or sum) of $y + 2$. Thus, $y + 2$ will be added before the sum is used as a divisor; otherwise, the numerator of the fraction would be divided by y and then 2 would be added to the total. The entire numerator is enclosed in brackets to establish that those operations must be performed before division. Both the square brackets and the parentheses, as well as braces, are signs of aggregation or fences. The preferred order for these signs is parentheses, square brackets, and braces:

$$\{[()]\}$$

Fractions containing square root signs can be set inline if the sign is converted to the exponent ½. The square root sign may also be set without the top bar, thereby allowing it to fit within a normal line of type. All three of the following expressions say the same thing.

$$\frac{a+b}{\sqrt{\dfrac{2a-12}{6}}} \qquad\qquad (a+b)/[(2a-12)/6]^{1/2} \qquad\qquad \frac{a+b}{\sqrt{[(2a-12)/6]}}$$

Mark a stacked fraction for inline presentation by creating a "line break" mark using the line in the fraction.

$$\frac{1}{2} \qquad \text{marked} \qquad \frac{1/}{2} \qquad \text{becomes} \qquad 1/2$$

Equations

Equations are statements that one group of figures and operations equals another. Equations can contain known quantities (expressed as numbers), variables (often expressed as *a, b,* and *c*), and unknowns (often expressed as *x, y,* and *z*). Equations always include an equal sign.

Displaying and Numbering Equations

Equations may be set inline, or they may be displayed (set on a separate line). If multiple equations in a text are displayed, they should be either centered or indented a standard amount from the left margin.

Displayed equations may be numbered if they will be referred to again in the document—the numbers provide convenient cross-references. If an equation is to be numbered, it must be displayed, but not all displayed equations are numbered. Generally, the equation number appears in parentheses to the right of the equation.

Equations are numbered sequentially either through the work or through a division of it. A system of double numeration (chapter number followed by a period and the equation number) can save time and labor if an equation number must be changed (perhaps because one equation was deleted). With double numeration, only the equation numbers in that chapter must be changed. The possibility that equation numbers may change during copyediting and revision is also a good reason for numbering only the equations that must be referred to, not all those that are displayed.

Let there begin the value of the $y_0, y_1, \ldots, Y_n$ of the function $y = f(x)$ at the $(n + 1)$ points $x_0, x_1, \ldots, x_n$. A unique polynomial $P(x)$ whose degree does not exceed n is given by the equation

$$P(x) = a_n x^n + a_{n-1} x^{n-1} + \ldots + a_0, \tag{2.2}$$

for which $Pn(x)$ at x_i must be satisfied by

$$P_n(x_i) = f(x_i) = y_i, \ i = 0,1,\ldots, n. \tag{2.3}$$

The conditions in (2.3) lead to the system of $n + 1$ linear equations in the a_1:

$$a_0 + a_1 x_i + \ldots + a_n x^n = y_i j = 0,1,\ldots, n. \tag{2.4}$$

Equation (2.3) is referred to again by its number. The parentheses identify the figures as an equation number. Some associations representing disciplines, such as the American Psychological Association, prefer that the text read "Equation 2.3" rather than (2.3).

Breaking Equations

Equations that are too long to fit on one line should be broken before an operational sign (+, −, ×, ÷) or a sign indicating relations (=, ≤, ≥, <, >). However, you should never break terms in parentheses. Equation (2.4) in the preceding example is broken before an equal sign. Use the line break mark to indicate a break in the equation.

$$a_0 + a_1 x^i + \ldots + a_n x^n \Big/ = y_i^i = 0,1,\ldots, n.$$

In a document with a series of related equations, the equations should be aligned on the equal sign, regardless of the amount of material to the right and left of the equal sign.

The mathematical formula for computation of the mean is

$$\bar{X} = \frac{\Sigma X}{N}$$

where

$$\bar{X} = \text{mean of the scores}$$

$$\Sigma X = \text{sum of the scores, and}$$

$$N = \text{number of scores.}$$

Punctuating Equations

Equations are read as sentences when they appear in prose paragraphs, with the operational signs taking the place of verbs, conjunctions, and adjectives. Thus, $a + b \geq c$, where $b = 2$, would be read aloud as a plus b is greater than or equal to c, where b equals 2.

Practice in punctuating equations varies. Some publishers omit punctuation such as commas and periods on the grounds that the marks are potentially confusing if read as part of the equation rather than as punctuation. They also argue that when the equation is displayed, the space around it signals its beginning and end, making punctuation redundant. Other publishers, however, follow the same rules for punctuating equations as for punctuating sentences. Displayed equations are punctuated as though they appear in sentences. Thus, they may be followed by commas or periods as the structure of the sentence requires. In the following examples, no colon appears after "by" in the first equation, but a colon appears after "as follows" in the second equation. The words that introduce the displayed equation are not followed by a colon unless they would be followed by a colon in the same circumstances if all the text were prose. The displayed equations do not have end punctuation; the space takes the place of commas and periods.

Let the polynomial be given by

$$P(x) = a_0, + a_1x + a_2x2 + \ldots + a_nx_n$$

where the coefficients a_1 and a_2 are to be determined.

If the center of a circle is at the origin and the radius is r, the formula can be reduced as follows:

$$x^2 + y^2 = r^2$$

Grammar and Punctuation

Mathematical copy, like prose copy, should follow accepted rules of grammar: subjects and verbs must agree, nouns take articles, and clauses contain subjects and verbs. You do not need to understand the following equation to edit the subject-verb agreement error, to place a comma before the nonrestrictive clause, and to conclude the sentence with a period.

There exists a unique polynomial $P(x)$ whose degree do not exceeds n which is given by $P(x) = a_nx^n + a_{n-1}x^{n-1} + \ldots + a_0$

A common error in mathematical copy is the dangling participle. Dangling participles occur in mathematical text for the same reason they occur in prose: the subject to be modified is absent from the sentence, often because the passive voice omits the agent. Revisions must insert the agent or delete the modifier.

Constructing interpolation polynomials, the inverse can be explicitly calculated.

By solving $x = 12 - y$, a contradiction can be obtained.

Solutions:

The inverse can be explicitly calculated with the construction of interpolation polynomials.

Solving $x = 12 - y$, we obtain a contradiction.

Solving $x = 12 - y$ yields a contradiction.

Copymarking for Typesetting

Copymarking clarifies type style (italic, roman, or bold), spacing, subscripts and superscripts, and ambiguous characters. Unknowns and variables in expressions and equations are set as lowercase italic letters. These are usually letters in the

English alphabet, but they may be Greek letters. If they are not typed in italics on the hard copy, they should be marked with an underline to denote italics. A slash through a letter indicates lowercase when the letter is typed as a capital. Numbers, symbols, and signs are set in roman type. Vectors are set in bold. (Vectors are quantities with direction as well as magnitude. They are sometimes set with an arrow over them.)

Operational signs ($+$, $-$, $\times$, $\div$) and signs of relation ($=$, $\leq$, $\geq$, $<$, $>$) are set with space on either side. The sign for multiplication is roman, not italic as for the unknown x.

Copymarking should clarify the intent when characters are ambiguous. For example, a short horizontal line may be a minus sign, hyphen, en dash, or em dash. It could even indicate polarity (a negative charge). Some, but not all, possible ambiguities are listed here.

0	zero	β	Greek beta
O	capital "oh"	B	capital "bee"
o	lowercase "oh"		
°	degree sign	e	exponent
1	number one	e	charge of an electron
l	lowercase "el"	Σ	summation
$\times$	multiplication sign	$\in$	element of
x	unknown quantity	m	abbreviation for meter
χ	Greek chi	m	unknown quantity

If the text contains just a few of these ambiguities, they may be marked as in any other copymarking. A clarifying statement may be circled next to the character. For example, you might write and circle "el" next to the letter that could be misread as the number one. Mark unknown quantities as italic if they are not typed in italic.

Even when the typist has been careful to use symbols, italics, and spacing correctly, marks can clarify the intent for the compositor and leave no doubt as to how the elements are to be set. For example, you might mark the superscripts and subscripts, even if they are typed correctly, to confirm that the positions should be retained in typesetting, and you might write and circle "minus sign" next to a minus sign so that it will not be mistaken for an en dash. The extent of your marking for clarification will depend, in part, on your company's practice and on your compositor's expectations, but it is better to err on the side of over clarification than to leave interpretation of elements up to the compositor and spend time—and money—correcting errors after typesetting. Check with the writer if you cannot tell what characters were intended.

Statistics

Statistics is a system of procedures for interpreting numerical data. The science of statistics governs the collection of data (as by sampling), the organization of data, and the analysis of data by mathematical formulas. Using statistics, researchers learn the significance of "raw" data, or figures that have not been analyzed. Researchers use the numbers to make predictions. Statistics is widely used in any empirical research, or research that relies on observation and experiment. Thus, it is used in the natural, physical, and social sciences and in engineering. You may copyedit material based on statistical analysis if you work in any of these fields and if you edit research proposals and reports. The results of product testing or field testing of documentation may also be analyzed statistically.

Two usage notes: the word *statistics* is singular if it refers to the system of interpreting numerical data; the word is plural if it refers to numbers that are collected. "Statistics is a science" is the correct usage because the reference is to the science. "The statistics are misleading" is correct as a reference to specific numbers. Also, the sciences (though not engineering) regard the word *data* as plural; thus, "data are" is the correct usage in science, in spite of what your ear may tell you.

If you work extensively with statistical material, you should learn more about it than this chapter will tell you. But this chapter will introduce some of the more common terms and symbols to enable you to copyedit.

Letters used as statistical symbols are italicized, whether they appear in prose or in tables, unless they are Greek letters. If they are not italicized or underlined in the typescript, they will have to be marked. Some common symbols and abbreviations and their meanings follow.

ANOVA	analysis of variance
r	correlation
df	degrees of freedom
F	F-ratio
μ (Greek letter mu)	mean
n	number of subjects
N	number of test results
p or ϕ (Greek letter phi)	probability
SD	standard deviation
$S; Ss$	subject; subjects
t-test	test of differences between two means

Equations are placed on the page according to the same rules that govern mathematical material. Thus, related equations are aligned on the equal sign, and operational or relational signs are set with space on either side (for example, $\mu = 92.55$, not $\mu=92.55$). Place a zero before a decimal in a quantity of less than 1, except for correlation coefficients (r) and probabilities (p), which are always less than 1. Form plurals of abbreviations by adding *s* only, with no apostrophe (Ss, not S's; IQs, not IQ's).

One other note: capitalize *experiment* or *trial* when these words refer specific tests. Thus, "The results of Experiment I revealed that…" but "the experiment."

Tables

Tables represent an efficient way to present quantitative data, enabling readers to refer quickly to specific numerical (or verbal) information. Tables often require close scrutiny by the copyeditor for correctness, consistency, accuracy, and completeness. A copyeditor should be alert to misspellings and other errors; inconsistencies in capitalization, spelling, and abbreviation; clarity of identifying information, such as whether the numbers represent percents or totals and whether the measures are meters or feet; correctness of arithmetic, such as whether totals are correct and whether percentages total 100. Apparent inconsistencies in numbers or other data may signal inaccuracy and should be checked. In addition, the copyeditor should check readability of the table, such as the amount of space between rows and columns and variations in type style (bold, italic, capitalization) to indicate different levels of headings. A reference to the table in the text should precede the table. Tables should be numbered sequentially in the chapter, and each table number should match the number used in cross-references. The title should accurately reflect the contents.

General Guidelines

Tables should be constructed to enable accurate reading and comprehension. You can find guidelines to aid you in copyediting tables in textbooks on technical writing and in some handbooks and style manuals. The following points briefly summarize these guidelines.

- Illustrations that are tabular are called tables; other illustrations, such as line drawings, graphs, and photographs, are called figures. The information in the rows and columns of tables may be quantitative, verbal, or even pictorial. The table title and number are identified at the top of the table, although informal tables that lack titles and numbers are also permissible. Some disciplines specify roman numbers for table numbers.
- Items compared in a table are listed down the stub (left) column, while points of comparison are listed across the top. This arrangement allows easy comparison of related items.
- Vertical and horizontal lines (rules) separating rows and columns are discouraged except for highly complex tables because the lines clutter the table with visual noise. White space is the preferred method for separating rows and columns. Too much space between columns, however, may result in inaccurate reading because the eye may skip to a lower or higher row. Tables do not need to be spaced to fill the same width as the text. Figure 10.1 depicts a table that requires vertical lines to distinguish different levels of information.
- To increase accuracy of reading across the rows of long tables, a space may be used between groups of five or so rows, or alternate groups of five rows may be lightly shaded.
- Columns of numbers should be aligned on the decimal.

No. 379. Toxic Release Inventory by Industry and Source: 1989 to 1993

[In millions of pounds. Based on reports from almost 23,000 manufacturing facilities which have 10 or more full-time employees and meet established thresholds for manufacturing, processing, or otherwise using the list of more than 300 chemicals covered. Only chemicals that were reportable in all years shown are compared so that data do not reflect any chemicals added or deleted from the list covered. The inventory was established under the Emergency Planning and Community Right-to-Know Act of 1986 (EPCRA)]

INDUSTRY	1987 SIC[1] Code	1989	1990	1991	1992	1993 Total[2]	1993 Air[3], point	1993 Air[4], non-point	1993 Water
Total	(X)	4,405.2	3,719.7	3,393.5	3,190.4	2,791.4	1,175.1	480.2	271.1
Food and kindred products	20	37.2	39.2	39.8	38.7	38.0	15.7	11.5	1.4
Tobacco Products	21	1.8	2.5	2.3	2.0	2.4	2.1	0.2	-
Textile mill products	22	32.2	27.2	25.2	21.9	20.4	15.3	4.8	0.3
Apparel and other textile prod.	23	1.4	1.3	1.4	1.6	1.1	1.0	0.2	-
Lumber and wood products	24	38.0	35.8	32.6	31.0	29.5	24.6	4.7	0.1
Furniture and fixtures	25	65.5	62.6	56.6	56.7	58.1	50.8	7.0	-
Paper and allied products	26	262.8	254.5	246.4	234.1	216.1	171.9	21.1	18.1
Printing and publishing	27	58.7	56.0	47.2	41.1	36.5	17.2	19.3	0.0
Chemical and allied products	28	2,093.3	1,629.5	1,546.9	1,543.0	1,308.4	320.9	148.1	234.1
Petroleum and coal products	29	99.2	86.9	79.9	84.2	74.5	21.8	35.3	3.3
Rubber and misc. plastic prod.	30	184.6	178.7	151.5	135.6	125.1	86.9	37.5	0.4
Leather and leather products	31	13.5	12.8	10.2	10.7	8.4	4.5	3.0	0.1
Stone, clay, glass products	32	37.2	31.2	29.8	26.1	26.7	15.4	2.6	0.2
Primary metal industries	33	522.9	476.7	424.7	348.6	328.6	106.0	30.6	6.8
Fabricated metals products	34	137.5	128.9	111.6	103.0	91.1	57.8	32.6	0.1
Industrial machinery and equip.	35	58.4	49.4	39.3	34.1	27.5	17.0	10.1	0.2
Electronic, electric equipment	36	100.6	82.5	67.7	53.2	39.6	28.8	9.9	0.3
Transportation equipment	37	206.0	176.6	150.8	137.2	135.5	94.9	38.9	0.1
Instruments and related prod.	38	52.5	44.3	39.7	33.3	26.6	19.9	5.8	0.8
Misc. manufacturing industries	39	29.3	26.2	20.8	18.9	17.2	11.4	5.8	-
Multiple codes	20-39	362.0	303.0	242.1	220.3	159.6	84.1	46.2	4.5
No codes	20-39	10.8	13.8	27.1	15.1	20.4	7.0	4.7	0.3

- Represents zero.   X Not applicable   [1] Standard Industrial Classification. see text, section 13.   [2] Includes other releases not shown separately.   [3] Stack.   [4] Fugitive

Source: U.S. Environmental Protection Agency, *1993 Toxics Release Inventory*, March 1995.

FIGURE 10.1   Table with Footnotes and Multiple Levels of Headings

Source: U.S. Department of Commerce, Bureau of the Census. 1995. *Statistical Abstract of the United States 1995.* 115th ed. Washington, DC: Government Printing Office, 235.

- Headings for columns should identify the measure ($, %, m) that is represented by the numbers in the column. The tables are cleaner (less noisy) if the abbreviation is not repeated for each entry. Another method is to insert the symbol or abbreviation before or after the first number in each column. This method reduces clutter in column heads.

  Notes to clarify information or to cite sources may be necessary, just as they are in prose text (see Figure 10.1). When the table as a whole comes from a single source, a footnote number often appears on the table title, and a note that cites author, title, and publication data appears at the bottom. The form of the note corresponds to the form used for the reference list.
- All tables must be referred to in the text.

Application: Copyediting a Table

Copymarking right on the table is sufficient for simple corrections. You can specify capitalization, deletions, and insertions on the table and supplement these marks with marginal instructions. When the changes are extensive, however, the table may have to be retyped.

Figure 10.2 illustrates a copyeditor's marking of a table for correction before printing. The table was created for a report on AIDS in correctional facilities. It shows the results of screening to determine how many inmates test positive for the HIV antibody, which would indicate that they have been exposed to AIDS. At the time of the screening, only four states had mass screening programs.

The copyeditor has marked several types of changes, not only in the familiar copyediting realms of correctness, consistency, accuracy, and completeness but also with an eye to visual readability.

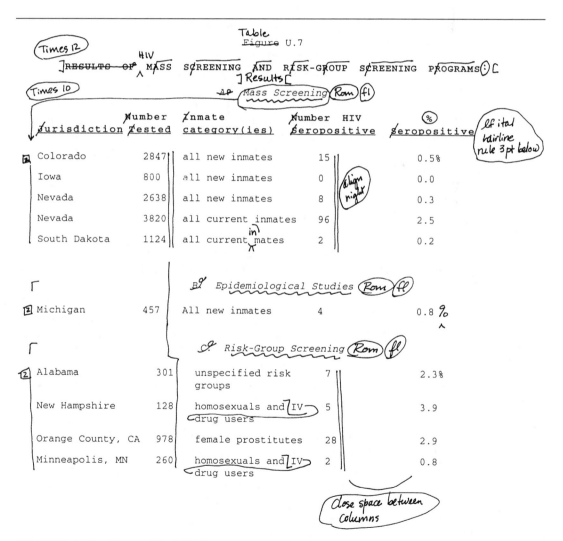

FIGURE 10.2　Copyedited Table

- **Correctness.** A check for correctness reveals several problems. The illustration was inappropriately labeled as a figure, but because the material is tabular, its title is changed to *table*. Also, the numbers in columns 2 and 4 are marked to align on the decimal, and the misspelling in the fifth row under "Mass Screening" is corrected.

- **Consistency.** Several changes are required for consistency. The column headings are marked for left justification. Capitalization in headings and in the third column is made consistent. And, because "number" is spelled out in the heading for column 4, the "percent" is also spelled out in the heading for column 5.

- **Accuracy/completeness.** Placing "HIV" at the beginning of the title gives a more accurate view of the subject matter than the generic "results." The selection of states might raise questions about completeness, but the prose part of the text verifies that only a few jurisdictions have screening programs. Also, the numbers are consistent enough to suggest that they are accurate.

  A copyeditor may question the middle category, "Epidemiological Studies," because the title names only the first and third categories, and the middle category contains only one study. The table may in fact be complete without this information. However, a deletion would be a content change and therefore beyond the expectations for copyediting. The copyeditor should query the writer before such a deletion, perhaps suggesting that the information on that single study be included within the text of the report where the table is discussed. Another alternative would be to change the title of the table to "Results of Screening Programs for the HIV Antibody in Inmates," which does not imply just two categories. However, a title change would also have to be approved by the writer, and all references to the table in the text and in preliminary pages would have to be changed accordingly.

- **Visual readability.** The copyeditor's biggest task is adjusting the visual aspects of the table to make it easy to read.

  title   The title is marked for lowercase letters with initial caps rather than all caps. Presumably, the titles of other tables in the report use lowercase letters. If the title style has not been chosen, lowercase letters are a good choice because they are easier to read than all caps.

  headings   Headings for the three major categories are aligned on a left margin that protrudes farther into the left margin than other data. Such placement is conventional for tables as well as logical for left-right reading. The headings are printed in boldface type rather than italicized for greater typographic prominence. The letters "A," "B," and "C" are deleted as irrelevant because they are not used in the text references and because the three side headings establish visually that there are three categories of information.

  column headings   Column headings are placed above the heading for the first category because they head the columns for all three categories. They are marked to set lightface italics rather than boldface. The rule distinguishes the column headings from the body of the table.

alignment   Numbers are marked to align on the imagined decimal. Columns are marked to maintain consistent indentations.

column spacing   The space between columns 4 and 5 is reduced so that the eye will not have to leap so far to get from one item to the next. In this case, the specific amount of space is left up to the compositor and will be based partly on how much space the column headings require.

typeface   The table is set in the same typeface as the rest of the text.

Figure 10.3 shows how the table will look when printed as marked by the copyeditor.

Table U.7

HIV Mass Screening and Risk-Group Screening Programs: Results

jurisdiction	number tested	inmate category(ies)	number HIV seropositive	percent seropositive
Mass Screening				
Colorado	2847	all new inmates	15	0.5%
Iowa	800	all new inmates	0	0.0
Nevada	2638	all new inmates	8	0.3
Nevada	3820	all current inmates	96	2.5
South Dakota	1124	all current inmates	2	0.2
Epidemiological Studies				
Michigan	457	all new inmates	4	0.8%
Risk-Group Screening				
Alabama	301	unspecified risk groups	7	2.3%
New Hampshire	128	homosexuals and IV drug users	5	3.9
Orange County, CA	978	female prostitutes	28	2.9
Minneapolis, MN	260	homosexuals and IV drug users	2	0.8

FIGURE 10.3   Printed Version of the Copyedited Table in Figure 10.2

Standards and Specifications

Standards and specifications describe products and processes according to measures of quality and intended uses. They affect the work of a technical editor in several ways: they may describe the details of type, structure, layout, and coding of the materials that the editor prepares and even the process of developing those materials; they may describe a product yet to be developed and serve as a tool for planning and designing the product's documentation; they may be documents to edit.

Standards for technical materials and processes are developed and published by government, industry, and national and international standards organizations. Some prominent developers of standards are the United States Department of Defense, the American National Standards Institute (ANSI), the Institute of Electrical and Electronics Engineers (IEEE), and the International Organization for Standardization (ISO). These standards cover many kinds of equipment and processes, including aircraft, computers, and quality assurance procedures. They may specify, for example, the size and brightness of screens for visual display terminals and the interfaces for microprocessor operating systems.

In addition to standards for equipment and processes, a number of standards, particularly in the military, govern publications. These specify standards for abbreviations, format, parts lists, readability, abstracts, and reports. In the United States, military specifications are identified in labels with the prefix MIL- and are commonly referred to as MIL-SPECS. Documents for all government agencies except the Department of Defense are governed by Federal Information Publication Standards. These specifications for publications will be particularly important if you work for the military, the government, or military and government contractors. The standards for SGML (Standard Generalized Markup Language—see Chapter 4) are defined in an ISO standard. The standards aim for a uniform quality and consistency in products that may be produced by various organizations. Products that meet those standards give consumers and other users some confidence in their quality. An editor checks publications for conformity to applicable specifications.

One set of standards particularly important to technical communicators is ISO 9000. This standard requires that all companies doing business in the European Economic Community maintain and follow a quality management system, documented in a quality manual. The quality management system could have the effect of encouraging a systematic process of document development that requires planning and review, a welcome prospect for editors who work in organizations that ask the editors to perform miracles at the end of document development when it is too late to do more than surface cleanup. In addition, editors may be responsible for maintaining the quality manual that documents the policies and procedures in a company.

In addition to the standards that apply across organizations, a product to be developed within a company will probably be described by specifications that establish product users, environment of use, purposes, requirements, and functions as well as details such as weight, size, materials, parts from other products, and interfaces. These specifications may also define resources for development, documentation needed, and a development schedule. Product documentation often begins on the basis of such specifications—and sometimes before the specifications become detailed. By developing or reviewing product specifications, editors can participate early in designing the documentation—contents, structure, publication medium, and other high-level features that determine usability and usefulness. The first versions of this documentation may even help to determine the way the product itself develops.

Because specifications precede the product, they may be vague or incomplete. The editor has a chance to check for completeness and accuracy of the information about the product to be developed. These checks support good planning at the beginning—much more efficient than efforts to correct problems at the end of the document development cycle.

Editing specifications also requires the ability to discover ambiguities in phrasing that might provide a loophole for a contractor trying to save money. John Oriel notes, for example, that the phrase "stainless steel nuts and bolts" requires only that the nuts be of stainless steel.[1]

Summary

Quantitative and technical material may look alien to copyeditors, but to respond by ignoring it is to overlook an important copyediting task and to miss an opportunity to develop a valuable skill. Your knowledge about punctuation and grammar and about marking type for a typesetter are pertinent to copyediting material full of quantities and symbols. You do not have to be an expert in the subject matter to copyedit this type of material, but you must be diligent about checking details, cautious about marking changes, and willing to inquire when in doubt. Chances are you won't be expected to copyedit complicated technical material until you gain some experience.

Note

1. John Oriel. 1992. "Editing Engineering Specifications for Clarity." *Conference Record: IPCC 92–Santa Fe*. New York: Institute of Electrical and Electronics Engineers, Inc.: 168–173.

Further Reading

Barry W. Burton. 1992. "Dealing with Non-English Alphabets in Mathematics." *Technical Communication* 39.2: 219–225.

Judith Butcher. 1992. "Science and Mathematics Books." Chapter 13 in *Copy-Editing: The Cambridge Handbook,* 3rd ed. Cambridge: Cambridge University Press. This book is oriented to British standards; it is thorough in noting what needs to be marked in scientific and mathematical text.

Barry Fisher. 1995. "Documenting an ISO Quality System." *Technical Communication* 42.3: 482–491.

Karen Judd. 1991. *Copyediting: A Practical Guide,* 2nd ed. Los Altos, CA: Crisp.

Jerry Kass. 1972. "Editing Mathematical Copy." *Journal of Technical Writing and Communication* 2(4); 307–330. This article is dated in its references to typesetting methods, but the editorial advice on mathematics is sound.

Joseph M. Kleinman. 1983. *List of Specifications and Standards Pertaining to Technical Publications.* Washington, DC: Society for Technical Communication.

David Purdy. 1991. *A Guide to Writing Successful Engineering Specifications.* New York: McGraw-Hill.

Ellen Swanson. 1979. *Mathematics into Type,* rev. ed. Providence, RI: American Mathematical Society. The authority of the professional society makes this reference the first choice of a mathematics editor.

University of Chicago Press. 1993. "Mathematics in Type." Chapter 13 in *The Chicago Manual of Style,* 14th ed.

Discussion and Application

1. Copyedit the sentences below to show the correct use of numbers, symbols, and abbreviations. You may need to refer to the discussion of abbreviations in Chapter 7 and to a list of abbreviations and a style guide. Prepare a style sheet if there are options.*

   a. When the electrode is fully in the spinal cord tissue, the resistance shoots up to 1000-ohms or more.

   b. There are two methods of applying a coagulating current. One uses a fixed time, eg. 30 sec, and varies the power applied, eg, 5-mA, then 10mA, 15mA, etc, up to a limit given by the manufacturer. The other method fixes the power, eg at 30 mA, and varies the time, eg 5 sec, then 10 seconds, 15 sec, etc.

   c. The spinal cord at the C1-C2 level is about 15-MM across and 10- to 12-mm from front to back, so the maximum lesion needed is 6 mm × 4 mm. Furthermore, a cylindrical electrode with a 2-mm uninsulated tip will provide a lesion somewhat barrel shaped; an exposed tip of 3-mm. is also used and will provide a lesion of about 4.5 × 3.0 mm.

   *Sentences in 1a, b, and c are adapted from "Neurolytic Blocks Around the Head" (first draft), by Samson Lipton, M.D. Used with permission of the author.

2. Copyedit the following sentences and equations to clarify punctuation, spacing, italicization of letters that represent unknowns, and grammar.

   a. Consequently an integer p is even if p^2 is even and is odd if p^2 is odd.

   b. Then, squaring we have:

   $$p^2 = 2q^2$$

   c. Hence: since 2 is rational, both p and q are even integers.

   d. Thus,

$$|4| = 4, \qquad |-4| = -(-4) = 4 \qquad |0| = 0.$$

   e. Using the *less than, equal to, greater than* symbols, these three possibilities are written:

$$a < b \qquad a = b \qquad a > b$$

3. Convert these stacked fractions for inline presentation.

   a. $\dfrac{1}{16}$

   b. $\dfrac{3}{a + b}$

   c. $\dfrac{x + 2}{2y}$

4. Mark this equation to show an appropriate line break.

$$(-x) = -(-x)^4 + 3(-x)^2 + 4 = -x^4 + 3x^2 + 4 = H(x).$$

5. Mark these equations to align them on the equal sign. Clarify other spacing as necessary.

$$(2x^2 - 2)(x^2 - x - 12) = 0$$
$$2(x - 1)(x + 1)(x - 4)(x + 3) = 0$$
$$x = -1, 1, 4, -3$$

6. Mark these sentences from a statistical analysis to show italics, spacing, and capitalization.

   a. The analysis revealed a significant effect of method in both experiments. In experiment I, $F = 8.09$ and $p = .001$. In experiment II, $F = 8.58$ and $p = .0007$.

   b. Results of a two-tailed t test ($t = .728$, $p < .20$) indicate that the performance difference between the two groups is not significant.

Chapter 11

Proofreading

Printing is expensive. It's risky to print directly from a copyedited document because the copyeditor's directions may not be incorporated accurately. One intervening process between copyediting and printing is proofreading. Proofs in printing are analogous to proofs in portrait photography: proof copy allows a final check before printing. Proofs show how the document will look and read with all the editing and graphic design directions incorporated.

The term *proofreading* is sometimes used loosely as a synonym for copyediting. The symbols for both processes are similar, and as online publication begins to dominate print, the distinctions between copyediting and proofreading may blur further. However, the processes as part of production in print differ in the purposes they serve, in the stage of production at which they occur, and in the placement of marks on the page. Copyediting prepares the text for printing; proofreading verifies that the copyediting specifications have been implemented. Copyediting takes place early in production; proofreading occurs later. On paper, copyediting marks are interlinear; proofreading marks are mostly marginal. Figures 11.3 and 11.4, later in the chapter, compare the copyediting and proofreading of the same document.

Copyeditors establish spellings and mechanics where more than one option is available and also impose correct grammar and punctuation. Copyeditors or graphic designers establish type style and size, typeface, and spacing. Proofreaders do not change these choices but verify that they have been incorporated into the document. Proofreaders introduce changes only where the copyeditor has overlooked errors.

Proofreaders compare a version of a document with an earlier version to establish that marked corrections have been made and instructions followed. Proofreading may occur at any point in production when the document is prepared in a new form. In fullscale printing, a typeset document usually appears first as a galley proof. (A galley proof is printed on a long sheet of paper; type size and style and line length are fixed, but the text has not yet been divided to form pages.) The galley is proofread against the typescript. Next, page proofs are printed, showing page breaks. The page proofs are compared with the marked galleys to determine that corrections have been made. Once page proofs are corrected, metal printing plates are made, and a blueline may be prepared. The blueline, a photographic print in blue characters of the pages as they will be printed and bound, is proofread to confirm that pages appear in the right sequence and that folds, if any, are correct. At each stage, a proofreader confirms that errors previously noted have been corrected and checks that new procedures (such as page breaks) have been completed correctly.

As each version of the document is replaced by a newer version, it becomes dead copy. When the galleys are being proofread, the marked typescript is the dead copy; when page proofs are being proofread, the galleys are the dead copy. The dead copy indicates the established text. Proofreading marks show where the proofs differ from the dead copy and what corrections need to be made before the next stage in production.

Editors often proofread, though large companies may hire people with the job title of proofreader. Editors who supervise proofreaders can better schedule and support the task if they understand the time and skill required to proofread well.

This chapter identifies goals of proofreading, symbols and their placement on the proof copy, and strategies for effective proofreading. It also considers the job of proofreading for online documents.

The Value and Goals of Proofreading

Three justifications for careful proofreading are accuracy, dollars, and credibility. Errors in numbers and names can confuse budgets, orders, sales, and management, as well as give faulty information. For example, is the part number 03262 or 03226? Have 20 or 200 parts been ordered? Is Andersen or Anderson in charge of arranging the meeting? Should the pressure be set to 22 psi or 32? Is the claim for $38 million or $3.8 million?

The cost of fixing errors increases exponentially at each stage of production. If a correction costs $1.00 at the galley stage, it may cost $6.00 at the page proof stage and $60.00 after metal plates for printing have been made. Each stage of production provides a chance to correct the document, but efficient proofreaders do a good job early and thus save money down the line.

Figure 11.1 illustrates some failures of proofreading. They may also be failures of copyediting, but because the texts have appeared in print and the proofreader has responsibility before print, they represent lapses in proofreading. A reader who recognizes these errors will probably laugh, gloat, and feel superior—counterproductive responses to a serious document from the writer or publisher's point of view. Spotting the one error that remains in a printed document, the reader will not stop to think about the fifty errors that the proofreader found. See how quickly you can spot the errors in the examples in Figure 11.1.

Readers may forgive an error or two, but if errors are too common, readers may dismiss the document as carelessly prepared and lose trust in the information. One editorial services company distributed a brochure advertising experienced "proffers" for proofreading. It is hard to imagine that the company got much proofreading business.

from a physician's advertisement	**LAURA WARD, M.D.** *Obstetrician & Gynecologist* *Diplomate American Board of Obstetrics and Gynecology* *Fellow American, College of Obstetricians & Gynecology* *Medical Director for Planned Parenthood*
from an investment firm's newsletter	Mr. Alberts rises early every weekday for a busy day of investing. After a 35-minute drive into town, Mr. Alberts starts his day pouring over the morning papers at a nearby coffee shop.
from a newspaper report of the Chernobyl disaster in the Soviet Union	The first detailed Soviet description of the Chernobyl disaster and its aftermath was given by Moscow Community Party chief Boris Yeltsin.... "We are undertaking measures to make sure this doesn't happen again," said Yeltsin, who was attending a Communist Party in Hamburg.
from a newspaper ad for a series of workshops for women	**Sexual Harassment - Petticost Wars - Time Management Men's Rections to The Women's Movement - And More**

FIGURE 11.1   Typos and Punctuation Errors in Print

Errors in spacing, type, and capitalization are less easy to spot than spelling errors, but such errors also detract from the quality of the publication. Shifts in spatial patterns and typography can distract readers or even give false signals about meaning. A spacing error may signal an omission. Figure 11.2 shows spacing and typography errors.

from an annotated bibliography	The metric system is increasingly the norm in U.S. indus- try. The system is economical…
from a technical report	A well-constructed message prototype for an emergency is important to the quick dissemination of information. The system and content of a message can have a dramatic effect on public response. Enough research has been conducted to discern a poor message from a good one.
from the printed minutes of an annual meeting	1. **Report of the Secretary/ Treasurer** The treasurer reported a balance of $9,978.90 in the checking account, but observed that the expenses exceeded income in this fiscal year. 2. **Report of the Journal editor** The editor presented a budget of $11,002 for 1998 for the journal. The Board advised her to investigate the feasibility of new page layout software.
from the headings for a project summary	**PROJECT:** Multi-Enzyme Reactor Design **SPONSOR:** State of Texas **PRIMARY INVESTIGATOR:** F. Senatore **Graduate Students:** A. Lokapur, P. Zuniga, T. Jih
from a book on word processing. The spacing error—no indentation for a new paragraph—indicates a content omission	…This is easy and the results far more convincing than simply adding one's name to someone else's style. the whole essay into the computer in the first place. Some enterprising…

FIGURE 11.2   Spacing and Typography Errors in Print

Proofreading Marks and Placement on the Page

Table 11.1 identifies proofreading marks for typos, punctuation, typography, and spacing. A comparison of these marks with those for copyediting (Tables 3.1–3.3) will reveal the similarity. The same marks are used in copyediting and proofreading for indicating changes in type and for correcting typos. However, they are placed

TABLE 11.1 Proofreading Marks—Spelling, Punctuation, Typography, Spacing

Proofreading Marks: Spelling

Margin	In Text	Meaning	Result	Comment
⟆	w⟆ords	delete	words	
⟆	to the the printer		to the printer	
i	ed⟨ting	insert character	editing	Place the caret at the point in the text where the insertion will go. In the margin, write the characters to be inserted.
the	to printer	insert word	to the printer	
#	in⟨the	insert space	in the	
⌒	edit ing	close up	editing	
⟆	typeface	delete and close up	typeface	
(*tr*)	typescript	transpose	typescript	
a	toler ance	substitute letter(s)	tolerance	Write just the letters to be substituted, not the whole word.
(*sp*)	%	spell out	percent	Spell the whole word if the spelling will be in question.
(*stet*)	proceed	"let it stand"; revert to the unmarked copy	proceed	Use dots under the words in the text that should remain the same and write "stet" in the margin.

Proofreading Marks: Punctuation

Mark all punctuation insertions in the text with an insertion symbol (caret) at the point where the punctuation occurs. The mark of punctuation appears in the margin as noted here. To replace one mark with another, show the insertion only, not the deletion.

Margin	Meaning	Margin	Meaning	Margin	Meaning
⋀	comma	:		colon	⌵"	quotation marks
⊙	period	|=|	hyphen	⌵'	apostrophe
(*set*) ?	question mark	⁄M	em dash	⸦ ⸧	brackets
; |	semicolon	⁄N	en dash	⟨ ⟩	parentheses

TABLE 11.1   *Continued*

Proofreading Marks: Spacing

Margin	In Text	Meaning	Result	Comment
lc	ʎTITLE	set in lowercase	title	
cap	title	set in capital letters	TITLE	
ulc	THE TITLE	set in upper- and lowercase letters	The Title	If only the first word is to be capitalized, write "init cap" for "initial cap." "ULC" means that all words are capitalized.
sc	document	set in small caps	DOCUMENT	Small caps have the shape of regular caps. They are not lowercase. Their most frequent use is in the AM/PM abbreviations.
ital	document	set in italic type	*document*	
rom	document	set in roman type	document	
bf	document	set in boldface	**document**	
lf	document	set in lightface	document	
wf	helvetica	wrong font	helvetica	Specify the font if you know it.
x	broken	reset a broken letter	broken	
	m2	set as superscript	m^2	Visualize the inverted caret pushing up.
	H2O	set as subscript	H$_2$O	Visualize this caret or "doghouse" pushing down.

differently on the page, and proofreading uses some spacing and typography marks that copyediting does not require.

To affirm that the proofs are correct and that production may proceed, the proofreader must review such visual features as alignment, spacing, and the clarity of individual letters. Thus, the proofreading marks include marks for alignment, space between letters and between lines, and broken or upside down letters. The copyeditor does not care so much about these visual characteristics of the document because the final form of the document will probably differ from the typescript.

TABLE 11.1 *Continued*

Proofreading Marks: Spacing

Margin	In Text	Meaning	Result	Comment
¶	sentence. ¶ A new...	start paragraph	sentence. A new...	
(run in)	sentence.⌐ ⌐ A new...	run lines together	sentence. A new...	
⊢─────⊣	end. ─────⊣ The line is	fill line	end. The line is	Use with awkward gaps at the ends of lines, especially with right justification.
(eq #)	to ᵛa ᵛbook	equal space between words	to a book	Checks within the text mark the spaces that should be equal.
⌐ ⌐	the ⌐first⌐ book	move type up	the first book	
⌊ ⌋	the second book	push type down	the second book	Use for words or letters within words.
↓	⌐Headings No space after.	push the line of type down	**Headings** no space after.	Use for whole lines of type.
( ) or (close up)	extra space (between lines)	close up the space between lines	extra space between lines	
⌐	word hangs out. ⌐Move it in.	move right	word hangs out. Move it in.	
⌐	⌐paragraph	move left	paragraph	
⌐⌐	⌐Title⌐	center	Title	
‖	‖margin is ‖ not straight ‖ on the left	align horizontally	margin is not straight on the left	
=	base͞line	align vertically	baseline	

Figure 11.4, the typeset version of the copyedited document in Figure 11.3, shows that most of the proofreading marks appear in the margin. The marginal marks alert the keyboard operator to check specific lines. The marks in the body of the text only show where to make the correction requested in the margin. In Figure 11.4, the caret in the second line signals that the marginal word, "calcium," should be inserted at that point.

When a line requires multiple corrections, they may be listed, left to right—in the order in which they appear—with a slash between the individual corrections.

MILK: NOT QUITE THE PERFECT FOOD

New Century Schoolbook
12, ulc

New Cent
10/12×25

Many people think that milk is nearly a perfect food. However, while calcium is a necessary mineral in the diet, many adults cannot tolerate even small amounts of milk. A scoop of ice cream or a glass of milk can cause abdominal cramps, gas, and sometimes, diarrhea for about thirty million Americans. These people lack an enzyme, lactase, which is needed to digest lactose, the sugar in milk. After about the age of two, many people gradually stop producing this enzyme. It is thought that fully 70% of the world's population has this condition. People of Asian, Mediterranean, or African descent are more likely to be lactase deficient while those of Scandinavian descent rarely lose the enzyme.

When lactose, the sugar found in milk and milk products, is not digested, bacteria in the colon use the sugar to produce hydrogen and carbon dioxide. The result is gas, pain, and a number of other symptoms that are sometimes mistaken for Lactose intolerance is not hte same thing as an allergy to the protein in milk or irritable bowel syndrome, although the three conditions may produce similiar symptoms. Because hydrogen gas is produced, a test that measures hydrogen in the breath can confirm lactose intolerance is the problem.

Since lactose digestion depends on the ratio of lactase to the amount of lactose containing food eaten, many people can drink small amounts of milk or eat some ice cream. The type of milk product can also make a difference. While yogurt contains more lactose than milk, it is often better tolerated because the bacteria in live cultures of yogurt break down the milk sugars. The same things appears to happen in hard cheeses such as cheddar or swiss.

Fortunately, there are some products on the market that can make life easier for those with lactose intolerance. Some foods are treated with the enzyme to break down the lactose into glucose and galactose. LactAid or Lactrase may be added to food 24 hours before consumption.

FIGURE 11.3   Copyedited Article for a Health Newsletter

The title in Figure 11.4 contains both a transposition and a capitalization error. Both are listed in the margin with a slash between the two circled instructions.

Corrections may be placed in either margin, but the marginal marks will be easier to match up with the body marks if they are in the margin closer to the error. Don't confuse the person making the corrections by placing in the left margin a note about an error on the right side of the line while placing in the right margin a note about an error on the left side of the line.

Where the corrections within the body of the text might be confusing, particularly when there are multiple corrections within the same area, you may write out the correct version in the margin, as well as marking within the body. Otherwise, it is sufficient to write just a single letter when there is an insertion or substitution.

Milk: Not Quiet the perfect Food

Many people think that milk is nearly a perfect food. However, while is a necessary mineral in the diet, many adults cannot tolerate even small amounts of milk.

A scoop of ice cream or a glass of milk can cause abdominal cramps, gas, and sometimes diarrhea for about 30 million Americans. These people lack an enzyme, lactase, which is needed to digest lactose, the sugar in milk. After about the age of two, many people gradually stop producing this enzyme. Fully 70 per cent of the worlds population have this condition. People of Asian, Mediterranean, or African descent are more likely to be lactase deficient while those of Scandinavian descent rarely lose the enzyme.

When lactose, the sugar found in milk and milk products, is not digested, bacteria in the colon use the sugar to produce hydrogen and carbon dioxide. The result is gas, pain, and other symptoms that are sometimes mistaken for allergy to the protein in milk or irritable bowel syndrome. Because hydrogen gas is produced, a test that measures hydrogen in the breath can confirm that lactose intolerlance is the problem.

Since lactose digestion depends on the ratio of lactase to the amount of food eaten containing lactose, many people can drink small amounts of milk or eat some ice cream. The type of milk product can also make a difference. While yogurt contains more lactose than milk, it is often better tolerated because the bacteria in live cultures of yogurt break down the milk sugars. The same things appears to happen in hard cheeses, such as Cheddar or Swiss. Fortunately, some products on the market can make life easier for those with lactose intolerance. Some foods are treated with the enzymeto break down the lactose into glucose and galactase. LactAid or Lactrase may be added to food 24 hours before consumption.

FIGURE 11.4   **Proofread Version of the Copyedited Document in Figure 11.3**

To write more than is necessary will make you look like an inexperienced proof-reader and will waste valuable time.

Usually, the compositor will be able to distinguish instructions from modifications of the text, but if there can be doubt, circle your instructions. The circled "run-in" that appears beside paragraphs 2 and 4 is an instruction to the compositor.

If there is more than one way to mark an error (for example, by transposing or substituting letters), choose the mark that will be easier to understand. Also try to anticipate how the compositor will apply the direction: if the probable method will be to delete an entire word and retype it, then your mark may reflect that process; if only one or two characters need to be replaced, mark single characters.

Use a colored pencil or pen to mark in proofreading so that the marks will be easy to see. Legible marks also increase the likelihood that the compositor will be able to follow your proofreading instructions. For purposes of cost accounting, the responsibility for each correction should be identified unless all the work is being done inhouse. The print buyer should not be charged for the printer's errors, nor should the publishing company absorb the expense of excessive changes by the writer after typesetting. If everyone involved in writing, editing, and designing the document is on the staff of a single company, it may be sufficient for the proofread-er to use a different color pencil for errors made by the printer and errors made by the company buying print. When the company contracts with people outside the company for services, a more detailed accounting is necessary. Each correction may then be accompanied by a circled mark to indicate responsibility: "pe" for printer's error, "ea" for editorial alteration, "da" for designer's alteration, and "aa" for author's alteration. An unmarked printer's error in the previous version of the document becomes the proofreader's error in a later version. Thus, if a typo was not corrected at the galley stage and remains in the page proofs, it is charged to the print buyer even if the printer made the error in the galley.

Generally, you will proofread to make the new version conform to the previous version. That is, you will not copyedit—because of the expense and because substantial changes will slow down production. However, if errors remain in the proof copy that were uncorrected in the dead copy, you will, of course, correct them rather than making conformity to the dead copy a higher aim.

Computers and Proofreading

Computers affect conventional proofreading in two main ways: they make the task easier, and when the publication is online rather than print, the division of responsibilities between copyeditor and proofreader is less rigid.

The spelling and grammar check features of word processing programs locate many typos and other lapses, such as failure to close parentheses. These checks will not locate all typos, however, and thus work as a first step in proofreading, not the final one. The computer cannot tell when the text should read "their" instead of "there," or "now" instead of "not." It cannot recognize omissions of content. Using the computer's checks to clean up the copy will make it easier to find the errors that the computer cannot catch.

Just as important as these checks at the end of production is the marking of text structures at the beginning of production. Text created with the styles and templates features of word processing and page layout programs or SGML or HTML markup will be visually consistent. A style defines typeface and size, style (bold, italic), and spacing for a structural part, such as level-one headings and paragraphs. A template is a collection of styles. Because the style definition incorporates all the choices, the writer, editor, or compositor can apply all the choices to a structural part with the one procedure of choosing its style. The advantage to the proofreader is less marking for variations in type and spacing in the final copy.

Electronic transmission of text also minimizes errors. Typesetting from the computer disks rather than from hard copy saves the expense of rekeyboarding and also minimizes the risk of introducing new errors.

When publication is electronic, copyediting is ongoing, and the various versions are not necessarily recorded for comparison with later versions. Thus the proofreading task is less distinct from copyediting. Some editors and proofreaders print the copy before online publication to look for errors on paper that they may overlook on the screen.

Strategies for Effective Proofreading

Most of us have spotted proofreading failures in printed documents and may therefore feel confident about our proofreading ability. Yet even the best of proofreaders will miss some errors. Proofreaders may get caught up in content and overlook surface errors. Reading quickly, we are likely to identify words by their shapes rather than by their individual letters, yet the word shape will remain essentially the same with a simple typo, particularly one that involves a narrow letter such as *i* or a missing member of repeated letters, such as two *m*s. The word shapes are identical for both the correct spelling of evaluation and its misspelling. "Evalvation" would be easy to overlook in proofreading.

[evaluation]   [evalvation]

Following these guidelines will increase your proofreading effectiveness:

1. **Take advantage of the computer.** Spelling and grammar checkers will catch many typos as well as some punctuation and capitalization errors. Styles and templates regularize type and spacing for parallel structural parts. Electronic transmission of text eliminates the need for repeat keyboarding. But don't depend on the computer for complete proofreading.

2. **Check the dead copy.** Proof copy can omit phrases or substitute words and still make sense. The dead copy is also an authority for spellings of names and other unfamiliar material as well as numbers.

3. **Pay special attention to text other than body copy.** Titles, tables, lists, and nonverbal text may contain just as many errors as body copy, but they are easy to overlook. Make a special point of proofreading the parts of the document other than paragraphs.

- **Titles and headings.** Look for typos; check for capitalization, type style, and typeface.
- **Illustrations.** Look for typos; check for spacing, alignment, keys, completeness, and color or shading.
- **Numbers.** Check for accuracy (reversal of numerals and errors in addition are common). Look for alignment on the decimal and for a match of numbers with references to them in the text.
- **Reference lists and bibliographies.** Check the spelling of authors' names and of titles; check for spacing, capitalization, and italics as specified by the appropriate style manual.
- **Names of people and places.** Check them against the dead copy; use reference books if the dead copy seems inaccurate.
- **Punctuation.** Check for closing of quote marks and parentheses.

4. **If you find one error in a line or word, go back over the line to look for other problems.** Mistakes cluster, perhaps at points where the keyboarder became confused or tired. Yet you feel as though you have "finished" a line if you find an error. In the following reference list entry, one proofreader noted the extra space and inappropriate italics—but forgot to check the date. The accurate date was 1996, not 1995.

    Tech nical Communi cation <u>43.2 (May 1995): 130–138.</u>

    [underlining changes italic type to roman]

    Technical Communication 43.2 (May 1996): 130–138. [corrected version]

5. **Review at least once solely for visual errors.** Review just for spacing, alignment, type style, and typeface with out the distraction of the text content. Some proofreaders turn the copy upsidedown to check spacing; others read right to left.

- **Typeface, type size, type style.** Type should not change arbitrarily.
- **Spacing above and below headings.** Are headings set solid over the body copy, or does some space separate headings from the body? Is consistent space before and after headings of the same level? Measure the type from the baseline of one line of type to the baseline of the next if you don't trust your eyes.
- **Centering or left or right justification.** Check all like headings (first level, second level) together to spot inconsistencies.
- **Indentation.** Check systematically, especially after headings.
- **Alignment.** Scan the left and right margins and any other aligned material.

6. **Check hyphenated words at the end of lines.** If the design calls for hyphenation at the ends of lines, check that word divisions conform to those in the dictionary, or at least that the divisions make sense. Note the difference between *thera•pist* and *the • rapist*.

7. **Use techniques to force slow reading.** Some errors will pop out when you skim, but most will not, especially because the eye recognizes and accepts word shapes even when some of the letters may be incorrect. Unless you can slow yourself down, you may read for content and miss errors.

- **Proofread with a partner.** One partner can read aloud from the dead copy while the other checks the proof copy. Partners can take turns to keep from tiring at one task. If you don't have the time to work this way for the entire document, try at least to use a partner for difficult text, such as columns of numbers and reference lists.
- **Read out of sequence.** Some proofreaders read backwards or read pages out of sequence to avoid getting caught up in the content.

8. **Use multiple proofreaders and proofreadings.** Publishers that value high quality insist on multiple proofreadings for any important document. Ideally several proofreaders, including someone other than the writer and editor, will read the document. Proofreaders who are familiar with the document or its content are likely to miss errors because they see what they expect to see. A single proofreader can increase productivity by reading the document for a specific feature (such as the visual features or word division), then rereading for another feature.

9. **Schedule to proofread when you are alert.** When schedules permit, proofread at the time when you most alert, such as first thing in the morning. When you are tired, interrupt proofreading to do other tasks and return to the proofreading refreshed.

Summary

Proofreading provides a check of the document after copyediting but before printing. Good proofreading helps to ensure a high-quality publication, notable for the apparent care with which it has been produced and the absence of distracting errors. Proofreading is skilled work.

Further Reading

Laura Killen Anderson. 1990. *Handbook for Proofreading.* Lincolnwood, IL: NTC Business Books. Especially good on working with type and understanding the printer's language.

Peggy Smith. 1993. *Mark My Words: Instruction and Practice in Proofreading.* 2nd ed. Alexandria, VA: Editorial Experts, Inc.

Discussion and Application

1. Compare the copyedited newsletter article in Figure 11.3 with the proofread version of the same document in Figure 11.4. In your own words, distinguish the types of corrections made on each document, and explain how and why the marks differ.

2. During the preparation of an annotated bibliography on technical editing, two collaborators missed the typos in the following words after repeated checks of the hard copy. (The spelling checker on the computer identified them.) For each word or phrase, hypothesize why the collaborators overlooked the error. That is, what is it about the spelling, word shape, or other feature of the word that encouraged the collaborators to read it as correct?

gullability   progams   responsibilites
indentification   edtors   comform
labortatory   embarassing   comunication
manuscritps

3. With regard to application #2, discuss how the fact that the collaborators were proof-reading their own document affected the quality of their proofreading.

4. A national insurance company mailed to shareholders a ballot for trustees and a book-let describing the qualifications of 27 nominees. Inside the printed booklet was a slip of paper entitled "Errata" that included these comments:

> Robert C. Clark is 45, not 55, as appears on page 5 of the Trustee booklet.
>
> William H. Waltrip is 52, not 62, as appears on page 6 of the Trustee booklet.
>
> Uwe E. Reinhardt's name incorrectly appears on the ballot as Uwe E. Remhardt.

What is the meaning of *errata*? Speculate on why the proofreaders missed these partic-ular errors. When the errors were discovered (after the booklet was printed), what were the company's three options? Why did the company choose to include the errata slip? What were the consequences of the proofreading failures to this company, in terms of money, time, and image?

5. Proofread the version of the document in the right column by comparing it with the ver-sion in the left column. Mark the typeset version to make it match the typescript version.

Ivory Trade Continues

Elephants at risk of extinction

Recent studies have established that ivory poaching has reduced the ele-phant populations in East Africa by half in less than a decade. The same story is repeated for the rest of Africa except parts of southern Africa where rigorous management has actually made it possible for the elephant num-bers to increase. Over much of the east and central Africa, elephants are so heavily poached, even in previ-ously secure sanctuaries such as Selous, Tsavo, and the Luangwa Val-ley, that it will not be long before the elephant is extremely rare or even extinct. There is little doubt that short-term profit-motivated poaching is responsible for the enormous decima-tion of the large herds of elephants. Conservation organizations of the world are now demanding immediate enlightened action, strong political will, and a high degree of interna-tional cooperation to avert a disaster.

Ivory Trade Continues

Elephants at risk of extintion

Recent studies have established that ivory paoching has reduced the elephant populations in East Africe by half in less than a decade. The same story is repeated for the rest of Afirca except parts of Southern Africa where rigorous managment has actually made it possible for the elephont numbers to increase. Over much of central Africa, elephants are so heavily poached, even in previously secure sanctuaries such as Selious, Tasvo, and the Luangwa Valley, that it will not be long before the lelphant is extremely rare or even extinct. There is little doubt that short term profit-motivated poaching is responsible for the enormous decim-ation of the large herds   of elephands. Conservation organizations of the world are now demanding imediate enlightened action, strong politicial will and a high degree of international cooperation to avert a disaster.

6. Proofread the right column to match the dead copy on the left; check especially for errors in type style, typeface, and spacing.

Ban on Ivory Imports Established

A moratorium on the importation of African elephant ivory was implemented through an announcement in the *Federal Register.*

A quota system for legal, regulated trade in ivory authorized under the Convention on International Trade in Endangered Species of Wild Fauna and Flora (CITES).

The United States, Western Europe, and Japan consume two-thirds of the world's "worked ivory."

Ban on Ivory Imports Established

A moratorium on the importation of African elephant ivory was implemented through an announcement in the Federal Register.

A quota system for legal, regulated trade in ivory was was authorized under the Convention on International Trade in Endangered Species of Wild Fauna and Flora (*CITES*).

The United States, Western Europe, and Japan consume two-thirds of the world's "worked ivory."

7. Correct spacing errors in this paragraph.

If the United States, Western Europe, and Japan   were concerned enough as parties to CITES to agree to the appeal and pro hibit importation of all ivory without exception,

the present enormous demand for ivory wou ld cease. IN   turn, poached ivory would become less lucrative. CITES, the   one instrument   of international   standards available, should impose a world wide ban on the ivory trade to stop the convention being used to channel hundreds of tons of illegal ivory into legal trade.

$Part$ 3

Comprehensive Editing

Chapter 12
Comprehensive Editing: Definition and Process

Chapter 13
Style: Definition and Sentence Structures

Chapter 14
Style: Verbs and Other Words

Chapter 15
Style: The Social and Global Contexts

Chapter 16
Organization

Chapter 17
Visual Design

Chapter 18
Illustrations

Chapter 19
Editing Online Documents

Chapter 12

Comprehensive Editing: Definition and Process

In Part 2 of this book, you learned ways to make a document correct, consistent, accurate, and complete. But a document may achieve all of these standards and still not work—readers may not be able to use it or comprehend it. Usefulness and comprehensibility depend on a document concept that matches the need for the document and on organization, visual design, and style that support not just comprehension but also the probable uses of the document by readers.

In comprehensive editing, the editor considers a document's concept and intended use, content, organization, design, and style. The purpose of comprehensive editing is to make the document functional for its readers, not just to make it correct and consistent. Other terms for this type of editing are *developmental editing, macro editing,* and *substantive editing.*

The job of comprehensive editing is challenging because it requires analysis and judgment. Although some guidelines will help you, decisions about style, design, and organization more often represent judgments than applications of rules. Furthermore, you have to think with the writer's mind as well as with the reader's and with your own and to imagine the document in use. While comprehensive editing will take more time and effort than basic copyediting, it gives you more options for making the document work.

In this chapter, you will compare the copyediting and the comprehensive editing of a single document to illustrate the differences between the two editing tasks. The chapter will walk you through the process of document analysis and goal setting that precedes comprehensive editing. The following chapters give specific guidelines for comprehensive editing.

Example: Copyediting vs. Comprehensive Editing

To compare the purposes, methods, and results of copyediting and comprehensive editing, let us consider the following scenario. You work in the communications department of a large medical center. Today you are editing a grant proposal, due in Washington in five days, worth $500,000 if it is funded. You also have an appointment with the graphics specialist to review the photographs for a slide show for which you are writing the script, to be completed in two weeks. The Director of Maintenance, with whom you have a cordial relationship, drops by with a memo (Figure 12.1), written for immediate distribution throughout the medical center. He asks you to look it over. Although you can spare only about 15 minutes, you agree to check the memo. Actually, even the 15 minutes is precious, but taking a break from the proposal will help refresh your mind so that you can think more clearly when you return.

The punctuation and mechanics of this memo need attention, but the work will be fun. Doing something on this level will offer relief from the intensity of the grant proposal. And you can perform a service for the Director of Maintenance with little effort.

To participate in this scenario, imagine that you are the editor assigned to the editing task. Using correct copymarking symbols, edit the document in Figure 12.1 for grammar and mechanics.

Your edited memo will look much like the marked-up version in Figure 12.2. On most points your editing will agree with the editing shown in Figure 12.2 because you are applying the same rules of grammar and spelling. Some editing for mechanics will depend on your assumptions about titles. For example, is "Maintenance Department" a proper name or a descriptive one? Is "manual" a part of a title or just a descriptor? Your answers to these questions will affect your decisions about capitalization. If you actually worked at this medical center, you would either know the answers to these questions or be able to check them quickly. Editing in a hypothetical situation, you either have to make an educated guess or make a note to check the information.

Your editing may vary from the sample in Figure 12.2 on some other capitalization questions. For example, you may have decided to capitalize "Service Call" each time it appears in order to emphasize the main topic of the document. Or you may have determined that the repeated capital letters distract readers. Either decision is legitimate, so long as you are consistent.

Some of your decisions require more thoughtful analysis. For example, in the last line of the first paragraph, the series ending in "hazards" is not clearly punctuated. "Safety" does not fit in the same class of things as wiring and fire (it is not a problem requiring a service call), so it must modify "hazard." But then, where does the series end, and where should you insert the "and" to conclude the series? You have two options:

...wiring, fire, and safety or security hazards.

...wiring, and fire, safety, or security hazards

TO: All Medical Center personnal

FROM: Tom Barker

DATE: January 2, 1997

SUBJECT: Service Calls

The Maintenance Department will initiate a Service Call system beginning Monday,January 6th, 1997, that is designed to provide an effective response time with equal distribution for all departments for work categorized as Service Calls. Service Calls are defined as urgent minor work requiring immediate attention, for example, loss of heat, air conditioning, water leaks, clogged plumbing, faulty electrical wiring, fire, safety or security hazards.

To initiate a Service call a telephone call to the Maintenance Department, (742-5438, 24 hours a day), is all that is required. A Service Call number will be assigned to the job and furnished to the originator if requested. This number may be used when referring to the status of the Service Call

During the hours of 5:00 p.m. through 8:00 a.m. daily, all day weekends and holidays, only one Maintenance Mechanic is on duty, who must insure total systems are oprational on his shift, in addition to accomplshing Service Calls. That will limit the amount of calls, but should not deter anyone from calling in a Service Call to insure it is accomplished when the manhours are available. If no one answers, please call again.

For urgent or emergency requirements, the PBX Operator must be notified to page Maintenance. Please do not use the paging system unless your request is justifiable.

Your cooperation in this matter would be greatly appreciated. Service Call requirements are outlined in the Policies and Procedures Manual. Requests exceding these these requirements will be classified as Job Orders, and are also outlined in the Policeis and Procedure Manual.

FIGURE 12.1   Unedited Memo on Service Calls

TO: All Medical Center personnel

FROM: Tom Barker, *Director of Maintenance*

DATE: January 2, 1997

SUBJECT: Service Calls

Excess capitalization is distracting.

The Maintenance Department will initiate a Service Call system beginning Monday, January 6th, 1997, that is designed to provide an effective response time with equal distribution for all departments for work categorized as service calls. Service calls are defined as urgent minor work requiring immediate attention; for example, loss of heat, or air conditioning, water leaks, clogged plumbing, faulty electrical wiring, fire, safety, and or security hazards.

The punctuation in the list makes the examples confusing. Surely "air conditioning" and "safety" are not problems requiring service calls. ("Loss" and "hazards" are.)

To initiate a service call a telephone call to the Maintenance Department (742-5438, 24 hours a day) is all that is required. A service call number will be assigned to the job and furnished to the originator if requested. This number may be used when referring to the status of the service call.

During the hours of 5:00 p.m. through 8:00 a.m. daily, all day and weekends on and holidays, only one Maintenance Mechanic is on duty, who must insure total systems are operational on his shift, in addition to accomplishing service calls. That will limit the amount number of calls, but should not deter anyone from calling in a service call to insure it is accomplished when the manhours are available. If no one answers, please call again.

A list of two items (daily hours and weekend hours) requires a conjunction

Use "number" with quantifiable amounts

For urgent or emergency requirements, the PBX operator must be notified to page Maintenance. Please do not use the paging system unless your request is justifiable.

Your cooperation in this matter would be greatly appreciated.

Service call requirements are outlined in the Policies and Procedures Manual. Requests exceding these these requirements will be classified as Job Orders, and are also outlined in the Policies and Procedures Manual.

FIGURE 12.2   **Memo from Figure 12.1 Marked to Show Copyediting**

The second option identifies "fire hazards," not "fire," as the problem requiring a service call. You can check with the maintenance department if necessary, but good sense tells you that fire, as opposed to a fire hazard, would require a call to the fire department.

Apart from these possible variants, however, the results of copyediting by any two competent editors will be similar. Furthermore, the edited document, as shown in Figure 12.3, will be similar in content, form, and style to the original.

What, then, has copyediting accomplished? First, because the document is correct, readers will take the message more seriously than they would if they recognize errors. Readers, fairly or not, evaluate the message on superficial text characteristics

---

TO: All Medical Center personnel

FROM: Tom Barker, Director of Maintenance

DATE: January 2, 1997

SUBJECT: Service Calls

The Maintenance Department will initiate a service call system beginning Monday, January 6, 1997, that is designed to provide an effective response time with equal distribution for all departments for work categorized as service calls. Service calls are defined as urgent minor work requiring immediate attention; for example, loss of heat or air conditioning, water leaks, clogged plumbing, faulty electrical wiring, and fire, safety, or security hazards.

To initiate a service call, a telephone call to the Maintenance Department (742-5438, 24 hours a day) is all that is required. A service call number will be assigned to the job and furnished to the originator if requested. This number may be used when referring to the status of the service call.

During the hours of 5:00 p.m. through 8:00 a.m. daily, and all day on weekends and holidays, only one maintenance mechanic is on duty, who must ensure that total systems are operational on his shift, in addition to accomplishing service calls. That will limit the number of calls, but should not deter anyone from calling in a service call to insure it is accomplished when the manhours are available. If no one answers, please call again.

For urgent or emergency requirements, the PBX operator must be notified to page Maintenance. Please do not use the paging system unless your request is justifiable.

Your cooperation in this matter would be greatly appreciated.

Service call requirements are outlined in the *Policies and Procedures* manual. Requests exceeding these requirements will be classified as job orders, and are also outlined in the *Policies and Procedures* manual.

FIGURE 12.3   Copyedited Memo from Figure 12.2

such as spelling and punctuation. Editing has increased the chances that readers will respond to the memo in the intended way rather than being distracted by errors.

Copyediting has also increased clarity. For example, in line 7 of Figure 12.2, the insertion of the conjunction *or* establishes that "loss" refers to "air conditioning" as well as to "heat." The insertion of the conjunction *and* in line 8 clarifies that "safety" modifies "hazards." These insertions clarify the information. The reduction of document noise (in this case excess punctuation and capitalization as well as errors) also makes the message clearer.

From the perspective of your work load, the task has taken little time and energy. If you used the computer to edit, you may have used the global change function to make the capitalization and spelling consistent, and you quickly deleted unnecessary punctuation and inserted necessary conjunctions. You can now return to your more important task—the proposal.

From the perspective of your relationship with the memo's writer, you have performed a service without challenging his competence. If you and he disagree on some specific emendations, you can confirm your choices by pointing to a handbook or style guide, or you can yield on issues when his choice is acceptable (if not preferable). If copyediting is correct, it should create a happy writer.

Comprehensive editing, on the other hand, looks beyond words and sentences to the way in which readers will read and use the document—to the context. Comprehensive editing should make the document more usable and comprehensible. While you will use judgment in comprehensive editing, you use informed judgment. That is, you edit with a full awareness of principles of style, organization, page design, and other factors of document design as well as with an awareness of the context.

Comprehensive editing will take more time than copyediting, so if you have the same commitment to other pressing projects described in the previous scenario, you may have to decline this project or negotiate for a lower level of editing.

The process of comprehensive editing is also more complex than the process of copyediting. Instead of reacting line by line to the text, you begin by assessing the document as a whole and interviewing the writer to determine how readers will use the document and to establish editing objectives. The actual editing is followed by an additional review. The discussion of the comprehensive editing process that follows will continue with the example of the memo on service calls.

The Process of Comprehensive Editing

Comprehensive editing is a multistage process that parallels the writing process. It begins with analysis and planning. The editing itself is likely to require several passes through the document, just as writing requires review and revision. It requires reviews with the writer to ensure that editing has not introduced content errors.

In analysis, the editor considers the document as a whole in its context—as it will be read and used. In well-managed publications companies or departments, the editor participates in the analysis and planning that take place before the writing begins. As a result, everyone involved in document development shares from the start a common understanding of its purpose and goals. But some comprehensive editing occurs after documents are developed—the editor enters the project at the end, as in the service call memo. Comprehensive editing at this point still requires analysis and planning. Just as brainstorming, researching, and outlining provide direction for a writer, analysis yields an overview of the editing task. This overview helps an editor work systematically toward goals that are consistent with the needs of the readers and the purposes of and uses for the document. Analysis discourages line-by-line reaction to errors and sentence structure. The line-by-line approach can work for copyediting, but it does not direct the editor's attention to content, organization, and style nor to the document in use. A sense of the whole document in its context and a concept of what the document seeks to achieve are necessary for the editor making judgments about content, organization, visual design, and style.

As editor, you are more likely to achieve the overall goal of improving the document's usability and comprehensibility if you edit with a plan. Before you begin to mark the page or edit on the computer screen, you should complete a four-part process that will result in a plan.

1. Analyze the document's readers, purpose, and uses to determine what the document should do and the ways it will be used.
2. Evaluate the document's content, organization, visual design, and style to determine whether the document accomplishes what it should.
3. Establish editing objectives to set forth a specific plan for editing.
4. Review the plan with the writer to work toward consensus.

These procedures may overlap with one another and may not be as linear as the list suggests. For example, a review with the writer to determine document readers, purposes, and uses (procedure 1) may satisfy the review of the editing plan (procedure 4). Because analysis requires you to be familiar with the document as a whole, you will have to read sample sections throughout the document to determine editing goals.

Once you have established the editing goals, you are ready to edit. Editing may require more than one pass through the document, and it will certainly include a review of the editorial decisions by others involved with the project. Throughout the process consultation with the writer and subject matter experts will be necessary.

The steps in comprehensive editing are discussed more fully in the following pages, and then the process is applied to the memo on service calls.

Analyze the Document's Readers, Purpose, and Uses

Comprehensive editing always begins with an analysis of who will use the document and for what purpose. Ideally, this analysis (and comprehensive editing) begins before the writing does, when the document is being conceived. This kind of advance planning directs the development of the document in an efficient and effective way.

It also prevents a conflict of wills when the editor's assumptions differ from the writer's. If the editing follows development of a draft, editing requires an evaluation of how well the document in its present form will achieve its objectives.

Analysis begins with questions. These are familiar questions—the ones writers ask before drafting documents:

- What is the purpose of this document? What should happen as a result of its use?
- Who will read it and why? Are there cultural variations in readers that will affect comprehension? Will the document be translated?
- What should readers do or know as a result of reading it?
- What do they already know about the subject?
- In what circumstances will they read it? (In good light or poor? Inside or outside? While doing a task?) What equipment or software do they have that might influence how they read and what they can see in the document?
- Will they read it straight through or selectively?
- Should they memorize the contents or use the document for reference?
- What will they do with the document once they have read it? (Throw it away? File it? Post it? Refer to it again? Print it from its online version?)
- What are their attitudes toward the subject of the document? (Cooperative? Unsure? Hostile?)
- What constraints of budget, equipment, or time influence the options for the document?

Only by analyzing what you want the document to do and by imagining it being used by readers can you make valid judgments about editorial emendations. For complex documents, you will need to interview the writer to determine answers to these questions. You can answer a number of the questions, though, by inference and from experience.

Evaluate the Document

The analysis represents a basis for evaluating the document. While the purpose of analysis is to determine what the document should do, the purpose of evaluation is to determine how well the document does it. An evaluation should systematically review these features as they support (or detract from) the document's purposes and uses:

- Content: completeness and appropriateness of information
- Organization: order of information; signals about the order
- Visual design: prose paragraphs, lists, or tables; paper size; screen display
- Style: writer's tone or persona; efficiency of sentence structure; concreteness and accuracy of words; cultural bias; grammar, usage, punctuation, spelling, and mechanics
- Illustrations: type, construction, placement

Establish Editing Objectives

Setting objectives may begin with an evaluation of what is ineffective in the original document, but it goes beyond a critical evaluation to identify what you want to accomplish in editing. If you observed in evaluation, for example, that the writer's style is full of nouns and weak verbs, the editing objective may be to substitute strong verbs for weak ones. If you determined that readers would be likely to read selectively but that the document provided inadequate identifiers of sections, your goal might be to insert headings for major divisions or better navigation buttons for online documents.

It may help to make notes of your editing goals, just as you take notes on what topics you want to cover and how you might organize them when you write. You can jot these notes on the original copy as you read it through, but collecting them in one place before you actually begin the editing will help to ensure consistency in the objectives.

The analysis, evaluation, and establishment of objectives let an editor know how to proceed with the editing and make all editorial emendations serve an overall concept of how the document should function. The objectives give focus to the editing and help the editor proceed efficiently, with few false starts.

Review Your Editing Plans with the Writer

After you have analyzed the document and established editing objectives, consult with the writer about the editing plans. You may save yourself the time of misdirected editing if you have misinterpreted editing needs or if the writer can offer additional suggestions before you begin. Furthermore, the review encourages cooperation and support. You will avoid surprising the writer and thereby avoid the negative response that surprise can provoke. The more substantial the changes, the more important is the review.

Complete the Editing

Top-down editing is generally most efficient. You start with the most comprehensive document features, the ones that will affect others. These features include content, organization, and visual design, all of which need to be consistent throughout the document. Bottom-up editing, beginning with sentences, can waste time and divert your attention from the comprehensive goals. You could start by correcting sentences for grammar and editing them for style and then proceed to content; however, you might ultimately decide to delete many of those edited sentences. On the other hand, if the surface errors distract you from the content, it may make sense to clean up the document first by correcting spelling and other errors. With clean copy, you can pay closer attention to more comprehensive document features. A strategy for perceiving the document structure may be to make the headings consistent—a bottom-up task that supports top-down editing.

You will probably not be able to achieve all your objectives working just once through the document. On your first pass, you might work on the content and organization. On the second pass, you can review the organization, edit for style, and make sure the document is consistent and correct. With a longer document, you can expect to make three or more editorial passes to complete all your goals. As you see the document take shape, you may revise some editorial objectives.

Evaluate the Outcome

What determines that the editing is good or right? The measures move from the text and conformity to standards of correctness to the context, where the measures are looser. As Chapter 2 explained, the best decisions about textual features reflect not just handbook standards but also what will happen to the document when it is distributed. In addition to readers and purpose, the context includes storage and disposal, budgets, related documents, and other variables.

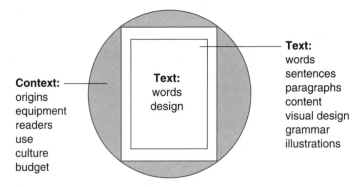

Comprehensive editing is good if it has produced a document that achieves the goals established in document analysis. The document is not simply different. The procedure itself is a good measure of quality. The editing has been based on a thoughtful analysis of document function and use and applies known principles of document design. The emendations are thoughtful and functional, not arbitrary. The editor can give a reason for each emendation, showing the logic of his or her decisions. The results can be tested, either objectively (as in a usability test with sample users) or by informed judgment.

Review the Edited Document with the Writer or Product Team

Both as a courtesy to the writer and as a check on the accuracy of your editing, you should give the writer a chance to review the edited document and to approve it or to suggest further emendations. Because you and the writer have already agreed on editing objectives, you will not surprise him or her with substantial emendations. You may need the writer to clarify ambiguous points. The writer may also have questions, and you need to be able to answer them intelligently. (See Chapter 20 for suggestions on how to confer productively with a writer.)

The outcome of the review should be either the writer's approval or identification of additional editing tasks.

Application: The Service Call Memo

Comprehensive editing of the service call memo will result in a document that differs from the copyedited memo. Figures 12.4 and 12.5, which appear later in the chapter, illustrate comprehensive editing. This discussion will show how those versions of the document emerged in the various stages of the editing process—analysis, evaluation of the content, establishment of editing goals, and evaluation of the outcome.

Analysis

Even this short memo raises some questions that may require some input from the writer. However, common sense will let us make some initial judgments about readers, purpose, and uses.

- **Readers.** This memo is addressed to "all medical center personnel." That label, however, only generalizes about readers. More specifically, the readers who will use the memo are people who will actually place service calls. Realistically, the readers are secretaries and nursing supervisors. Secondary readers may be people who need to know the procedure but who will not actually place the calls, such as department managers who will direct a support staff person to place the call. Readers will generally be receptive to the instructions, because they want to do their jobs correctly; they will be even more receptive if they can see how the new system will benefit them. But if they have had trouble in the past with the maintenance department or if they are constantly bombarded with procedure changes, they may be impatient or unreceptive.
- **Purpose.** The main purpose is to give information and instructions. For the primary readers (those who will place the service calls), the primary function is to instruct. Thus, the style and design of the document should make it easy for readers to determine what to do. A secondary purpose is to encourage cooperation and goodwill. Perhaps the procedure has changed because the previous one alienated some departments. Whatever the situation, the procedure will work best if readers want to and can cooperate.
- **Use.** Readers will not use the instructions immediately upon receiving them by mail. Rather, they will skim the memo to get its gist and then file it for reference when they need to place a service call. Thus, the document should be designed for filing, easy reference, and selective reading.

Evaluation

A critical evaluation of the memo should yield the following observations:

- The sentences and words are unnecessarily long.
- The instructions are buried in the passive voice.
- Specific information (such as the hours in which the policies are in effect) is difficult to find and interpret. Nothing attracts attention to the subject of the memo.

- The memo shows the value to the maintenance department of the change, but a "you" attitude is missing; cooperation could be better motivated if the service itself rather than the justification for it were stressed. The proposed date for implementing the new policy may occur before readers get the memo, which may exasperate them.
- Excessive capitalization creates distracting document noise.

Establishment of Editing Goals

Based on the evaluation, the following objectives would be reasonable for the service call memo.

- **Content.** The information seems complete, but emphasize some information and delete some information to fit the memo on one page. Propose a later date of implementation.
- **Organization.** Begin with an overview of the process; then give specific instructions. The information at the beginning will establish the purpose and significance of the instructions.
- **Visual design.** First, restrict the memo to one page to prevent some instructions from getting lost and to force simplification of the instructions. Second, create a memo of transmittal to explain the document if necessary, but place the instructions that will be filed on a separate single page. Third, hole-punch the left side of the instructions on the assumption that readers will file the instructions in the *Policies and Procedures* manual and that the manual is in a three-ring binder. Finally, use white space, boldface headings, and lists to make information accessible.
- **Style.** Use the language of instructions—verbs in the imperative mood.
- **Illustrations.** Do not add any illustrations. Because no equipment (other than the telephone) is necessary and no forms will be filled out, illustrations will probably not clarify the procedure.
- **Mechanics.** Establish consistent capitalization style for "service call," "maintenance department," and the manual title.

Evaluation of the Outcome

Figure 12.4 shows the results of comprehensive editing based on the established objectives. Although the message is the same as the message in Figure 12.3, this document will enable readers to read selectively and to follow instructions, the stated purposes of the document. The edited version shows a new date of implementation, but the editor must confer with the writer before such a change. The editor does not have liberty to change a policy, but thinking about the document from the reader's perspective, the editor can recommend such a change to maintain goodwill with readers.

TO: All Medical Center personnel

FROM: Tom Barker, Director of Maintenance

DATE: January 2, 1997

SUBJECT: **Service calls: New procedure**

Beginning Monday, January 13, 1997, a new procedure for placing service calls will take effect. This procedure will allow the maintenance department to respond to requests from all departments quickly and effectively.

Shorter sentences make comprehension easier.

What is a service call?

Headings in question form aid selective reading.

Service calls are for minor work requiring immediate attention. Examples:

List form lets readers skim the examples. It also clarifies that "fire" and "safety" modify "hazards" rather than being problems that require service calls.

    loss of heat or air conditioning
    water leaks or clogged plumbing
    faulty electrical wiring
    fire, safety, and security hazards

The reference to the manual relates to the definition of a service call. Reorganization groups related information.

See the *Policies and Procedures* manual (page 18) for service call requirements. Requests exceeding these requirements are classified as job orders.

How do I place a service call?

Verbs in imperative mood direct a reader to act.
A new piece of information, the request for identifying information, clarifies the three-step procedure.

Call 742-5438.

Identify the nature of the problem and the location.

Request your service call number. Use this number if you need to check on the status of your service call.

Space sets off and calls attention to the second-level heading, "for emergencies."

For emergencies

Ask the PBX operator to page maintenance. Please request paging only when you have a true emergency.

When may I call?

You may place service calls 24 hours a day.

From 5:00 p.m. through 8:00 a.m. weekdays and all day weekends and holidays, only one maintenance mechanic is on duty. He or she can respond to a limited number of calls during those times. However, you can still place a service call on nights or weekends to be completed when the mechanic is available.

Thank you for your cooperation.

FIGURE 12.4   The Memo After Comprehensive Editing

This version shows some emendations that the objectives did not specify. Because the memo fits on one page, a separate memo of transmittal is unnecessary. Some new information has been added after all, following "How do I place a service call?" Besides placing the call and requesting the number, the caller will have to identify the problem and location. The definition of service calls that previously appeared at the end is now incorporated in the introduction with other definition material. Finally, in the last section, the statement that service calls could be placed 24 hours a day contradicted the purpose of the memo to encourage users to place their calls during normal working hours. The sentence has been restructured.

The editing plan did not anticipate each emendation, but it provided direction to make editing efficient and purposeful. Other goals became apparent as the editing progressed. These goals could be incorporated into the overall plan.

You may have imagined a different document. You can probably think of ways to improve the document in Figure 12.4. Or you may believe the version in Figure 12.5 would be easier to read. Differences in editing result from different initial assumptions. If you have assumed (in contrast to the analysis outlined here) that readers will read primarily for comprehension and that they will read most thoroughly when they remove the memo from the envelope, you will probably have planned for paragraphs rather than the highly formatted version in Figure 12.4. The highly formatted version assumes selective reading at the time a service call is placed. Because so many judgments have been involved in this editing and nothing comparable to a style guide tells what form a memo giving instructions on placing service calls should take, any two editors might edit in quite different ways. Careful analysis will help keep subsequent judgments sound. Consultation with the writer will reveal assumptions that are not apparent from the document itself.

Determining Whether Comprehensive Editing is Warranted

Comprehensive editing requires more commitment to a project than does copyediting. The choice to edit comprehensively has significant consequences for the editor's time as well as for the document. Furthermore, the greater the editorial intervention, the greater the risk of changing the message. The choice will depend on several criteria.

- **Limits in your job description.** You may not have a choice. Your supervisor or the writer may set limits on the extent of editing you can do. If you are charged only with correcting grammar and spelling and making mechanics consistent, you will stop there even if you can see how different organization or format might improve it. You can always present an argument for a higher level of editing, but sometimes you must be satisfied with doing less than you could.
- The one exception to this general guideline to edit according to your job description is a situation where there is potential danger to someone's health and welfare or violation of laws or ethics. You have an ethical responsibility to make sure

TO: All Medical Center personnel

FROM: Tom Barker, Director of Maintenance

DATE: January 2, 1997

SUBJECT: Service calls: New procedure

On Monday, January 13, 1997, the Maintenance Department will begin a new service call system. This system will let us respond to all departments quickly and effectively.

Service calls are for minor work requiring immediate attention, such as loss of heat or air conditioning, water leaks, clogged plumbing, faulty electrical wiring, and fire, safety or security hazards.

Imperative verbs and "you" in this paragraph help readers identify themselves as actors. Elimination of passive voice clarifies who is to do what.

To place a service call, telephone the Maintenance Department (742-5438), 24 hours a day. A service call number will be assigned to the job. You may request this number and use it when checking on the status of your service call.

Use of "we" and "you" throughout personalizes the memo and is consistent with the reality that this procedure involves people working together.

From 5:00 p.m. through 8:00 a.m. daily, and all day on weekends and holidays, only one maintenance mechanic is on duty. He or she can respond to a limited number of service calls during these times. However, you can still place a service call on nights or weekends to be completed when the mechanic is available. If no one answers, please call again.

In an emergency, notify the PBX operator to page Maintenance. Please do not use the paging system unless your request is justifiable. We will appreciate your cooperation.

Service call requirements are outlined in the *Policies and Procedures* manual (page 18). Requests exceeding these requirements will be classified as job orders, which are also outlined in the *Policies and Procedures* manual.

FIGURE 12.5   **Memo Edited for Mechanics and Style but not for Format and Organization**

the document will not result in harm to a reader. If you are only a copyeditor but can see that a safety warning ought to be added or that a warning about risks should be emphasized, you should make the recommendation anyway.

- **Time.** If you are pressed by more important projects and the task has not been previously scheduled, you will have to stop when you have no more time.
- **Importance of the document.** Comprehensive editing is more appropriate for important documents than casual ones. The memo on service calls is important because safety is an issue. It may be important as well if productivity in the maintenance department is low because callers are abusing the service call

system. Other criteria of importance are money and number of users. If the document will be used by thousands of people or if it accompanies a product that accounts for a large percentage of your company's revenue, it's worth comprehensive editing.

- **Document anonymity.** Messages that represent an anonymous voice of the company—whose author doesn't particularly matter—can generally be edited more comprehensively than can personal statements, such as editorials or essays. The memo on service calls, though it is signed, is essentially anonymous; it is not a personal statement but rather a procedure. On the other hand, an article for a professional journal reflects an individual's point of view and requires more cautious editing. A middle ground between copyediting and comprehensive editing is editing for style but not for content, organization, or visual design. The memo on service calls, for example, would be easier to follow in paragraphs, its original form, if it used imperatives rather than passive voice. Figure 12.5 is an example of a document edited for style but not for form, content, or organization. It accomplishes the objectives of shortening sentences and words, making the instructions friendlier and more reader oriented, and clarifying the information. It pays less attention to the objective of facilitating selective reading.

Be sure to determine what outside criteria may limit your right to edit comprehensively, either before you begin or after you have evaluated the document and set objectives. And don't exceed the limits without permission or other good cause related to safety and ethics.

The Computer in Comprehensive Editing

The computer is a good tool for comprehensive editing because of the ease with which the editor can use the computer to insert and delete for content changes, cut and paste for reorganization or style changes, and reformat. The styles feature of word processing programs saves time in formatting consistently. The computer also enables editors to obtain clean copy so that they can see the effect of their editing without the distraction of cross outs and paste-overs. Also, a quick spelling check and correction prior to comprehensive editing can eliminate some superficial distractions.

Yet many editors prefer to work with hard copy in planning the editing and working with organization because the hard copy gives them a sense of the whole document. The limited amount of text that shows at any one time on a computer screen invites line-by-line editing rather than editing based on understanding of the whole document. Once you are familiar with the whole, online editing may be more efficient than hard-copy editing.

Hard copy also is necessary at certain points in the editing process if the document is reformatted, because the screen design does not exactly reflect page design. What appears satisfactory on the screen may not appear satisfactory on paper. Thus, seeing the text on hard copy can help editors make good editorial decisions about format as well as about content and organization.

On the other hand if the document is electronic, such as an online help system, it will require you to review it in the same form that readers will use it.

Summary

Comprehensive editing makes a document more usable and comprehensible. An editor bases editorial decisions on a clear concept of the document's intended purpose and uses rather than on personal preferences. Comprehensive editing requires the editor to imagine the document being used in a context and not simply to apply handbook rules.

The comprehensive editing process requires analysis of the document's readers, purpose, and uses; evaluation of the document; establishment of specific editing objectives; and consultation with the writer about the plans before the editing takes place. After the editing is completed, the writer still has review privileges.

Analysis should be top down: it should first consider the document in use and then features that influence comprehension and usability, including content, organization, visual design, and style. Top-down editing can also help an editor avoid wasting time. Editing for grammar, punctuation, and mechanics can follow.

Further Reading

Sam Dragga and Gwendolyn Gong. 1989. *Editing: The Design of Rhetoric*. Farmingdale, NY: Baywood.

Carolyn D. Rude and Elizabeth O. Smith. 1992. "Use of Computers in Substantive Editing." *Technical Communication* 39.3: 334–342.

Discussion and Application

1. The following definition of open heart surgery is prepared for patients and their families. It is intended to answer a frequent question in a way that permits patients to study and ponder the answer. It may save the surgeon time, and it may answer questions that patients neglect to ask during consultation. The definition is printed on an 8 ½ × 11-inch page and folded in half for a two-page look.

   Analyze the document's readers and possible uses in more depth, evaluate the document, and establish editing objectives for content, organization, style, visual design, and possible use of illustrations. Do not edit. As you analyze, focus on readers and the document rather than on the writer. Use "readers" and "documents" rather than "the writer" as the subjects of your analytical statements. Try to anticipate what questions the readers may ask and the ways in which they will use brochure.

If your instructor directs you to, write a letter to the writer proposing the editorial emendations. Indicate the concept of the emendations you propose, and request the writer's response and suggestions.

What Is "Open Heart Surgery"?

By Donald L. Bricker, M.D.

This question is often asked perhaps more of patients who have experienced "open heart surgery" than of physicians. It is even posed in an argumentative fashion in some circumstances, and the author has been called more than once to arbitrate as to whether a given surgical procedure was or was not really "open heart surgery." The confusion surrounding the use of the term is quite understandable, since the term "open heart surgery" was coined over two decades ago and is very vague today when applied to the large area of cardiac surgery which it may be used to describe.

Perhaps the term was coined originally because the heart surgeon was concerned with methodology which would allow him to correct congenital heart defects which actually entailed opening the cardiac chambers for repair. Yet, this nosological consideration for procedures performed within cardiac chambers overlooked the true area of common ground which set cardiac surgery apart in terms of magnitude and risk. This area of common ground was simply the need to relieve the heart of its physiological burden while operating on it. In other words, heart operations of great magnitude are best grouped together by the necessity of providing an external mechanical support system to substitute for the function of the heart and lungs in pumping and oxygenating blood. Any heart operation, therefore, which requires that the heart either be stopped for the procedure, or undergo such manipulation that it cannot perform its designated function, would fit this classification. The external mechanical support system referred to is, of course, the "heart-lung machine." What is today implied by the term "open heart surgery" in common medical parlance, then, is any cardiac operation requiring the use of the "heart-lung machine."

What does the "heart-lung machine" do? First, let us exchange that term for "cardiopulmonary bypass" to aid in our understanding. Basically, cardiopulmonary bypass removes the heart and lungs as a unit from the circulatory system and temporarily bypasses them while performing their function. To accomplish this, blood is diverted from the heart by tapping into the great veins delivering blood from the upper and lower extremities. This blood flows by gravity into an oxygenating device which performs the lungs' function of adding oxygen and dissipating carbon dioxide. This blood is then pumped back into the circulatory system through a convenient artery, usually the aorta, the great artery that comes immediately from the heart. It can be seen then, that with this system functioning, appropriately placed surgical clamps on the venous and arterial sides of the heart and lungs would totally isolate them. This allows the surgeon to stop the heart if he wishes and perform his operation in a precise and unhurried fashion. Of course, the cardiopulmonary bypass unit pump-oxygenator, or if you insist, "heart-lung machine," cannot do this job indefinitely, but sufficient time is safely at hand with today's equipment that the surgeon need not hold the concern he once did for the time factor. Improvement in this equipment has, more than any other factor, led to the successes we routinely enjoy today. One question frequently asked is about the blood supply to the heart and lungs themselves during this period of "bypass." Their blood supply is markedly reduced since they are removed from the circulatory system, but since they are at rest, oxygen requirements are minimal, and for the duration of most procedures no problems are posed.

In conclusion, "open heart surgery" has come to be a less than literal term and in common usage implies a cardiac surgical procedure requiring use of cardiopulmonary bypass. It is this factor which sets the operation apart from other operations on the heart. If your operative procedure was or is to be done under cardiopulmonary bypass, rest assured you have undergone or will undergo "open heart surgery."

Reprinted by permission of Donald L. Bricker, M.D.

2. Locate a brief document or section of a document that may benefit from comprehensive editing. This document could be a letter, short instructions, flyer, brochure, announcement, or chapter from a text. Assume that you have been assigned to edit the document. Using the procedure for comprehensive editing described in this chapter, analyze, evaluate, and set objectives for editing your document. If your instructor requests, bring the document to class and share your analysis, evaluation, and objectives with the class orally.

Chapter *13*

Style: Definition and Sentence Structures

Writers have many choices about words and their arrangements. Sentences may be long or short; words may take form as nouns, verbs, or adjectives; verbs may be expressed in the active or the passive voice. The rules of grammar govern these choices, and more like them, in only limited ways. The choices are matters of style. Style in this sense differs from mechanical style—choices about capitalization, hyphenation, and numbers—and from computer styles (format).

People use subjective terms in describing style. We may say that a document is dense, clear, or wordy (terms that reflect the reader's response to the content); or we may call the style formal, stuffy, pretentious, informed, casual, or warm (terms that reflect the image the writer projects). These impressions are created by specific components of language—words and sentence structures. Style in documents is analogous to style in clothing or music. For example, the western style in dress is the cumulative effect of components—denims, plaid shirts, boots, and belts with big buckles; the formal style in male dress is created from tuxedos, tucked shirts, bow ties, and polished shoes. Different styles in music, such as classical and jazz, are created by choices of notes, rhythms, and instruments. As with dress and music, styles in writing are appropriate according to contexts.

Creation (and modification) of a style, whether in dress, music, or language, depends on knowing component parts and options for arrangement. To make a document less stuffy or pretentious or to make it informal or casual, an editor may shorten sentences and word length and convert verbs in passive voice to active voice.

Style influences the reader's response to the content. A style that seems right for the situation and projects the image of a competent person encourages a positive response. Style, through sentence structures and word choice, also affects comprehension.

This chapter continues with a fuller definition of style. It then presents guidelines about sentence structure. Chapter 14, also on style, emphasizes words, and Chapter 15 explains social and global issues in style. This chapter will be more meaningful to you if you have reviewed the components of sentence structure (noun, verb, complement, clause) in Chapters 8 and 9.

Definition of Style

Style in language is the cumulative effect of choices about words, their forms, and their arrangement in sentences. These choices create a writer's persona (voice and image) and affect comprehension. Although the purpose of editing for style is to increase overall effectiveness, the work is done at the word and sentence levels.

Writer's Persona and Tone

The choices about words and sentence structures project an image of the writer—a persona—that influences a reader's response to the content of the document. Even an image of an objective and detached person has a corresponding style. The writer's persona reveals his or her attitude toward the subject and readers. The writer may seem to be serious, superior, disdainful, indifferent, or concerned. This attitude creates a document's tone, analogous to a speaker's tone of voice. An inappropriate tone can create noise in a document that distracts from meaning or even prevent the reader from hearing the meaning. For example, a computer manual that creates the image of writer as pal looking over the user's shoulder may irritate the user by the presumption. The familiarity of "you" as preferred in American user manuals may seem presumptuous in England.

Style and Comprehension

Style affects a reader's comprehension. Readers may be distracted from meaning by a style that is inconsistent with their expectations, such as an excessively formal style in a user manual or a casual, flippant style in a research report. Certain structures and word choices are easier to comprehend than others. Readers usually understand subject-verb-object structures better than inversions, concrete terms more readily than abstract ones, and positive expressions more readily than negative ones. Editors aim for reader comprehension because a universal purpose of technical communication is to inform. Editing for style is editing for meaning.

Because choices about style are wide-ranging, the bases for editorial emendation are less certain than they are for grammar and punctuation. Inverted structures, abstract terms, and negative expressions may usually interfere with comprehension, but sometimes they have their own meaning and should be left alone. Editors can be arbitrary or wrong; they can clarify but they can also distort meaning. The ear (what sounds better) monitors choices in a good but limited way. Editing for style requires awareness of readers and purpose, knowledge about the subject matter of the document and about principles of communication, good reading, and judgment.

Example: Analysis of Style

The effectiveness of a document depends not just on style but also on accuracy and completeness of information, organization, format, correctness, and consistency. To distinguish the effect of style, let us look at two versions of the same paragraph that differ in style but not in any other features. The paragraphs illustrate the ways in which words and structures create a style and affect a reader's response and probable action.

The example is the first paragraph of a proposal for a grant to fund some research on the effect of job complexity in sheltered workshops. Like all proposals, this one tries to persuade the funding agency that the research will be worthwhile and that the person who proposes it is capable of completing the project. To understand the paragraph apart from the whole proposal, you should know that sheltered workshops are places where people with handicaps complete tasks comparable to those in any industry but in settings where the pressures are not so great. The assumption has been that simplifying their jobs will improve their performance. The question this proposal will raise is whether the tedium that results from simplification hurts performance.

Original Paragraph

The field of job complexity in industry has a varied history encompassing as it does philosophy, economics, social theory, psychology, sociology, and a myriad of other disciplines. In the rehabilitation field, the job simplification model has been the mainstay in sheltered workshops for retarded persons almost since their inception. As in industry, the reasons for adopting the job simplification model are varied, but it is the author's contention that in the workshop the reasons are philosophical and technological rather than economic.

In evaluating the paragraph for possible editing, the editor might ask: Does the idea seem important? Does the writer project competence and other desirable attributes? Does the style seem right for the audience and purpose? If the answer to these questions is "yes," the sentences should not require editing for style.

If you have any negative responses to this paragraph, either to the idea or to the writer, try to identify them specifically and pinpoint the features of the style that created your response. To develop your subjective response into something useful, try to correlate the response with objective triggers.

Some readers have described this writer as windy and inflated. That image would be counterproductive in a proposal. Inflation gives the impression that a writer has to expand a trivial idea to make it seem important. If the writer invites a negative impression through style, the proposal concept and method will have to be proportionally stronger to persuade the people at the funding agency to approve this project rather than a competing one.

To emend an inflated style, an editor looks for stylistic features that create it. This writer uses redundant categories, words that restate a concrete subject with an abstraction, such as "field." The style is heavy on nouns and light on verbs. All the verbs are *to be* verbs, inconsistent with a proposal writer's goal of projecting a person who can act and make things happen. Identifying the choices that inflate the paragraph prepares for editing by giving direction.

Edited Paragraph

The issue of job complexity in industry encompasses philosophy, economics, social theory, psychology, sociology, and technology. In rehabilitation, sheltered workshops for retarded persons have used the job simplification model almost since their beginnings. As in industry, the reasons for adopting the job simplification model vary, but in the workshop the reasons are philosophical and technological rather than economic.

Although the editing began with some specific observations, the editing process itself revealed even more instances of an inflated style. The category word "field" in the first sentence is inaccurate, so "issue" was substituted. Maybe on first reading you slip by the "myriad of other disciplines," but having observed the writer's inflation, you may question whether any issue can encompass a "myriad" (innumerable, ten thousand) of disciplines beyond the five named. This phrase is a sophisticated "et cetera" but not a meaningful one. In the second sentence of the original, "their inception" could be interpreted to modify "retarded persons" rather than "workshops" because of its placement. In that sentence, some phrases were rearranged to put key words in the subject, verb, and object, and "beginnings" was substituted for "inception." "Are varied" becomes "vary" to put the action of the sentence in the verb rather than in an adjective. Three of the *to be* verbs are now action verbs. The writer's self-reference is deleted; the context and phrasing reveal the statement to be a thesis rather than a fact.

What difference has editing made? The image of the writer is more subtle, but for the problem statement of a proposal it is better to be invisible than to get in the way of the meaning. The edited paragraph is also shorter than the original, but conciseness has been a happy consequence of editing to make sentences reveal their meaning rather than an end in itself. Eliminating inflation, using precise terms, and placing key words in the subject, verb, and object have focused attention on the problem that needs to be solved.

Analysis of style helps an editor evaluate the content. Some clarification of meaning, especially the substitution of "issue" for "field" and the deletion of the "myriad" phrase, have resulted from an analysis of style. One content addition in the first sentence, "technology," results from comparison of the list in the first sentence with the descriptors in the final sentence. Modifying word choices and sentence structures is risky because it can change meaning. Careful reading has preceded all of these changes to prevent errors.

Guidelines for Editing for Style

Although editors cannot depend on rules about style, they can use guidelines that have been established by experience and by research. These guidelines parallel the guidelines for effective writing style. Analysis for style may consider context, sentence structures, verbs, and other words.

context	Make style serve readers and purpose; consider the possibility of an international audience and the possibility of discrimination.
sentence structures	Use structure to reinforce meaning.
verbs	Convey the action in the sentence accurately and forcefully.
nouns, adjectives	Choose words that are accurate, concrete, and understandable.

These guidelines cannot be used as rules and may be inappropriate in a specific situation. Like all guidelines, they point but do not prescribe. Using them requires judgment.

Context: Make Style Serve Readers and Purpose

Context refers to the situation that surrounds the text. It refers to readers, document uses, physical conditions, other related documents, and more (see Chapter 2). Editors must make word and sentence structure choices that are appropriate for the context. Just as no one style in dress or music is suitable for all situations, neither does one style work for all technical documents. In the example on sheltered workshops, the justification for editing at the word and sentence levels was the need for the proposal to compete with other proposals. In contrast, a style for a computer manual for new users would have been less formal. One edits a scientific paper and article for a popular magazine according to different criteria. Editing work that will be translated into other languages requires knowledge of the language and culture. Contextual and social issues of style are discussed more fully in Chapter 15.

Sentence Structures: Use Structure to Reinforce Meaning

Like many structures, sentence structures are hierarchical. The independent clause, which makes a sentence grammatically complete, is the strongest part of the sentence. It is the structural core of the sentence. The structural core is the logical place for the main idea of a sentence. (See Chapter 9 for a review of these terms and identification of basic sentence patterns.)

Other structural signals about meaning are the patterns of coordination and subordination. An idea expressed in a structurally subordinate part, such as a dependent clause, phrase, or modifier, seems inherently less important than an idea in an independent clause. If two ideas are related and equal in importance, coordination and parallel structure may reinforce that relationship. When the main idea is buried in a dependent clause or phrase, or when parallel ideas do not have parallel structure, comprehension is more difficult because the reader has to overcome misleading structural signals about meaning.

The relationship between structure and meaning is the basis for this principle of editing for style: *make structure reinforce meaning.* Several guidelines apply this principle.

Place the Main Idea of the Sentence in the Structural Core

The main idea should appear in the structural core of the sentence, the subject and verb of the independent clause. Consider the following sentence:

The <u>course</u> of the twentieth century <u>produced</u> a cancer death rate that rose parallel to the advances in technology.

The core sentence is underlined, "course . . . produced," but the meaning appears elsewhere. The sentence is not about the course of the twentieth century and what it produced at all but rather about how the rising cancer death rate paralleled advances in technology. "Course" and "produced" are the least important words in the sentence, but they take the strongest structural part. The topic of the sentence differs from the grammatical subject. The meaning does not appear in the core sentence, and the structure does not reinforce meaning. Style interferes with comprehension. Two options for revision are these:

In the twentieth century, the <u>rise</u> in the cancer death rate <u>paralleled</u> the advances in technology.

In the twentieth century, the death <u>rate</u> from cancer <u>rose</u> parallel to the advances in technology.

Both revisions guide readers to the main idea using structural signals.

A good editorial strategy is to look for grammatical subject of the sentence to see if it is identical with the topic of the sentence (that is, whether the structure matches what the sentence is about). Then look for the verb. The following sentence may seem "wordy," but the reason is that the subject and verb include redundancies that lead to confusion.

The department's policies related to standards of behavior must be firmly maintained and affirmed.

One wordy phrase is the subject, "policies related to standards." What is the topic, "policies" or "standards"? "Policies" is the grammatical subject, but the real issue is standards, not the policies that define them. The department wants good behavior. Standards define behavior, and policies define standards. The policies are one level of abstraction further from the desired behavior. The end of the sentence is also redundant, especially with the adverb. (To "firmly affirm" seems like an overstatement that paradoxically weakens the insistent tone of the sentence.)

> The Department's standards of behavior must be maintained.

Sentences that open with the words "there are" or "it is" often waste the core on a simple declaration. As a result, the main idea of the sentence may be buried in a dependent clause. Furthermore, such openers delay the significant part of the sentence.

> It is often the case that a herniated disc ruptures under stress.

> Often, a herniated disc ruptures under stress.

> It is possible to apply for the scholarship by completing either of two forms.

> You may apply for the scholarship by completing either of two forms.

> There are two expenses to be justified.

> Two expenses must be justified.

In each of these pairs, the core sentence in the revised version gives more information. The sentences are more efficient not just because they are shorter but also because the key idea resides in the core.

"There are" and "it is" may be preferable to alternatives, especially when they introduce a list. "Exist" rarely substitutes effectively for the *to be* verb.

> There are three reasons for this problem. [effective use of the "There are" opener because
> it prepares readers to expect a discussion of the three reasons]

> Three reasons exist for this problem. [substitution without a gain in effectiveness]

The purpose of the sentence is to prepare readers to hear the three reasons; the first version does just that. Before you delete "there are" in order to use the sentence core efficiently, consider the whole situation.

Use Subordinate Structures for Subordinate Ideas

When the sentence pattern is complex (a main clause plus a subordinate clause), the pattern itself communicates the relationship of main and subordinate ideas. If structure and meaning conflict—that is, if the structure affirms one relationship but the words affirm another—comprehension will be more difficult. Or the words may lead the reader to one interpretation while the aim was another.

A letter of application for a job as bank teller included the following sentence:

> I was elected treasurer of a social fraternity, which enabled me to collect and account for dues.

The situation helps an editor determine which facts are subordinate. "Was elected" affirms leadership ability and the confidence of peers. Yet a person hiring a bank teller may be interested more in whether the applicant can handle money accurately. For this reader, then, the information in the subordinate clause is more important than the information in the main clause. An appropriate stylistic revision emphasizes the skills that matter in this context:

> As elected treasurer of a social fraternity, I collected and accounted for dues.

The following sentences might shape a reader's desire to purchase an old home in need of repair:

> Although the beams show signs of dry rot, the house seems structurally sound.

> Although the house seems structurally sound, the beams show signs of dry rot.

The second version of the sentence leads a reader to conclude that the purchase is risky. The idea of structural soundness is subordinate to the idea of damage.

Use Parallel Structure for Parallel Items

Parallel structure can reinforce relationships between items in a series or a compound sentence. All the items should be the same part of speech (nouns, verbs, participles) or same type of phrase. In a compound sentence, both clauses should use the active voice (or the passive voice); a mixture requires a shift in the way the reader processes the information.

> We can help to keep costs down <u>by learning</u> more about the health care system, <u>how to use</u> it properly, and <u>by developing</u> self-care skills.

The faulty parallelism in this sentence makes it difficult to determine whether there are three ways to keep costs down (as the punctuation suggests) or two (as the parallel phrases "by learning" and "by developing" suggest). Two revisions are possible:

> We can help to keep costs down <u>by learning</u> more about the health care system, <u>by using</u> it properly, and <u>by developing</u> self-care skills.

> We can help to keep costs down <u>by learning</u> more about the health care system and how to use it properly and <u>by developing</u> self-care skills.

Parallel structure, with appropriate punctuation, encourages the correct interpretation. Which is the better sentence of the two revisions? The answer depends on the middle phrase, regarding use. Reasoning suggests that learning how to use the health care system in itself will not keep costs down, but using the system correctly will. This argument favors the first version, in which "using" is a separate strategy rather than something to learn. Furthermore, the first version is easier to read because the series of three is a simpler structure than the double compound in the second version (the two parallel phrases plus the compound object of the preposition, "learning").

The repetition of the pronoun *by* is optional, but it can signal relationships, and you should be consistent in repeating it or not.

Sentence Arrangement

English sentence structure allows for different arrangements of parts, but for clarity and cohesion (the connection of sentences), some patterns of arrangement work better than others. The beginnings and endings of sentences matter the most.

Place the Subject and Verb Near the Beginning of the Sentence

A writer can anticipate that readers, even those who skim, will pay attention to the beginning of the sentence. An introductory modifying phrase is fine, but a delay of the subject and verb keeps readers wondering what the sentence is about. Likewise, separating subject and verb makes it hard for readers to keep them together in their minds. The core of the sentence belongs near the beginning, with subject and verb close together.

Stylistic problems tend to cluster, and the example that follows reveals problems in addition to the delay of subject and verb, but the goal of placing subject and verb near the beginning of the sentence leads to other improvements as well.

> In discussions with staff members in other divisions, <u>observations</u> in support of the budget assessment by the manager <u>were made</u>. [The subject is delayed until word 9, and the verb is delayed until words 19 and 20.]

> <u>Staff members</u> in other divisions <u>concurred</u> with the manager's budget assessment.

Arrange Sentences for End Focus and Cohesion

Sentence structure not only clarifies the meaning of individual sentences but also connects one sentence to the next, so that the larger text structures (paragraphs, sections) make sense. These connections between sentences develop from the relationship between the end of one sentence (predicate) and the beginning of the one that follows (subject). In the subject and predicate sentences follow a pattern of known-to-new information. The subject states familiar information. The predicate, in commenting on the subject, adds new information. This information, now known or familiar, becomes the subject of the next sentence. This pattern of known-to-new is called the "known-new contract," and it explains cohesive ties between sentences.

The paragraph you have just read illustrates the known-new contract and the cohesive ties that result from arranging sentences to place known information in the subject and new information in the predicate.

[1] *Sentence structure* is a familiar or known topic because it is the subject of the chapter. A new word, *connects,* appears in the predicate.

[2] *Connect* from the predicate of sentence 1, now a known term, becomes the subject, *connections,* in sentence 2. Key words in the predicate of sentence two are *end* and *beginning, predicate,* and *subject.* These words are echoed in the modifying phrase at the beginning of sentence 3.

Sentences 4, 5, and 6 work together to elaborate on sentence 3. "Known to new" of sentence 3 appears explicitly in sentence 7.

[1] <u>Sentence structure</u> not only clarifies the meaning of individual sentences but also <u>connects</u> one sentence to the next, so that the larger text structures (paragraphs, sections) makes sense. [2] These <u>connections</u> between sentences develop from <u>the relationship</u> between the <u>end</u> of one sentence <u>(predicate)</u> and the <u>beginning</u> of the one that follows <u>(subject)</u>. [3] In the <u>subject and predicate</u>, sentences follow a pattern of <u>known to new information</u>. [4] The subject states familiar information. [5] The predicate, in commenting on the subject, adds new information. [6] This information, now known or familiar, becomes the subject of the next sentence. [7] This <u>pattern of known to new</u> is called the "<u>known-new contract</u>," and it explains <u>cohesive ties</u> between sentences.

Because readers expect the predicate to add a substantial comment about the subject, good writers (and their editors) structure sentences to place important information at the end. Sentences generally should not dwindle into modifying phrases or weak predicates, such as those that merely declare that something "is important." Instead, the predicate should add new information that the next sentence will comment on. Structuring sentences to emphasize the end is called *end focus.*

The introductory modifying phrase or clause is a common pattern in English sentences because the information that tells when, why, or how is rarely the most important information. Placing the modifier at the beginning establishes the context for the information in the subject and predicate and leaves the end of the sentence free for the significant comment of the predicate. The modifier may provide a transition to the previous sentence; it is not the place to introduce new information.

The sentence that follows wanders through a long beginning and then drops off in the predicate into an almost meaningless statement ("is apparent"). The sense of empty words is increased by the lack of an agent (who is to seek and develop?) and a weak verb (is). The first revision shifts the modifying phrase to the beginning, uses a human agent as subject, and ends with an important point. The second revision avoids the ambiguity of "in the future," which may seem to justify postponing action, and substitutes "have the opportunity," which may better encourage action right now.

> Room for aggressively seeking and maintaining new markets for existing products and developing new products in the future is apparent.

> In the future, we must aggressively seek and maintain new markets for existing products and develop new products.

> We have the opportunity to seek and maintain new markets for existing products and to develop new products.

Prefer S-V-O or S-V-C Word Order

The subject-verb-object (S-V-O) or subject-verb-complement (S-V-C) pattern is the most common one in English. Its familiarity makes it easy to understand. Inversions of the pattern require extra mental processing. The sentence you just read is an example of S-V-O order:

> Inversions of the pattern require extra mental processing.
> S   V   O

A subject complement is either a substitute for a subject or a modifier, as in this sample of S-V-C order:

> A complement is a substitute.

Inversions of the order, because they use patterns that are less familiar to readers, call attention to themselves and possibly away from meaning.

> The jurors will no favors grant.
> S   O   V

> Favored by committee members is the plan to renegotiate his contract.
> C   V   S

Sentence Length

Sentence length influences readability and the sophistication of content. Length can make sentences seem to drag or to sound like a child's book. A reader's impression of length depends not just on the number of words but also on other choices that make sentences seem long, such as stacked phrases, depersonalization, separation of subject and verb, and negative constructions. These choices may actually encourage writers to construct long sentences.

Adjust Sentence Length to Increase Readability

If sentences within a document all contain roughly the same number of words, the reading will become monotonous. The rhythm will lull the readers rather than keeping them alert. Variety in sentence length may be desirable for its own sake, but variety also can be used to emphasize key points. A very short sentence surrounded by longer ones will draw attention and thus influence a reader's perception of the significance of the content.

Long and complex sentences are generally more difficult to understand than short and simple ones because they require a reader to sort and remember more information and more relationships. Yet they can be easier to understand than a series of short sentences just because they do establish relationships and help with interpretation. For example, a complex sentence (with a dependent and an independent clause) could show a cause-effect relationship. The first two sentences in this paragraph do just that. By contrast, the following series of simple sentences have a childlike quality:

> Long and complex sentences are hard to understand. Short and simple sentences are easy to understand. Long and complex sentences require a reader to sort information and relationships. They require a reader to remember more information than short sentences.

Length is relative to readers. Adult readers who know the subject appreciate the information they receive from complex sentences. But readers who have limited backgrounds on the subject will need to absorb the new information in smaller and simpler chunks. Also, length describes more than just a number of words. Sentences constructed from a series of phrases become too long more quickly than do sentences constructed with several clauses. If the sentence core gets lost in a series of modifying or prepositional phrases, the sentence is too long for easy comprehension.

Use People as Agents when Possible

One good way to create stuffy reading that puts readers to sleep is to keep people out of the sentences. Sentences full of abstractions seem longer than sentences in which human beings perform actions. Human agents engage readers.

The following sentences from an employee policy statement aim to clarify the terms of vacation eligibility. Employees become eligible for vacation after they have worked a year, but these sentences raise the question of what constitutes a year if an employee misses work time because of an accident or illness. The long sentence and separation of subject and verb increase reading difficulty, but employees will also struggle to see themselves in the action because they are never named. Edited version B takes advantage of headings and bulleted lists to increase readability, but it also uses human agents, shorter sentences, and closer connections between subject and verb.

Original

After the successful completion of the first 90 days of employment, <u>time lost</u> as a result of an accident suffered during the course of employment as recognized by the Workers' Compensation Board, <u>shall be added</u> to time worked during the vacation year to qualify under the provisions of vacation leave. Also, after the 90-day period, time lost as a result of a non-occupational accident or illness not to exceed 520 hours shall be added to the time worked during the vacation year.

Edited Version A *(sentence core near the beginning; human subjects)*

After <u>you</u> <u>complete</u> 90 days of employment, the <u>work time</u> you lose because of accident or sickness <u>may count</u> as time worked for purposes of calculating vacation eligibility. If the time lost results from a work-related accident that the Workers' Compensation Board recognizes, the time you lose will be added to the time you worked for counting when you have worked a year. If the time lost results from a non-occupational accident or illness, up to 520 hours of the time you lose will be added to the time you worked.

Edited Version B *(editing for style plus format)*

After you complete 90 days of employment, the work time you lose because of accident or sickness may count as time worked for purposes of calculating vacation eligibility.

- **Work-related accident.** If the Workers' Compensation Board recognizes that your accident occurred as you were completing work duties, the time you lose for the accident counts toward time worked.
- **Non-occupational accident or illness.** Up to 520 hours that you lose because of an accident or illness that is not related to work counts toward time worked.

Prefer Positive Constructions

Negative constructions will make sentences seem longer than they are because they increase reading difficulty. A double negative requires extra steps of interpretation. These steps slow down reading and increase the chance of error in comprehension.

It is <u>not</u> <u>un</u>common for employers to require writing samples from applicants.

Readers have to use "not" to cancel "un" before they understand that writing samples are common in job applications. They would comprehend more quickly if the sentence began, "It is common . . ." or "Employers commonly. . . ."

How long does it take you to figure out the meaning of the following sentence?

The <u>elimination</u> of disease <u>doesn't</u> guarantee that we <u>won't</u> die according to genetic timetables.

The three negatives ("elimination," "doesn't," "won't") are especially troublesome because the concept of "genetic timetables" is difficult. Though the following versions alter the emphasis, they are easier to understand.

Even if we eliminate disease, people may still die at the same age they do now because of genetic timetables.

Elimination of disease doesn't guarantee a longer life. Genetic timetables, rather than disease, may establish the lifespan.

Simple negative constructions in two clauses also load extra interpretation responsibilities on readers:

It is <u>not</u> possible to <u>reduce</u> inflationary pressures when the federal government <u>does not reduce</u> its spending.

These edited versions are easier to understand:

Inflation will continue if the federal government keeps on spending at the same rate.

Inflation will decrease only if the federal government reduces its spending.

In this next version, the negative construction remains in one clause, for emphasis. Even elimination of one negative construction, however, makes the sentence easier to understand.

Inflation will not decrease unless the government reduces its spending.

Positive language has a psychological benefit, as well as being easier to understand. In the next example, the company that resolves a complaint may negate its efforts to help if its letter ends with a reference to the problem:

If you ever have any more problems with our company, do not hesitate to call.

The edited version anticipates a positive future relationship:

If our company can serve you in the future, please feel free to call.

Summary

Style, the effect of word choices and sentence structures, either aids or interferes with comprehension. Style is not just a decoration but rather is a matter of substance. Style also affects a reader's attitude toward the document. The writer's persona can encourage or alienate readers.

Using sentence structure effectively has the power to clarify meaning. The sentence core—the subject and verb—announces what the sentence is about and what is important. Subordinate and parallel structures signal relationships of ideas. The beginning and end of the sentence help readers connect one sentence and its content to the next.

The range of choices in words and sentence structures is so wide that rules cannot prescribe the choices, and even guidelines must be used wisely—as guides, not as limits. Reading carefully and knowing the context for use can help an editor apply these guidelines.

Further Reading

Walker Gibson. 1969. *Persona: A Style Study for Readers and Writers.* New York: Random House.

Daniel Jones. 1998. *Technical Writing Style.* Boston: Allyn & Bacon.

Richard A. Lanham. 1991. *Revising Prose,* 3rd ed. Englewood Cliffs, NJ: Prentice-Hall.

Joseph M. Williams. 1997. *Style: Ten Lessons in Clarity and Grace,* 5th ed. New York: Addison-Wesley.

Discussion and Application

1. Edit the following sentences to use the sentence core effectively and to place subject and verb together near the beginning of the sentence.

   a. The focus of the test for Experiment II was tangential to and not a direct approach to leadership.

   b. The major framework of her essay involves presenting a discussion of health care funding.

   c. Some studies have revealed that there has been a small increase in mastitis cases involved with BST-supplemented cows.

   d. The expected results of the use of BST supplements is an increase in the profitability of the dairy producer operations.

   e. The reason why video vignettes are less used than lecture in teaching ethics is because of the higher cost and less available resources.

2. Edit the following sentences to reinforce meaning by use of parallelism and subordination.

   a. The student health center offers a variety of services such as physician appointments, mental health, health education programs, and lab and X-ray work.

   b. The report considers factors important to students in choosing a medical facility, ratings of services at the student health center, reasons why students do not use the center, and offers suggestions for increase in student usage.

   c. To become a mutual fund shareholder, an investor places an order with a local securities dealer or by contacting the fund sales staff directly.

   d. Even if callers refuse to state their name, messages can still provide helpful information. For example, some people call to report a change of a meeting, to file a complaint, or any other kind of message.

   e. There are two main elements that a tennis racket should deliver—power and control.

3. *Describe the different interpretations these two versions of the same sentence might invite:*

   a. Although the region is engaged in extensive tourist promotion, the historical site is accessible only by a secondary route.
   b. Although the historical site is accessible only by a secondary route, the region is engaged in extensive tourist promotion.

4. Discuss the suitability of the persona and style of the following paragraph assuming that the passage appears in a manual for volunteers assisting a probation officer. Discuss the suitability of the persona and style assuming the passage appears in a law textbook. How do context and readers influence your description of the style? What particular stylistic features make the passage appropriate or not for the two contexts?

   > Probation is a method of disposition of a sentence imposed on a person found guilty of a crime. It is a court-ordered sentence in lieu of incarceration. The offender remains in the community under the supervision of a probation officer for a predetermined period of time. If compliance with the terms and conditions of probation set by the court is made by the offender, he or she is discharged from the court's jurisdiction and the debt to society is considered paid. If compliance is not made, another method of sentence may be imposed which may include incarceration.

5. Sentences from the previous paragraph are reproduced here. The beginnings of revisions follow the sentences. With one or two classmates, continue editing the sentences with these aims: Use a human agent (person, offender, judge, probation officer) where you can. Use the subject and verb for important words. Place the subject and verb close together, near the beginning of the clause.

   You may use two sentences in place of one or combine sentences so long as you don't change the meaning.

   a. Probation is a method of disposition of a sentence imposed on a person found guilty of a crime. It is a court-ordered sentence in lieu of incarceration.
   When a person is found guilty _____, the court may _____.
   b. If compliance with the terms and conditions of probation set by the court is made by the offender, he or she is discharged from the court's jurisdiction and the debt to society is considered paid.
   If the offender _____, the court _____.
   c. If compliance is not made, another method of sentence may be imposed which may include incarceration.
   If the person_____, the court _____.

6. Discuss the consequences of your editing of the previous sentences on clarity, length, and tone. What has editing achieved? What has been sacrificed? What readers would prefer the edited sentences? What readers might prefer the first version?

7. Discuss your process of analysis and editing in #5. What issues did the group members need to resolve before agreeing on the words and arrangement of sentences? What were the uses and limitations of the guidelines for sentence structure?

8. Consult a professional journal and the style manual in your field. Do they give guidelines to authors for style? If so, summarize the guidelines for sexist language, use of passive voice, and other issues that they emphasize. In class share these summaries to determine where the disciplines agree and disagree.

Chapter *14*

Style: Verbs and Other Words

Once you understand the effect of sentence structure on style and comprehension (see Chapter 13), you will have a good basis for editing words. Good structures depend on good choice of subjects and verbs in particular. In order to convey meaning, these parts of the sentence must be accurate and specific. Good choices of words also add energy to writing and make it more readable.

This chapter continues the discussion of style begun in Chapter 13 with an emphasis on verbs and concrete nouns.

Verbs: Convey the Action in the Sentence Accurately

Sentences often falter stylistically because of their verbs. Writers can easily identify subjects and use nouns; more effort is required to figure out what the nouns *do*. Evaluating verbs is a necessary step in determining whether a sentence conveys its meaning. When you edit for style, consider verbs after you evaluate context and sentence structure.

Build Sentences around Action Verbs

Readers approach sentences wondering, "Who did what?" or "What happened?" The answer to the question should appear in the verb. When the action that should be conveyed by the verb is lost in imprecise substitutes or in other parts of speech, including nouns and adjectives, sentences fail to communicate effectively. Editors can help writers find their verbs.

In the following sentence, the verb does not tell what the subject does:

> A one-year <u>warranty</u> <u>was placed</u> on the tape deck through March 1998, guaranteeing that all parts and labor would be covered during this time.

The fact that the warranty "was placed" is less important than how long it lasts and what it covers. "Placed" is a weak, imprecise verb. A reader has to insert mentally the answer to the question of what the subject (the warranty) did. Putting the action in the verb, the structure where readers expect to find the answer to their question, would clarify the meaning.

> This tape deck's one-year warranty <u>extends</u> through March 1998 and <u>covers</u> all parts and labor.

The new verbs, "extends" and "covers," are stronger than the original verb, "was placed," because they tell more specifically what the warranty does. The action in the original sentence is buried in a past participle ("would be covered") in the case of the second verb and is only implied in the case of the first ("through March 1998"). The revision increases reader comprehension and reduces possible misinterpretation by eliminating the word "guaranteeing." A "warranty" and a "guarantee" mean different things in a legal sense.

This sentence also hides the action by using an imprecise verb.

> The <u>stream of air</u> that escapes the larynx <u>experiences</u> a drop in pressure below the vocal folds.

An editor examining the sentence core might ask: Can air "experience"? Does this verb tell what the subject does? The verb works with the subject grammatically but not logically. An editor can spot the hidden verb in the noun "drop."

> The <u>stream of air</u> that escapes the larynx <u>drops</u> in pressure below the vocal folds.

If you can help writers find their verbs, the main action will appear in the sentence core, and structure will reinforce meaning.

Choose Strong Verbs

Many writers draw on a limited repertoire of verbs. Comparatively weak verbs describe everything that happens in their writing:

is (or other variations of *to be:* are, was, were, will be)

have (has, had)	make	involve	provide
add	give	concern	become
deal with	do	reflect	use

Nothing much happens in these verbs. These all-purpose verbs work in many contexts because they describe many possible actions. As an editor, you won't eliminate all uses of these verbs, for occasionally they may be the best choice. But if such verbs predominate, you will have to hunt for the real meaning and substitute the appropriate verb. Consider the following example:

> The report <u>will deal with</u> the third phase of the project.

Although this sentence tells the reader the general subject of the report, the reader might have learned more had the verb been more specific. Will the report describe, state results, evaluate, or give instructions for the project's third phase?

The paragraph in the left column below uses weak verbs (*hold, are, have*). The second paragraph uses stronger verbs (including infinitive forms of verbs) to convey more action and decisiveness. You would probably prefer to read a report with verbs such as those in the paragraph on the right. The strong verbs were implied in the nouns—called *nominalizations.*

The City Council will <u>hold</u> a meeting for discussion of the possible expansion of the basketball arena. If Council members <u>are</u> in agreement about the need, they <u>will</u> <u>have</u> a consulting firm <u>proceed</u> with investigation of the feasibility.	The City Council will <u>meet</u> to <u>discuss</u> whether to <u>expand</u> the basketball arena. If Council members <u>agree</u> about the need, they will <u>hire</u> a consulting firm to <u>investigate</u> the feasibility.
Verbs: hold, are, have, proceed	*Verbs: meet, discuss, expand, agree, hire, investigate*

Avoid Nominalizations

One way to weaken verbs is to "bury" them in nouns or adjectives. A verb turned into a noun is a nominalization. Suffixes, such as *-tion, -al,* and *-ment,* convert verbs into nouns, as the following list illustrates:

Verb	Suffix	Nominalization
admire	-tion	admiration
agree	-ment	agreement
transmit	-al	transmittal
depend	-ence	dependence
rely	-ance	reliance
solder	-ing	soldering

Nominalizations do serve valid purposes in writing. They become problems only when they obscure what is happening, as when the action a sentence seeks to communicate is disguised as a thing. The form conflicts with meaning and therefore makes comprehension more difficult. Furthermore, because sentences full of nominalizations are wordy and lifeless, they discourage reading.

Verbs can become adjectives, too, in their participle form (with the addition of the suffixes *-ing* and *-ed*).

Some <u>magazines</u> <u>are</u> more specialized than others by dealing with just one topic, such as science or art.

The verb is the weak *to be* verb, "are." The writer threw in "dealing with" as well, sensing the need for a more specific action. The real action, however, appears in the past participle, "specialized."

Some <u>magazines</u> <u>specialize</u> in just one topic, such as science or art.

Well-chosen verbs will clarify meaning and enliven prose with action.

Prefer Active Voice

Voice refers to the relationship of subject and verb. In the active voice, the subject performs the action represented by the verb. The subject is the agent of action.

The board <u>reached</u> a decision.
    subject   verb

The subject of the sentence, "board," performs the action identified by the verb. In the passive voice, the subject receives the action identified by the verb. The subject is the passive object of action, or the recipient of action.

A decision <u>was reached</u> by the board.

The subject of the sentence, "decision," does not do the reaching. Rather, it receives the reaching. Don't confuse passive voice with past tense or with weak verbs. A sentence in passive voice always has these components:

a *to be* verb
a past participle

Some statements seem passive because nothing much happens in them, but they are not in passive voice unless they include a *to be* verb and a past participle.

Reasons to Prefer Active Voice

The use of the verb "prefer" in this heading is intentional. Active voice is preferable in many situations, but editors should not arbitrarily convert passive voice sentences to active voice. You need to know the reasons why it is preferable—usually—and why it is not—sometimes.

- **Adds energy to writing.** Active voice conveys directly that people do things or that things happen. In passive voice, the emphasis is on the result, the thing rather than the action.
- **Establishes responsibility.** Active voice is also preferable in many situations for ethical reasons. Sometimes writers use passive voice because they do not want to reveal who the agent of the action was.

    A decision <u>was reached</u> that you should be fired.

The writer of that sentence may not want the reader to know who made the decision. The passive voice protects the agent from identification.

Passive voice can mask responsibility for future actions as well as for past ones. Consider the group that plans to place a microwave oven in the company lunchroom. One person asks about the effect on workers who wear pacemakers. "A sign will be posted," says one. By whom? Who will take responsibility for posting the sign? A sentence that does not declare responsibility is less likely to shape subsequent action than one that tells who will do the task. "The supervisor will post a sign" or "I will post a sign" gives a person a task and thereby increases the chance that the task will be performed. If responsibility is not assigned, the members of the group may all assume that someone else will do the job.

Reasons to Prefer Passive Voice

The arguments for active voice are persuasive, so the reasons for choosing passive voice—in some situations—must be strong.

- **The agent is insignificant or understood.** Sometimes it doesn't matter who has done or will do something. The proper emphasis, then, is on the recipient of the action rather than on the agent.

    Computer <u>chips</u> <u>are made</u> of silicon. [passive voice; emphasizes the object of the action of making]

The purpose of this sentence is to identify the material that forms computer chips, not who makes them. The agent is irrelevant (unless context tells us otherwise).

    <u>Manufacturers</u> <u>make</u> computer chips of silicon. [active voice: introduces an irrelevant agent]

The context may make the agent of action clear and negate the need to name the agent in the sentences. A policy statement directed to supervisors, for example, may establish in an opening paragraph or heading that the statement identifies policies for supervisors. To write in active voice throughout the policy statement would require the repetition of "supervisors" as the subject in many of the sentences. This repetition would therefore subordinate the policy, and the action that results from the policy, to the agent.

- **Readers expect passive voice.** A publication or an organization may establish passive voice as the preferred style. Some of the sciences, for example, maintain the convention of passive voice sentences, with the writer's voice, or at least the "I," invisible. The reason for minimizing the "I" is to convey a sense of the writer's objectivity—the facts rather than the interpreter predominate. Thus, a science writer may write

It was determined that...

rather than

We determined that...

Though a number of studies have challenged the presumed objectivity of the passive voice, you should respect an organization's expectations in your editorial decisions. Readers will be distracted by variations from the norm and may discredit the findings of a writer who does not use the language of the community and therefore seems not to belong to it.

- **Creates cohesion and focus.** Because of the known-new pattern of sentences, the subject of one sentence generally restates an important idea in the predicate of the preceding sentence. (See Chapter 13.) Passive voice gives writers the option of using as the subject of a sentence a recipient rather than an agent of action. Stronger cohesive ties and focus may result. In the following pair of sentences, the first sentence focuses on readers: the second one focuses on style. If the point of the sentences is to suggest the effect of style on readers, the first revision, using passive voice to place "readers"in the subject position of the sentence, does a better job of achieving that focus.

    Style affects a <u>reader's comprehension</u>. <u>Readers</u> may be distracted from the meaning by a <u>style that is inconsistent</u> with their expectations. [passive voice to create a cohesive tie between the subject of the second sentence and the predicate of the previous sentence]

    Style affects a <u>reader's comprehension</u>. A <u>style</u> that is inconsistent with their expectations may distract readers from meaning. [active voice to emphasize the topic of style]

Use Concrete, Accurate Nouns

Readers depend on accurate words for the details as well as for forming concepts. If the words are inaccurate or difficult to understand, the reader must either apply extra mental effort to substitute the correct word or "learn" incorrectly. Writing that features abstract nouns, phrases or pairs rather than single words, and complex rather than simple words may be hard to understand. Yet, since changing a word changes meaning, editing of words must be cautious.

Concrete words evoke one of the senses—sight, sound, taste, touch, odor, motion. Words that help us "sense" the meaning are easier to understand than are abstractions. Technical writing is often more concrete than philosophical writing because the subject matter is objects and actions (though it can be philosophical and abstract as well).

All-purpose nouns are the close cousin of all-purpose, weak verbs. They often fail to convey a precise meaning, as these examples illustrate:

areas	aspects	considerations
factors	matters	

Try to imagine what the following sentence means:

These aspects are important considerations for this area.

Abstractions can be functional when they introduce a list. The concreteness appears in the list that follows the category.

Some abstract nouns appear in the next section in phrases and redundant categories. Their translations offer more concrete nouns.

Prefer Single Words to Phrases or Pairs and Simple to Complex Words

When a writer unnecessarily complicates information, readers work harder to comprehend.

Phrases vs. Single Words

Some writers try to sound sophisticated and knowledgeable by using phrases in place of single words and multisyllabic words when simpler words are more accurate. A restroom may be referred to as a "guest relations facility," and a hammer as a "manually powered fastener-driving impact device." Such a style may please a writer, but it rarely pleases a reader. In *1984*, George Orwell called such circumlocutions "doublespeak." The National Council of Teachers of English (NCTE) gives out doublespeak awards for the "best" examples each year. The military and government have won with these creations:

Phrase	Translation
unlawful or arbitrary deprivation of life	killing
permanent pre-hostility	peace
violence processing	combat
collateral damage	civilian casualties in nuclear war
frame-supported tension structure	tent

Military and government writers are not the only inventors of phrases when a word would do. A hospital described "death" as "negative patient care outcome." In finance, a "negative investment increment" means "loss."

These examples illustrate that a negative subject may motivate inflation of language: the phrases are euphemisms for unfortunate or tragic outcomes. But even without this motivation, writers often wander around a specific subject without identifying it. A style that inflates and abstracts loses clarity. If the subject matter involves potentially dangerous mechanisms or chemicals, such violations of clarity may even cause harm to the readers.

Multisyllabic Words

Multisyllabic words are more difficult to understand than their one-syllable synonyms; they also take up more room on the page and take longer to read. They may create the appearance of pretense rather than of sophistication. Readers appreciate the simpler version.

Multisyllabic Word	Synonym
utilize	use
effectuate	do
terminate	end

Won't readers respect the writer who is able to use the multisyllabic word more than the writer who uses the single-syllable synonym? This is a question you may be asked by a writer who objects to editing for simplification of words. The answer is not easy. This guideline does *not* advise you as editor to substitute imprecise generic terms for specific technical ones. It does *not* insist that you eliminate jargon. It does *not* advise you to edit all documents to the same simple reading level. However, it does advise you that writers rarely gain respect on the basis of an ability to use inflated words. They are respected because they have gathered and interpreted data in a credible way, or because they have solved a problem, or because they have helped a reader perform a task accurately. If writers rely on multisyllabic words because their substance is weak, readers will not be impressed. Readers appreciate being able to move through a document without artificial barriers of extra syllables.

Redundant Pairs

If writers can't focus on the exact subject, verb, or modifier, they may insert two or more, hoping to cover all the possibilities. A writer may announce the "aims and goals" of a meeting, for example, not to distinguish between aims and goals but to avoid choosing just one of the words. A computer manual may instruct a user to "choose or select" a command without really meaning that the user has an alternative. The result is wordy, unfocused writing that may confuse readers because *and* and *or* signal multiple possibilities. In your initial analysis of a writer's style, look for pairs joined with the connectors *and* or *or*. The pairs may state legitimate alternatives, or they may indicate definitions in apposition. However, if they are merely synonyms, they are redundant.

Redundant Categories

Redundant categories are abstract restatements of concrete words. The abstraction puts the more concrete word into a category. A weather announcer, for example, may predict "thunderstorm activity." The person planning whether to carry an umbrella to work that day imagines thunderstorms, not "activity," which is not something from which one needs the protection of an umbrella. Yet, because of the redundant category, the thunderstorm in the sentence is merely a modifier. The weather announcer could more directly predict thunderstorms.

The category is redundant if the specific term establishes the class.

Joe expects to set up a business in the Los Angeles community.

Los Angeles, by definition, is a "community," so the word is redundant.

Joe expects to set up a business in Los Angeles.

Redundant categories demote key terms from noun to modifier, as well as adding extra words.

Redundant	Edited
a career in the medical profession	a career in medicine
hospital facility	hospital
time period	time
red in color	red
upright position	upright
money resources	money; resources
field of industry	field; industry

Application: Editing for Style

In Chapters 13 and 14 we have been looking at sentences and words in order to define the components of style. The brief proposal that appears in Figure 14.1 illustrates the cumulative effect of stylistic choices at the word and sentence levels. It also illustrates the effect these choices have on the reader's comprehension and attitude. Like most documents, its needs for editing are not limited to style. You should spot at least one grammar error, and the insertion of headings could aid both comprehension and access. However, the analysis will focus primarily on style.

Analysis

Analysis should reveal specific editing goals so that the editing can be purposeful and systematic. It will begin with general responses based on awareness of the context and work toward specific goals.

- **Context and content.** Readers (the people who can determine whether to grant the funds or not) are likely to read the proposal and wonder: How will the money be spent? Is this expense worthwhile? Are the proposers capable of doing what they propose? Readers need to be persuaded that the proposed expense represents the best use for their funds. They would prefer not to read the proposal twice or more to find out what it is about. Yet the proposal is confusing on first reading. The final sentence compliments the writer and document rather than anticipating what a reader will need to know or do at this point.

[1]This request to the United States Geological Survey is in reference to the $15,000 allocated by the Office of Coal Management to the Ames District for use in hydrologic assistance. [2]At this time and stage of access and interpretation of existing data on record in the form of computer storage and publications in print that we may not be aware of is our main concern. [3]Due to the U.S. Geological Survey having vast storage of and access to this data we would like to suggest the available funds be used in the following two areas if possible.

[4]The first area may be handled by the Geological Survey district office in Wilson due to accessibility and central locale to all literature and data sources. [5]By compiling this data a comprehensive interpretation of surface water, i.e., quantity, quality, salinity etc. for site specific coal leases or areas immediately adjacent those leases can be provided to the Ames District hydrologist. [6]Thus, due to time constraints, time may be spent on analyzing these interpretations and conducting on-site calculations.

[7]Secondly, another area we foresee as a positive and very useful endeavor is the expertise that can be provided by the Water Resource Division of the Geological Survey in Mountainview.

[8]Because the subdistrict office and the White River Resource Area office are both located in Mountainview, we may obtain their help in the form of infrequent consultations, informal review of tract analysis and field reconnaissance on a one time basis of any lease area lacking available hydrologic data.

[9]The foregoing should provide adequate justification for requesting the U.S. Geological Survey's assistance.

FIGURE 14.1   Original USGS Proposal

- **Sentence structure.** The sentences are long, and it is difficult to find the sentence core. Verbs are weak ("is" and "would like" in the first paragraph). The "due to" construction (sentences 3 and 4) substitutes a prepositional phrase for a clause.
- **Verbs.** In addition to overreliance on weak verbs, the writer uses passive voice frequently (sentences 4, 5, and 6), making it difficult to determine who is to do what, as well as creating a dangling modifier in sentence 5.
- **Words.** The word "area" in sentences 4 and 7 is ambiguous. Is it a geographic area or a subject area? The mention of specific sites ("Wilson," "Mountainview") suggests a geographic area, but "area...is...expertise" in sentence 7 suggests a subject area. The writer also creates some ambiguity with redundant pairs and categories ("time and stage of access and interpretation of existing data on record in the form of computer storage and publications in print" in sentence 2, "positive and useful endeavor" in sentence 7). Some modifiers are excessive ("immediately adjacent" in sentence 5, "available...data" in 8).

Based on the analysis, the following editing goals may be established.

1. Shorten sentences and emphasize the sentence core by placing it earlier in the sentence.
2. Use action verbs. Prefer active voice; clarify responsibility if passive voice remains.
3. Make terms concrete. Delete unnecessary repetitions.
4. Leave the reader with a good impression.

Figure 14.2 shows the edited copy.

Because the editing was substantial, making the marked copy hard to read, the editor revised using the computer. The word processing program created the comparison of the two versions that you see in Figure 14.3. Underlines show additions, and the strikethroughs show deletions. This comparison function of the computer enables editors to keep track of editing they complete on the computer without marking everything by hand.

Evaluation and Review

The editing of the proposal began with analysis of the proposal as it will be used by readers. The analysis follows the process described in Chapter 12. It is a necessary

---

This request to the United States Geological Survey (USGS) refers to the $15,000 allocated by the Office of Coal Management to the Ames District for use in hydrologic assistance. We are concerned about our lack of access to existing data in publications and computer databases. Because the USGS has access to this data, we suggest that the available funds be used in the following two ways.

The USGS office in Wilson has access to all literature and data sources. It could compile and interpret data on surface water (quantity, quality, salinity, etc.) for site specific coal leases or areas adjacent to those leases. The Ames District hydrologist could then spend his or her time analyzing these interpretations and conducting on-site calculations.

Second, the Water Resource Division of the USGS in Mountainview could provide expertise both through the subdistrict office and the White River Resource Area. We could obtain their help through infrequent consultations, informal review of tract analysis, and initial field survey of any lease area lacking hydrologic data.

Please consider our request carefully and contact us at extension 388 if you need further information.

FIGURE 14.2   Edited USGS Proposal

[1]A Strong verb replaces "is."

[2]Subject and verb come early in the sentence. Publications are listed before computer databases to clarify that "computer" does not modify "publications."

[3]A phrase is converted to a clause for easier comprehension.

[4, 5]The core sentence comes early in both of these sentences for easier comprehension. Active voice replaces passive voice. Verbs are stronger ("compile," "interpret").

[6]Active voice replaces passive voice to establish who will do what.

[7, 8]Core sentences come early. Unnecessary modifiers are deleted. Abstractions ("area," "endeavor") are minimized.

[9]Instead of complimenting themselves, the writers offer the reader further assistance.

[1]This request to the United States Geological Survey ~~is in reference~~ (USGS) refers to the $15,000 allocated by the Office of Coal Management to the Ames District for use in hydrologic assistance. [2]~~At this time and stage of access and interpretation of existing data on record in the form of computer storage and publications in print that we may not~~ We are concerned about our lack of access to existing data in publications and computer databases. Because the USGS has access to this data, we ~~be aware of is our main concern.~~ [3]~~Due to the U.S. Geological Survey having vast storage of and access to this data we would like to~~ suggest that the available funds be used in the following two ~~areas if possible~~ ways.

[4]~~The first area may be handled by the Geological Survey district office in Wilson due to accessibility and central locale~~ The USGS office in Wilson has access to all literature and data sources. [5]~~By compiling this data a comprehensive interpretation of surface water, i.e., quantity, quality, salinity etc.~~ It could compile and interpret data on surface water (quantity, quality, salinity etc.) for site specific coal leases or areas ~~immediately adjacent those leases can be provided to the Ames District hydrologist.~~ [6]~~Thus, due to time constraints, time may be spent on~~ adjacent to those leases. The Ames District hydrologist could then spend his or her time analyzing these interpretations and conducting on-site calculations.

[7]~~Secondly, another area we foresee as a positive and very useful endeavor is the expertise that can be provided by~~ Second, the Water Resource Division of the ~~Geological Survey in Mountainview.~~ [8]~~Because~~ USGS in Mountainview could provide expertise both through the subdistrict office and the White River Resource ~~Area office are both located in Mountainview, we may obtain their help in the form of~~ Area. We could obtain their help through infrequent consultations, informal review of tract analysis ~~and field reconnaissance on a one-time basis of any lease area lacking available,~~ and initial field survey of any lease area lacking hydrologic data.

[9]~~The foregoing should provide adequate justification for requesting the U.S. Geological Survey's assistance.~~ Please consider our request carefully and contact us at extension 388 if you need further information.

FIGURE 14.3   Computer Comparison of Original and Edited Copy, USGS Proposal

first step in editing for style because the style guidelines do not prescribe changes and must be used with awareness of meaning and purpose. Because some of the sentences are so ambiguous, an editor should consult with the writer before sending the proposal forward.

Some content and format questions arise as a result of stylistic editing. For example, proposals usually include specific budgets. The reader may wonder how much of the $15,000 will go for each purpose and how much each data search and consultation will cost. Headings or numbers could help to identify the two proposed uses of the money. The editor may raise these questions with the writer and suggest additional data. But this level of editing exceeds editing for style and should be approved before it is done. Even if this proposal is edited only for style, it will be easier to understand and therefore more persuasive.

One consequence of editing for style has been to shorten the proposal. The revised version contains 185 words compared with 265 in the original. But the marginal explanations for editorial emendations in Figure 14.3 do not point to conciseness as an end in itself. Often the best way to achieve conciseness is to use strong subjects and verbs early in the sentence.

Summary

The arrangement of words in sentences gives them meaning by showing relationships, but if the words themselves are imprecise or incorrect, the meaning will be uncertain. Editors need to pay particular attention to verbs, because they should communicate the most important information about the subject and because good verbs enliven writing. Effective writing also requires specific nouns and modifiers. Because changing words changes meaning, editors approach style through meaning.

The method of comprehensive editing described in Chapter 12 applies to editing for style. This method requires consideration of the document as a whole and in the context of its readers and purpose and will help you focus on meaning. Before you edit for style, confirm that you are expected to edit for style and not just for grammar and typos. Edit with respect for the power of word and sentence choices to change meaning and to affect the reader's response.

Further Reading

Walker Gibson. 1969. *Persona: A Style Study for Readers and Writers.* New York: Random House.

Daniel Jones. 1998. *Technical Writing Style.* Boston: Allyn & Bacon.

Richard A. Lanham. 1991. *Revising Prose,* 3rd ed. Englewood Cliffs, NJ: Prentice-Hall.

Joseph M. Williams. 1994. *Style: Ten Lessons in Clarity & Grace,* 5th ed. New York: Addison-Wesley.

Discussion and Application

1. The following sentences do not use verbs effectively. Analyze each sentence to determine the source(s) of the problem (action not in the verb, weak verb, passive voice, nominalization), and then edit to use verbs more effectively. Maintain the writer's meaning; prepare a query to the writer if you cannot determine the meaning.

   a. Prolonged use of the battery can cause it to become drained of its energy.
   b. The crimper on the alfalfa mower breaks the stem every inch to allow the fluids in the stem to be released.
   c. *(from instructions for playing tennis)* Place your legs in a bent position with your toes pointing outward at an angle of 45 degrees.
   d. Because there is a trend toward fewer and larger farms, it will cause an increase in the demand for machinery, decreasing the demand for farm laborers.
   e. The report shows a recommendation toward simple, cost-effective advertising with the aid of either an advertising agency or an account executive from a media service.
   f. Further research needs to be entailed into the project.

2. Distinguish passive voice sentences from those with weak verbs or past tense. (Look for the *to be* verb and past participle to identify the passive voice sentences. Verify your identification by determining whether the subject performs or receives action.) Convert the passive voice sentences to active voice.

   a. The report was written collaboratively by three engineers.
   b. The report was informative but too long.
   c. The engineers have sent the report to the editor.
   d. The report has been shortened by three pages.
   e. This method of writing and editing is effective for us.

3. Describe the sentences below in terms of core sentence, use of verbs, and voice. Do they all mean the same thing? Which version is "better" stylistically? What are the bases for deciding the answer to that question?

   a. There must be thorough preparation of the specimens by laboratory personnel.
   b. Laboratory personnel must prepare the specimens thoroughly.
   c. The specimens must be prepared thoroughly by the laboratory personnel.
   d. Preparation of the specimens by laboratory personnel must be thorough.

4. Compare the following sentence pairs to determine how the revision for style has also inappropriately changed meaning. Define the change in meaning. Suggest a way to edit the first sentence in the pair to improve style without changing meaning.

   a. Technical writers are now finding themselves in roles in product design and managing the production as well.
   Technical writers now find roles in product design and production management.
   b. The problem involves derivation of objective methods for evaluating the effect of adriamycin on the heart.
   The problem derives objective methods for evaluating the effect of adriamycin on the heart.

5. Locate examples of "doublespeak," and conduct a doublespeak awards contest in your class.

6. Select a passage from a technical document, such as a computer manual, or from a textbook on a technical subject. Analyze the style, using the terms and principles discussed in this chapter and in Chapter 13. Then, taking note of the presumed readers and purpose for this document, evaluate the style. What are its strengths? What, if any, editing goals would be appropriate?

7. Analyze your own writing style. Look for placement of subject and verb in sentences, use of verbs, active and passive voice, and concreteness of nouns.

8. Vocabulary: If you are not confident about explaining the meaning of the following terms, check the glossary or review the chapter: *nominalization, passive voice, past tense.*

Chapter **15**

Style: The Social and Global Contexts

Guidelines for sentence structures and words, such as you studied in Chapters 13 and 14, approach style textually. Editors also need to consider social and cultural issues of style. Values, customs, experiences, and ways of thinking all contribute, along with the words and sentence structures, to interpretation. The chances for failures in communication increase whenever there are differences between writers and readers—differences of gender, age, race, health, language, ways of thinking, and culture. Good editors take their cues for good style from the context in which documents will be read and used as well as from the grammar, sentence structure, and word choice.

This chapter considers two contextual issues in editing for style: the possibility of discriminatory language and the globalization of the workplace. These considerations take an editor outside of the text to the places where the text is used, but both also bring the editor back to the text to make the textual choices work in the context of document use.

The Language of Discrimination

Word choices, as they relate to human subjects, may imply bias against groups on the basis of their gender, age, race, religion, culture, politics, or physical or mental disability. Rarely do writers intend discrimination. They may find incredulous the idea that the phrase *the disabled* is more negative than *people with disabilities*, or that the term *handicap* or the generic *he* may offend some readers. How the writer (and editor) feel, though, is less important than how the readers feel and how word choices will affect their attitudes and comprehension. Whatever your personal feelings about particular phrases or pronouns, you edit for neutral, unbiased language for two good reasons:

- Language that appears discriminatory to readers creates such significant noise in the document that it may block comprehension. Readers will never hear the intended meaning if discriminatory language interferes.
- Most professional associations and journals have policies to discourage discriminatory language, particularly as the language relates to gender. You edit the document to conform to these policies, just as you edit reference lists and punctuation to conform to the accepted form.

These reasons supersede your own feelings about the significance of word choices. They also reflect a professional respect for the people who will read the words that you edit.

Unnecessary demographic information can make language discriminatory. For example, the sentence "Dr. Alice Jones, paralyzed from the waist down, was named dean of the College of Home Economics" introduces information that is irrelevant to her appointment and to her ability to do the job. That information about her paralysis would be appropriate only in a human interest story. Information that indicates race, age, gender, religion, politics, or physical attributes is rarely appropriate in professional situations because it does not relate to professional credentials. It discriminates by implying that a given age, gender, race, or physical status may affect competence or by conveying surprise that the achievement is inconsistent with the demographic characteristic.

Application: Discriminatory Language

The paragraphs in Figure 15.1 appear in an architect's program for a visitor's park design. The program is a report with recommendations for design based on analysis of the facility function, use, and setting. It guides the designer by establishing a concept of the facility as well as specific goals and criteria for design. In these paragraphs, the writer addresses the issue of visitors who have disabilities and how the park should be designed to accommodate them.

The message of the paragraphs is positive: the architect is concerned not just with physical accommodations, such as wheelchair ramps, but also with the visitors as whole persons. However, the language contradicts the message by making "the handicapped" an abstraction, and by separating these visitors from others. The language can shape the designer's response to the message by subtly suggesting that "the disabled" are significant primarily in terms of their handicaps. Though the message discourages differentiation of handicapped and other visitors, the language permits it.

An analysis of the style will reveal nominalizations and weak verbs as well as some discriminatory language. Passive voice is appropriate to the extent that the designer is understood throughout the program to be the agent of implementing the concepts. Yet overuse of passive voice may reduce readers' access to and acceptance of the ideas. Figure 15.2 shows how this report might be edited for style.

Handicapped people have expressed that they do not need or desire segregated outdoor activities. They prefer not to be singled out, but instead appreciate efforts made to accommodate their special needs. A sensitive approach without differentiation is preferred, and demonstrations of sympathy should be avoided.

The disabled possess different learning styles which tend to focus on using the sensory perceptions to the greatest extent possible. Depending on their particular disability, they utilize their hands, eyes, and ears to perform as informational transmitters. To enhance the overall experience for the disabled visitor, and encourage his participation, all kinds of sensory experiences should be incorporated into exhibit and facility design.

Activities should be created which will enable participation by the physically disabled. Easy access to areas in the way of ramps, minimal inclines, and railings should be provided. Developing designs and activities that accommodate the handicapped population can provide them with a more enjoyable and secure experience.

FIGURE 15.1   Original Visitor Park Report

~~Handicapped~~ *with disabilities* people have expressed that they do not need or desire segregated outdoor

activities. They prefer not to be singled out, but instead appreciate efforts made to accommodate

their special needs. A sensitive approach without differentiation is preferred, and demonstrations

of sympathy should be avoided.

People with disabilities use their ~~The disabled possess different learning styles which tend to focus on using the~~ sensory

perceptions to ~~the~~ *a* great~~est~~ extent ~~possible.~~ *in learning* Depending on their particular disability, they ~~utilize~~ *use*

their hands, eyes, and ears to ~~perform as~~ *transmit* informational ~~transmitters.~~ To enhance the overall

experience for ~~the~~ disabled *s* visitor, and encourage ~~his~~ *their* participation, all kinds of sensory

experiences should be incorporated into exhibit and facility design.

Physical disabilities require ~~Activities should be created which will enable participation by the physically disabled.~~ Easy

access to areas in the way of ramps, minimal inclines, and railings ~~should be provided. Developing~~

designs and activities that accommodate ~~the handicapped population~~ *visitors with disabilities* can provide them with a

more enjoyable and secure experience.

FIGURE 15.2   Edited Visitor Park Report

Editing for a Nonsexist Style

The generic *he* is now shunned by most professional associations. Research also shows that readers associate male images with the male pronoun far more often than they associate female images. One study required subjects to complete sentence fragments such as "Before a pedestrian crosses the street,… " and "When a lawyer presents opening arguments in a court case,…." Subjects also had to describe their images of the people in the sentences and give them names. People who used *he* in completing the sentences gave the subject a male name five times more often than a female name, and they imagined the subject as a man four times more often than as a woman. By contrast, people who used *they* or *he or she* selected a male name only twice as often as a female name and imagined the subject as a woman as often as a man.

The male bias in imagery caused by the generic *he* is especially inappropriate in evaluative situations. The following guidelines for employee evaluation encourage in a subtle (and probably unintentional) way the identification of outstanding employees as male. Supervisors are asked to classify employees into one of three rankings: needs improvement, good, and outstanding. The definitions of performance levels for the first two descriptions contain no gendered pronouns, but the description of the outstanding employee contains four references to males.

> The employee is clearly superior in meeting work requirements, and he consistently demonstrates an exceptional desire and ability to achieve a superior level of performance. His own high standards have either increased the effectiveness of his unit or set an example for other employees to follow. This rating characterizes an excellent employee who consistently does far more than is expected of him.

Professional associations have developed guidelines for avoiding sexist language. The National Council of Teachers of English (NCTE) suggests these alternatives for terms that include "man":

Sexist Language	Alternative
mankind	humanity, human beings, people
man's achievements	human achievements
the best man for the job	the best person for the job, the best man or woman for the job
man-made	synthetic, manufactured, crafted, machine-made
the common man	the average person, ordinary people
chairman	coordinator, presiding officer, head, chair
businessman	business executive or manager
fireman, mailman	firefighter, mail carrier
steward and stewardess	flight attendant
policeman, policewoman	police officer

The NCTE advises these alternatives for masculine pronouns:

1. Recast into the plural.
2. Reword to eliminate unnecessary gender problems.
3. Replace the masculine pronoun with *one, you,* or *he or she,* as appropriate.

Style for a Global Workplace

Businesses increasingly define themselves globally. Manufacturing and markets extend beyond the national boundaries of the company. With this global expansion comes the need for communication through letters, email, faxes, and reports. In addition, manuals for products such as computer hardware and software may require translation for international audiences. Both for business correspondence and translation of manuals, knowing the customs and the culture aids in communication. Furthermore, planning for eventual translation before writing the original facilitates both the procedure and success of translation.

International Correspondence

Iris Varner and Linda Beamer identify categories of differences in cultures that affect communication: ways of thinking and knowing (linear and logical or dualistic); attitudes toward achieving and activity; attitudes toward nature, time, and death; sense of the self; and social organization. They also distinguish "high context" from "low context" cultures, with a high context culture depending more on contextual information than on words to communicate meaning. These differences become apparent in various forms of communication, such as the business letter.

Whereas linear, activity-oriented Westerners favor a direct style of correspondence with the emphasis on the business transaction, people in Asian countries may be indirect about the transaction and try to establish a relationship with the correspondent. The first paragraph of a Western business letter conventionally states the business directly and purposefully, whereas a business letter from Japan or from another Asian country may begin with a reference to the seasons and conclude with good wishes for the recipient's family. The Western style might be described as hard and direct, whereas the Eastern style is more poetic. Miscommunication occurs when the Eastern reader is offended by the apparent indifference of the Western writer to the setting and the person or when the Western reader is frustrated by the apparent vagueness of the Eastern writer.

Various cultures, Western or Eastern, have different expectations for formality, such as whether it is appropriate to address the recipient by a first name. Languages place different meanings on words and colors. These cultural differences are displayed in the organization and format of the business letter as well as in the style.

If your work as editor requires you to review correspondence and reports for international readers, you will need to look beyond handbooks of grammar and style to edit well. Learning about the customs of the culture will help you advise writers about communicating with sensitivity and with awareness of how readers in different settings may respond. The books on international communication listed at the end of this chapter provide a starting point.

Globalization

Globalization refers to developing products, including their documentation, for international users. If a product is usable in many cultures, it does not require as much adaptation (localization) for individual cultures.

Globalization begins at the planning and concept stage for products and is not something added on at the end, nor does it simply exist in product documentation. The product development team should include people who know the cultures in which the product will be distributed. Planning for international users is audience analysis on a global scale. This analysis considers not just the tasks that users will perform or concepts they will learn but also their styles of learning, equipment available, and cultural values and customs. Because multiple cultures expand the number of variables, globalization aims to reduce the amount of culture-specific bias in the products and documentation. It limits the needs for translation and localization. Where globalization reaches its limits, translation and localization take over.

Translation

Instructions for products distributed internationally must be understood by people using different languages. Two ways to achieve that goal are to bypass language by using visual instructions and, more commonly, to translate the English version into the languages that customers use.

Using Visual Instructions

When the instructions are simple and the languages of potential readers are numerous, pictures may substitute for words. For example, emergency procedures for air travelers are visual rather than verbal. Pictures, however, may be as ambiguous as words, and, like words, they reflect culture. Because some reading patterns are right to left or top to bottom rather than the Western left to right, marking of the sequence of frames by number may be necessary. Colors, symbols, and dress may have different meanings in different cultures. And, pictures do not allow for explanations or alternative courses of action. If visual instructions will work, however, a company may be able to bypass translation. Editors should be called on to edit visual as well as verbal instructions, and a usability test should be mandatory (with users from different cultures) if safety is involved.

Writing to Facilitate Translation: Minimize Ambiguity

Planning for translation before writing and editing requires knowledge of how translation occurs and of specific strategies to facilitate translation. Writers and editors can make choices in their native language that will make translation more accurate and minimize ambiguity.

The first step in translation is often a computer translation. The program can match English words with their equivalents in other languages and parse sentence structure, but machine translation by itself is not enough to convert a manual written in North America for use in another culture. The users in different countries vary according to communication and learning styles as well as in the way they respond to visuals. Machine translation achieves about half the work of translation before a human translator takes over.

To increase the efficiency and accuracy of the translation, writers can try to eliminate the ambiguity that comes from unfamiliar words, complex structures, and culture-specific metaphors. Errors in grammar and punctuation compound the problems of translation. The following guidelines can aid in writing and editing a document destined for translation.

- **Use short sentences and substantial white space.** Many languages use more characters to say the same thing that an English sentence says. If the translation will preserve format, including page and screen breaks, you may have to edit to shorten sentences so that when they are translated they will still fit on the screen or page.
- **Avoid jargon and modifier strings.** The principles that produce clarity are even more important when the text will be translated. There may not be equivalents in the translation language for unfamiliar words and jargon. Strings of modifiers raise questions about what modifies what.
- **Create a glossary of product terms.** Definitions help translators find equivalents for unfamiliar or product-specific terms.
- **Eliminate culture-specific metaphors.** Sports and political metaphors are common within a culture, but not all cultures play the same sports, and political systems vary. Varner and Beamer observe that military metaphors are common in Western countries—*strategies, opposition, price wars, planning attacks,* and *digging in* have become the language of competition, but their military roots make Americans seem aggressive (p. 54). Even so apparently neutral a metaphor as "typewriter keyboard" may not be meaningful in Asian countries, where hundreds of characters take the place of much shorter Western alphabets.
- **Avoid acronyms and abbreviations.** An organization is a social construction; unless an organization is international, its acronym may not be recognized in countries other than the host country. The translation program (like spelling checkers) will not recognize many acronyms. Abbreviations pose similar problems. Technology terms, like CPU and RAM, known by their acronyms better than by the words they stand for, are exceptions to this guideline.
- **Avoid humor and puns.** Much humor is at the expense of something valued by someone else, and puns depend on specific meanings of languages. Both are difficult to translate well and risky to use.

Localization

Localization refers to adapting material for the local culture. At the very least it means converting numbers and measurements, dates, and spellings to usage in the culture in which the document will be used. Even among English speaking countries, usage varies. For example, in Great Britain, "localization" would be spelled "localisation," with the *s* taking the place of the *z*, and currency would be expressed in terms of pounds, not dollars. Because Canadian and U.S. dollars have different values, numbers would either have to be converted or the amount qualified according to whether it is U.S. or Canadian. Differences expand when the language changes from English. Simple translation of words does not completely localize the material, although it is a first step.

More extensive localization could require modification of examples and illustrations to reflect customs of dress, interpersonal interaction, attitudes toward gender and age, corporate structure, and ways of thinking.

In Figure 15.3, Nancy Hoft offers numerous categories for "cultural editing."

Anticipating these needs to localize, you may avoid as many references as possible to these cultural differences when developing the document.

Researching Social and Cultural Information

Even a person who is sensitive to and respectful of differences will not intuit all the ways in which language may evoke unintended responses. Acquiring information about any unfamiliar audience requires research.

- Dates and date formats
- Currency and currency formats
- Number formatting
- Accounting practices
- List separators
- Sorting and collating orders
- Time, time zones, and time formats
- Units of measurement
- Symbols (in English, some symbols are / and &)
- Telephone numbers
- Addresses and address formats
- Historic events
- Acronyms and abbreviations
- Forms of address and titles
- Geographic references
- Technology (electrical outlets, computer keyboards, printer page size capability
- Legal information (warranties, copyrights, patents, trademarks, health- and safety-related information

- Page sizes
- Binding methods
- Illustrations of people
- Many everyday items (refrigerators, trash cans, post office boxes)
- Hand gestures
- Clothing
- Architecture
- The relationship of men and women in the workplace
- The role of women in the workplace
- Popular culture
- Management practices
- Languages
- Text directionality
- Humor
- Color
- Communication styles
- Learning styles (the relationship of the instructor and the students)

FIGURE 15.3   Categories for Cultural Editing

Source: Nancy L. Hoft. 1995. *International Technical Communication: How to Export Information About High Technology.* New York: Wiley, 129–130. Copyright © 1995. Reprinted by permission of John D. Wiley & Sons, Inc.

Some sources are readily available: printed materials in the genres in which you will work, international employees of your company, professors and international students, and customers. The sources listed at the end of this chapter provide numerous helpful explanations and details. For products for international distribution, a consultant in international communication or the country in which your products will be distributed may be hired. Asking questions at the beginning of document development will help prevent problems, but because you won't be able to anticipate all the questions to ask, the review of the draft should include a review for globalization and localization.

Summary

The meaning of written communication resides not just in words and their arrangement but also in the cultural and social experiences that readers bring to the text. Part of editing for style, then, is considering the values and expectations of readers, especially when readers may differ from the writer. Two values that enhance editorial effectiveness are respect for differences and sensitivity to the experiences of readers. Research into the values and customs of intended readers is necessary to supplement what an editor can sense intuitively. In addition, anticipating translation and localization, editors can aim for words and structures that will minimize ambiguity.

Further Reading

Deborah Andrews, ed. 1996. *International Dimensions of Technical Communication.* Arlington, VA: Society for Technical Communication.

Francine Wattman Frank and Paula A. Treichler. 1989. *Language, Gender, and Professional Writing: Theoretical Approaches and Guidelines for Nonsexist Usage.* New York: Modern Language Association.

Nancy L. Hoft. 1995. *International Technical Communication: How To Export Information About High Technology.* New York: Wiley.

Iris Varner and Linda Beamer. 1995. *Intercultural Communication in the Global Workplace.* Chicago: Irwin.

Discussion and Application

1. Edit the definition of an outstanding employee that appeared in the "Language of Discrimination" section (page 246) to remove references to gender.

2. The following paragraph incorporates language that is common and familiar in discussing business in North America. Underline the metaphors of war and sports. Discuss the difficulties in translation such metaphors might create.

A department committee will meet for strategic planning to identify goals for surviving in light of the new policies of retrenchment. We will have to dig in to fight the misunderstanding of management about the department needs and goals. If our current funding is reduced, we will not be able to play on a level playing field with the competition. It appears that the vice president is trying to do an end run around our department manager.

3. Look for examples of writing in which cultural metaphors and values are embedded. Weekly news magazines, reports from corporations, and fundraising letters could be sources.

Chapter 16

Organization

Organization affects a user's performance and a reader's comprehension. In brief, when the document is used to complete a task, an editor looks for organization by task. When the document is used to learn, an editor looks for document structure that matches the content structure.

Organization is a powerful aid (or hindrance) to performance and comprehension and requires an informed and systematic approach to editing. Because organization is so closely linked to meaning, editing for organization also may reveal content gaps or inconsistencies.

This chapter defines the two broad uses of technical documents: performance and learning. These purposes roughly correspond to user-based organization and content-based organization, though most documents will use a mixture. The chapter presents guidelines for editing for organization and applies these guidelines to the editing of a proposal. It also considers paragraph organization. The chapter emphasizes the sequence of information as it influences performance and comprehension. Chapter 17, "Visual Design," shows how to provide selective access to information and how to reveal the structure visually.

Organization for Performance: Task-Based Order

Many problems in software documentation occur because writers organize according to program function rather than according to the tasks the users need to perform. For example, the programmer for software to create a database for a bail bond company wanted to organize the manual according to the screens, each one of which records different types of information. However, the user normally begins with the eighth screen in the program, "entering personal data," and thus would have been frustrated by wading through seven categories of instructions before finding the ones needed to get started. The editor advised organizing according to tasks the user would perform in the probable sequence of use. The user of the program does not care as much about program functions as about getting the job done. The sequence of screens in the program is invisible and irrelevant to the user.

The editor's role in organizing according to task should take place at the planning stage, because reorganization after the manual is drafted is a huge project that somewhat repeats the efforts of organizing to begin with. The task-based order begins with a task analysis: identification of tasks that users need to do. The writing should proceed around this task analysis. For complicated procedures, the task analysis includes major tasks plus individual tasks. These major tasks define the divisions of a long manual. For example, all the tasks related to setting up and naming files will appear in one chapter of the manual. The plan for the major divisions and the items within them will save editorial and revision time later.

In addition to helping plan the task-based order, the editor will check it during development and when a draft is complete. The editor can look for grouping of related tasks, sequence, and completeness according to the task analysis. The editor can check for whether there are too many or too few steps within a section (which could become confusing), in addition to editing for style, visual design, illustrations, and organization.

Eventually some trial of the instructions will probably be necessary. The editor may try the product while reviewing the instructions to supplement reading. The best usability test, however, examines representative users in their work settings using the product to perform their work tasks. Any failures or frustrations point to gaps in the instructions or problems with the product itself.

Reference manuals are sometimes organized alphabetically so that readers can search for specific items. Alphabetical order assumes familiarity with the product and with the terms that describe it. It does not aid comprehension because it does not show logical relationships of information. A good index can accomplish the reference task in a manual that is organized by related procedures.

Organization for Comprehension: Content-Based Order

The structure of a document influences how well readers understand content and how easily they read. A review of structures may also reveal gaps and inconsistencies in the information. Thus, editing for organization is also a way to edit for content.

One purpose of a technical document is to give information so that readers can learn or use it (or both). The principles of learning, developed by researchers in cognitive psychology and instructional technology, apply to documents. A number of text features, including organization, visual design, and sentence structure, can influence learning. This chapter applies learning theory specifically to document and paragraph organization.

The Schema Theory of Learning

The most widely accepted theory of learning is schema theory. Basically, the theory proposes that people learn and remember information by perceiving patterns, or *schemata* (singular, *schema*), for concepts and facts. We organize information in order to understand and remember it.

The word *schema* comes from the Greek word for *form*. You are probably familiar with one variation of the word as used in the term *schematic diagram*, which outlines the form of an object without representing it realistically. You might find such a diagram pasted on the back of your refrigerator to show the wiring pattern. Schemata resemble schematic diagrams.

Even an abstract and simple concept such as "give" has a structure in memory. *Give*, in any context, implies certain components, including a *giver, recipient,* and *gift.* These components remain constant even when the specific examples of giving change. (See pages 101–103 of the selection by Rumelhart and Ortony in the list of readings at the end of the chapter for an elaboration of this example.)

Each remembered schema becomes a pattern or template against which new information can be matched. These templates help us learn new information because we can relate the new to old information. When we use a schema to understand new information, we may modify the existing schema and make it more complex, or we may create a new schema. Thus, the schema for *give* may develop over time. A child's concept may be limited to the experience of receiving gifts for a birthday. The adult's concept, based on broad experience, is more complex and may include giving of service and love as well as tangible gifts. Nevertheless, the basic components of giver, recipient, and gift remain constant.

The schema theory explains why analogy is useful in teaching. The analogy relates new information to familiar information; it taps into an existing schema to modify it or to develop it with the additional information presented. The theory also explains why learners with expertise on a subject can learn new material on the subject more quickly than can equally intelligent learners with limited knowledge of the subject. The experts can draw on more schemata, and more complex schemata, to learn new information.

Organization and Comprehension

Schemata are hierarchical. That is, they have major components and subordinate components. People learning something new first seek the macrostructure, or major points. Then they fill in the microstructure, or the details. This is called top-down processing. To phrase this concept in terms that are familiar to writers and editors, most people learn from general to specific, or from the concept to the details. That is why documents begin with an introduction that establishes context or problem and research question or plan. Writers and editors appreciate the readers' need to develop a concept (the macrostructure) within which to order the details (the microstructure). A schema should be available early.

A document's organization affects how well and how easily readers learn the information contained in it. The major and subordinate points in the document should match the major and subordinate structures of the content. The document structure helps readers place the details in the hierarchy. Likewise, a faulty structure can cause a reader to identify a minor point as a major point and thus to "learn" the new information incorrectly. Readers may compensate for faulty structure by rereading and by imposing the correct order mentally. Each time the text demands that they do so, however, the chances increase that readers will comprehend incorrectly.

In summary, the overall structure of the document influences the reader's comprehension of the content. So do the reader's prior learning and other text characteristics, but at the very least we may say that *structure reinforces meaning*. A collection of facts has little meaning without a structure and a context. The reader's perception of meaning is affected by the order in which the information is presented (the sequence) or the categories by which it is organized (as in hypertext, in which sequence may not be significant). In addition, the emphasis given to various parts through the organization helps readers establish hierarchies of importance.

Analysis as a Means of Understanding

Good informative documents are organized in predictable ways, according to templates for document structures that readers have learned through experience with reading. Their patterns of organization match the patterns readers know. Print documents generally have three main parts: a beginning, a middle, and an end. The beginning (often called an introduction) presents the concept by identifying the topic and by placing it in its context (background, purpose, significance). The middle analyzes the overall topic by identifying component parts and develops the topic with details. The end, like the beginning, considers the topic as a whole, by summarizing, drawing conclusions, or anticipating applications of the information. Thus the beginning-middle-end pattern corresponds to a whole-part-whole structure.

All patterns of development, or all ways in which you can arrange data in the middle of a document, are based on the principle of analysis. That is, the whole of the topic is analyzed (broken down) into its component parts. If the document contains instructions, the whole (the overall task) is analyzed into major tasks and steps. If the document describes a mechanism, the whole mechanism is analyzed into functional parts. If the document is a feasibility study, the project is analyzed into criteria that will influence the decision, such as cost and availability of equipment.

Analysis is a means of understanding the whole, not an end in itself. Analysis gives meaning and coherence to a subject by revealing the structure of the information. However, to have meaning, a subject cannot be a series of isolated facts or objects; rather, the parts must cohere into a whole. Thus, the end of a document reassembles the material to force readers to look again at the topic as a whole. Instructions are an exception in the sense that they often have no formal conclusion; however, the completion of a task provides the reader with the sense of closure and completeness that a printed conclusion provides in a report. The beginning and end of a document, then, look at a topic as a whole, while the middle considers the component parts as a means of clarifying, supporting, or enabling the whole.

As an editor who can manipulate organization to improve clarity and usability, you need to consider both the overall structure (beginning, middle, and end) and the structure within the middle section. To relate this observation to the cognitive theory just discussed, we may say that the beginning and the end focus on the macrostructure of a concept, item, or task. The middle clarifies the macrostructure by naming components of the whole. The details within each section of the middle form the microstructure.

Principles of Organization

Language is so complex, and documents are so varied in type and purposes, that we can only generalize about what constitutes good organization. The organization that works in one situation may not work in another. You cannot rely on rules in editing for organization any more than you can rely on rules in editing for style. You can, however, apply principles of organization, using them with good judgment and remaining flexible enough to apply them differently in different situations.

This section of the chapter presents some principles that will help you improve a document's clarity and effectiveness by reorganizing it. The principles suggest guidelines for editing for organization. The order in which the principles are presented also suggests a process for applying them. That is, if you consider the principles in the order that they are presented, you will work efficiently. Briefly, when editing for organization, you should do the following:

1. Follow pre-established document structures.
2. Anticipate reader questions and needs.
3. Arrange information from general to specific and from familiar to new.
4. Apply conventional patterns of organization.
5. Group related material.
6. Use parallel structure for parallel sections (chapters, paragraphs).

Let us examine each guideline in more detail.

Follow Pre-established Document Structures

Someone other than the writer may establish the structure of a document. For example, an organization that publishes an RFP (request for proposals, inviting proposals for providing a service or solving a problem) may specify what components should appear on what pages. The RFP for the proposal in Figure 16.3 specified that an abstract and a work schedule appear on the title page of the proposal. A manual in a series of manuals follows a structure established for the series (document set).

Certain organizational patterns are widely accepted for documents in different disciplines. Readers who know these conventions expect documents to follow them. For example, a research report in a scientific journal typically follows the pattern of problem statement, literature review, methodology, results, and discussion. These terms are likely to be section headings. Many business executives, especially those who manage according to management by objectives (MBO), expect that business plans will address goals (broad aims), objectives (specific aims), strategies (means of achieving the goals and objectives), and evaluation procedures (means of measuring whether the goals and objectives are achieved).

Because of the structure established by an organization's specifications or a discipline's conventions, readers expect that they will find certain kinds of information in certain places. Variations from established structure will distract readers from the content and from their purpose for reading. Furthermore, failure to follow conventions may cost the writer some credibility.

Thus, your first task in editing for organization is to compare the structure of the document to be edited with any external patterns to which it must conform. Ideally you and the writer will do this step in planning for document development. When the document is complete, this comparison will help you spot missing parts as well as ineffective structure. For example, if you are editing a management plan that lacks evaluation procedures, you will advise the writer to complete that section.

Anticipate Reader Questions and Needs

You will have to get into the intended reader's mind to assess likely questions and needs for information. Readers ask predictable questions from a document, including "What is it?" "What is this about?" "Why is it important?" "Who is affected by it?" "How do I get started?" These are versions of the investigative journalist's questions: who, what, when, where, why, how, and so what? As editor you can use these questions to review whether probable reader questions have been covered in the order in which readers are likely to ask them. You can be almost certain that if the document defines or describes an unfamiliar concept or object, the first question will be "What is it?" The predictability of the question explains why introductions often include definitions.

Anticipating user needs will also help in organizing hypertext. A decision about whether to create one long document with links to places within it or to create a series of short segments in separate files with links among them depends in part on whether readers are likely to read in a linear sequence or selectively. It also depends on whether users are likely to print the entire document. It will be easier to print one long document than to print numerous sections, such as chapters.

Arrange from General to Specific and Familiar to New

Because people learn top down, by relating details to concepts and to familiar information, readers need general information in an introduction to provide a concept. Readers who encounter the details before the concept may struggle with the details (because there is no context for understanding them), or they may form their own concept or schema, which may not be correct.

As editor you evaluate the beginning of a document for information that orients readers. Writers may inadvertently omit information that is familiar to them but that readers need. You can look more objectively at the introduction from a reader's point of view. It is easiest to do this if your level of understanding matches the intended readers' level. (Not being a subject matter expert has some advantages.) If you know more than readers probably will, you can still anticipate basic reader questions and evaluate opening paragraphs to see whether the questions are answered.

All readers learn new information best if it is prefaced with familiar information. The familiar information may be the problem that led to the feasibility study or proposal, or prior research on a subject, or common knowledge. This familiar information places the new information to follow in a context and links the new with the old. In the terms of learning theory, the familiar information lets readers draw on a schema that the new information will modify.

The amount of information to precede details will depend partly on the document's purpose. Instructions do not generally need to begin with detailed conceptual information. Readers, at least readers in North America, are more eager to get on with the task than they are to analyze processes and principles. The general information in instructions may be as minimal as a definition of the scope of the task (what a reader will learn to do), expected competencies, and equipment and materials needed. The introduction may also include a brief description or explanation of relevant equipment and processes if these will help the reader complete the task accurately, efficiently, and safely.

When the reader's task is comprehension rather than performance, the general information needs to orient the reader not just to the topic but to its background and significance and to relevant principles. The amount of general information in such documents will depend on the expertise of the reader. Beginners are likely to need more explanation and background than are experts; and people who will use the information casually need less orientation than do people who will use the information in research.

The analysis of document readers and purpose described in Chapter 12 will help you assess the circumstances of use of the document you are editing and make editorial decisions accordingly.

After you review overall document structure according to pre-established structures and probable reader questions, you review the introductory material for the whole and for each major section. Does it orient the reader to the subject and purpose? Does it begin with information that is familiar to the intended reader?

Use Conventional Patterns of Organization: Match Structure to Meaning

Readers have learned schemata or patterns of organizing information, such as chronological, spatial, comparison-contrast, and cause-effect. These patterns reflect generic relationships of information rather than externally imposed patterns such as the scientific report and MBO patterns described previously. Using one or more of these patterns in a consistent and predictable way will probably increase the ease with which readers understand.

A chronological structure is appropriate for narratives, instructions, and process descriptions. A spatial order (for example, top to bottom, left to right, in to out) is useful when the text describes a two- or three-dimensional object or space. Comparison and contrast can follow point-by-point or item-by-item patterns depending on what is to be emphasized. A document organized to illustrate causes and effects should consistently move either from causes to effects or from effects to causes. All of these patterns help readers perceive meaning because they match the document structure with content structure and show how parts relate.

In a report to aid decision making, such as a feasibility study, the middle section addresses each of the issues that affect the decision, such as cost, legal restrictions, available workforce, and competition. The topics in such a report are arranged by importance and by relationship to one another. All financial topics, for example, would be grouped together. As you consider the organization of these topics, you can also assess whether they are complete. That is, are there other topics that will bear on the decision that are not discussed?

The order of importance pattern is more abstract than the others. Who is to decide what is more important than something else? The problem—the reason for the document to begin with—should give cues. If the problem was environmental, the investigation on that impact is more important than the research on costs or equipment. As an editor, you need to be particularly cautious about rearranging material to emphasize one point at the expense of another. Your perception about what is important may not be the same as the writer's. The writer, usually the subject matter expert, should have final say about rearrangement to suggest degree of importance.

All of these patterns—chronological, spatial, comparison-contrast, cause-effect, order of importance—are subordinate to the overall pattern of general to specific, as discussed previously. As editor, you first ensure that the general information precedes the specific; then check that patterns are used effectively in developing the parts. Many documents use more than one pattern. For example, a chronological narrative may also use comparison-contrast within the sections on various time periods. One pattern, however, should predominate.

Group Related Material

This is a familiar concept for writers and editors. An editor tries to keep paragraphs focused on a topic rather than wandering, to avoid mixing unrelated information, and to develop lists meaningfully rather than randomly. Editors have to be concerned with grouping because the order in a draft may reflect a writer's mental associations rather than a coherent structure.

The list in Figure 16.1 illustrates the problem resulting from failure to group. The list appeared in the first draft of an employee handbook for a video rental store; its purpose was to delineate duties of the manager. The writer's free association is evident from the listing. For example, does item 15, "filing on people," refer to filing in a file cabinet (as in item 16) or to filing in courts to pursue writers of bad checks (as in item 14)? The lack of organization in the list would cause problems for managers trying to develop a concept of their job and to identify and recall specific duties.

Duties: Manager

1. waiting on customers
2. selling memberships
3. inventory of tapes, control cards, etc.
4. inventory of store
5. employee meetings
6. keeping employee morale up
7. accounts payable and receivable
8. computer input
9. computer reports—late, end of week, end of month, etc.
10. daily deposits
11. totaling time cards
12. payroll—total system done in store
13. checkbook balanced
14. hot check system
15. filing on people
16. filing system
17. ordering/merchandising
18. budget for store
19. prebooking
20. customer orders and order system
21. P. O. P. ordered and regulated
22. receiving
23. mail outs (free movie postcards)
24. static customer mail-outs
25. tape repair
26. cleaning VCRs
27. having anything that is broken fixed right away—the longer it sits the more money tied up
28. brown book—bookkeeping record
29. monthly report book
30. payroll book
31. schedule for employees
32. keep store appearance bright, eyecatching, clean, fun; movies playing by customer counter and children's movies in children's section at all times
33. keep coop checked on and updated
34. taking care of customer complaints or problems
35. taking care of employee problems
36. ordering computer supplies
37. ordering office supplies
38. keep plenty of paper supplies on hand
39. reshrinkwrap tape boxes that are getting worn out
40. Commtron bill
41. video log (introducing and explaining it to customers)
42. machine inventory every Wednesday—better control over what you have in the store; just in case one is stolen the time span isn't very long
43. promotions and marketing
44. know what's hot and new and learn the titles and actors
45. know your stock and what you have on hand—better control of movies and VCRs

---

FIGURE 16.1 **Unedited List of Duties of Video Store Manager**

One task in editing this list, then, is to group various tasks into categories, such as finances (budget, accounting, and bookkeeping), personnel, marketing, customer relations, maintenance, and inventory/supplies. Figure 16.2 shows an initial grouping based on these categories.

For your information, "Filing on people" (item 15) does refer to filing charges against customers who fail to return the movies they have rented. "P.O.P." (item 21) is an abbreviation for "Point of Purchase" and refers to displays supplied by the wholesaler. "Coop" (item 33) refers to a cooperative advertising arrangement between the movie distributor and the retail store. "Commtron" (item 40) is the major supplier of movies to this store. This information was not obvious from the document; queries to the writer were necessary.

Figure 16.2 shows how the list might look after one editorial pass. The editing is not yet complete, but the grouping will help the editor when reviewing the list with the writer.

Duties: Manager

1. **Finances**
   accounts payable and receivable
   daily deposits
   checkbook balanced
   hot check system
   budget for store
   brown book—bookkeeping record
   Commtron bill
   filing system
   computer reports: late, end of week, end of
      month etc.
   monthly report book
   filing on people

2. **Personnel**
   employee meetings
   keeping employee morale up
   payroll—total system done in store
   schedule for employees
   taking care of employee problems

3. **Inventory/Supplies**
   inventory of tapes, control cards, etc.
   inventory of store
   ordering computer supplies
   ordering office supplies
   keep plenty of paper supplies on hand
   machine inventory every Wednesday
   computer input

4. **Marketing**
   selling memberships
   P. O. P. ordered and regulated
   ordering/merchandising
   mailouts (free movie postcards)
   static customer mailouts
   keep store appearance bright...
   keep coop checked on and updated
   know what's hot and new and learn the titles
      and actors
   know your stock and what you have on
      hand—better control of movies and VCRs
   prebooking
   promotions/marketing

5. **Customer Relations**
   waiting on customers
   customer orders and order system
   taking care of customer complaints and
      problems
   video log—introducing and explaining it to
      customers

6. **Maintenance**
   tape repair
   cleaning VCRs
   having anything that is broken fixed...
   reshrinkwrap tape boxes that are getting
      worn out

FIGURE 16.2   Manager's Duties Grouped after One Editorial Pass

The grouping will probably reveal some missing duties. For example, will the manager train new employees? It may raise questions about whether the duties are really managerial duties. Might a subordinate handle orders? Grouping may also encourage clarification of items. For example, is "promotions and marketing" (item 43) a category term, or does it refer to a specific duty? Thus, editing for organization, using the guideline about grouping, should help the editor and writer make the content more complete and accurate as well as easier to comprehend.

After the content is complete and meaningfully organized, the editor can work on consistency (for example, "mail-outs," "mail outs," or "mailouts"?) and parallelism of phrases. (The example in Figure 16.4 will illustrate the importance of grouping related ideas in paragraphs.)

Research has shown that memory decays after readers are asked to remember more than seven items, plus or minus two. That limit on memory is one reason why phone numbers have seven digits. If your groups contain more than seven items, as in the list of managerial duties in Figure 16.1, consider regrouping to shorten the list. The video store manager will have a better chance of learning and remembering the 45 duties grouped into six major categories with subduties than if the list remains 45 items long. Likewise, users performing a task can comprehend and remember six major steps more readily than 45 separate tasks. The grouping creates the hierarchical structure for the task schema.

The principle of restricting the number of chunks of information relates to prose as well as to instructions and lists such as the one in Figure 16.1. Chunks may be the major sections in a report, proposal, or chapter as revealed by level-one headings. If such a document has many more than seven sections, readers may lose a sense of how the individual sections relate to one another and to the whole. Excessive efforts to reveal the structure of the information with divisions of text may actually diminish the coherence.

The principles of grouping and chunking also explain the effective organization of hypertext and of menus in software applications. Top level categories reveal the main tasks or main organizational divisions to be featured. Secondary items are then grouped within these main categories, with the grouping revealing how the parts are related to each other and to the whole. The depth of the structure—number of levels of menu choices or categories of information—depends in part on the complexity of the information but also on how much a reader can grasp in a chunk. Some grouping will be preferable to undifferentiated sections.

Use Parallel Structure for Parallel Sections

The principle of parallelism at the sentence level is important for larger structures, too, such as paragraphs, sections of chapters, and chapters. The structure can clarify the content relationships of sections. For example, if you are editing a progress report that is ordered by tasks, each section probably will proceed from work completed to work remaining. The editor of this book looked for a pattern in each chapter of introduction, principles and theory, application or guidelines, and summary.

Paragraph Organization

Paragraphs, like whole documents, make more sense if they are well-organized. The subjective descriptor for good organization in a paragraph is flow. A paragraph that flows enables a reader to move from sentence to sentence without halting or rereading. In more objective terms, what creates "flow" (or cohesion) is the linking of sentences and repetition or variation of key words.

Linking Sentences

Sentences must connect to each other to make sense in a paragraph. Abrupt shifts in topic will puzzle readers. A workable pattern for those connections is to tie the beginning of one sentence to the end of the previous sentence. The subject offers the familiar information whereas the predicate offers the new information. The new information in the predicate of one sentence becomes familiar and can then be the subject of the next sentence. That linking helps to create flow or cohesion.

You can see the pattern of linking in this diagram of the key words of the subjects and predicates for three sentences in the previous paragraph:

subject	predicate
pattern for connection	tie beginning…to end
subject/familiar	predicate/new
new information	familiar/subject

Sentences don't have to rigidly follow this pattern, but a sentence should echo a key point of the one that precedes it.

Repetitions and Variations

In addition to links, repetitions or variations of key words keep a paragraph focused on a topic. Variations can be restatements in different terms, examples, opposites, and pronouns. They can also be effects or causes. In the paragraph on linking, with the heading announcing "linking" as the topic, there are two repetitions of "linking," but there are also these variations of the concept:

variation of "linking"	type of variation
connect	restatement in different terms
connections (twice)	restatement
abrupt shifts	opposite
tie	restatement
subject/familiar–predicate/new	example
"flow," coherence	effect

These terms keep readers focused on the topic of linking. A subtopic is the idea of making sense, announced in the first sentence. The second sentence varies this concept with the opposite, "puzzle." "Cohesion" in the final sentence is a restatement.

Application: The Problem Statement for a Research Proposal

This section applies the guidelines for editing for organization to increase comprehension. It also illustrates the process of analyzing documents described in Chapter 12. Figure 16.3 shows the problem statement from a proposal for research on the bulb onion. The proposal was written by a graduate student competing for research funds distributed by the graduate school of a university. The graduate school distributed an RFP and promised to fund some—but not all—of the proposed projects.

The writer's goal is to persuade the proposal evaluators that his project is feasible and worthy of being funded. To demonstrate feasibility, he has to show that

Isozyme Variability of Allium cepa Accessions

Statement of the Research Problem

Genetic variability is essential in the improvement of crop plants. Until recent years most of the single gene markers used in higher plant genetics were those affecting morphological characters, i.e., dwarfism, chlorophyll deficiency, or leaf characteristics. Molecular markers offer new possibilities of identifying variations useful in basic and applied research. Proteins are an easily utilized type of molecular marker. Protein markers code for proteins that can be separated by electrophoresis to determine the presence or absence of specific alleles. The most widely used protein markers in plant breeding and genetics are isozymes.

Electrophoretic studies of isozyme variation within a plant population provide information on the genetic structure (Sibinsky et al. 1984) without depending upon morphological characters, which are easily influenced by the environment. Genetic studies of isozymes have been conducted on more than 30 crop species (Tanksley and Orton 1983). Considerable variation can exist among plant populations as well as among individuals within a given population. Domestic and wild barley accessions were assayed and substantial differences within and between accessions were found (Kahler and Allard 1981). Significant variation in allelic frequency and polymorphism has been detected in lentil collections (Sibinsky et al. 1984) as well as in Zea mays (Stuber and Goodman 1983).

Researchers in the proposed project are particularly interested in analyzing isozyme variability within the bulb onion (Allium cepa L.). The bulb onion is a major horticultural crop in Texas with a cash value of $75 million in 1982 (Tx. Veg. Stats. 1983). As onions are a major crop, techniques are continually being explored to facilitate the breeding of superior varieties. However, a major factor limiting advances in breeding is the identification of selection criteria, i.e., genetic variation. Electrophoresis as discussed above provides a tool for selection. Electrophoretic techniques for analyzing isozyme variability in onions have been established (Hadacova et al. 1983; Peffley et al. 1985). Isozyme markers have many applications including introgression of genes from wild species, identification of breeding stocks, measurement of genetic variability, determination of genetic purity of hybrid seed lots, and varietal planting and protection (Tanksley 1983).

This research intends to explore isozyme variability of Allium cepa accessions with the intent of identifying molecular markers useful in onion breeding and genetics.

FIGURE 16.3   Unedited Problem Statement for a Research Proposal

he has the facilities and skills to complete the work and that he can finish within the given time. To demonstrate worth, he has to show that the research will yield significant information. (As a technical editor, you know these things because you are familiar with typical technical documents, such as proposals. If you do not know the purposes and methods of proposals, you consult a technical writing handbook or textbook. You also consult the RFP to determine criteria for these grants.) The problem statement reproduced here is mostly concerned with demonstrating worth.

The proposal evaluators are not expert in the subject of plant genetics. They are professors from various departments in the university. They are intelligent and well educated but probably uninformed about onion research. Furthermore, they will read quickly through a stack of proposals and will not have the luxury of rereading and mulling over meanings. (As a technical editor, you discover this information through queries to the writer or possibly to the contact person specified on the RFP.) As editor, you can probably act as a good stand-in for the intended readers. That is, unless you have a good background in the study of plant genetics, the material will be new to you, as it will be for some evaluators. Your responses, therefore, will be useful in assessing likely responses from the evaluators. Your analysis of the context is complete enough at this point for you to assess the document itself.

Guideline 1 for the editing for organization will not apply here because you are looking at just part of the proposal. But guideline 3—arranging information from general concept and familiar information to specific details and new information—will be very important.

Read the problem statement through. Note where you must reread sentences or look back in the text to verify assumptions. Also note where the meaning seems especially clear. These notes will guide you in determining editing goals and suggesting emendations. After you have noted both the confusing and the clear places, you can set some editing goals. You can determine what you would need to do as editor to clarify the problem and its significance for intelligent but uninformed readers.

Unless you are a plant genetics expert, you probably got lost early in the problem statement. The first sentence is easy enough because most educated readers know that crop scientists improve crops through genetic manipulation. You may have been relieved when you got to paragraph 3, where you can relate to the idea that onions are a major cash crop (even if you are surprised by this information).

If you stumbled over other parts, you may be thinking, as many editors do, that defining terms, either parenthetically or in a glossary, will solve the problem. Indeed, the numerous technical terms can slow down nonexpert readers significantly. If you follow the principles and processes of editing described in Chapter 12, though, as an editor you may rightly choose to think first about organization and completeness of information. Just as readers process a text top down, you will edit top down. Your first task should be to define the concept of the research problem. Definitions may be an option later. If you are not convinced, look up a few of the terms that are repeated frequently: "electrophoresis," "isozyme variation," and "accessions." Chances are that the definitions do not clarify the significance of the problem statement. One reason is that the predictable reader question, "so what?" (or "what is the significance?") is not really answered except indirectly in the information about cash value of the crop.

Reorganizing may make the concept and terms clearer. Guideline 3 suggests that you place general or conceptual information before specific. Thus, the concept of the research (the need and purpose) should appear in the first or second paragraph. While considering organization, you may also have noted that the discussion of electrophoresis in paragraphs 1 and 3 is interrupted by the introduction to paragraph 3 ("as discussed above" is a clue). Thus, an editing goal may be to group the sentences about electrophoresis together (guideline 5). Additional goals may include the definition of terms, correction of grammar and punctuation (especially the dangling modifier in paragraph 2), and other sentence- and word-level emendations.

To prepare for a thoughtful and systematic reorganization, begin with guideline 3: place general information before specific. To identify the general information (concept), look back through the problem statement to see if you can answer the basic question: What will the proposed research do that previous genetic research has failed to do?

The first paragraph contrasts two kinds of gene markers: those affecting morphological characters (the kind formerly used in research) and molecular markers (the kind to be used in the proposed research project). Why are molecular markers better? To answer that question, you may need to ask why morphological markers are limited. The answer to that second question is buried in a dependent clause in paragraph 2: "…which are easily influenced by the environment." If the morphological characters are easily influenced by the environment, the data generated by using them may be unreliable. These markers may measure environmental effects rather than genetic variation. This statement is implicit in the first paragraph, but it is not stated. Thus, this key information becomes available to uninformed readers only upon rereading (and mental reordering of the information).

This kind of questioning and probing for the central ideas that you do as an editor parallels a reader's questioning. Your editing task is to make the concepts so clear that a reader won't have to probe. Once you can answer the basic questions, you should have the conceptual information that should appear early in the document. This process of asking questions about concept is central to top-down editing. If the questions cannot be answered from the existing text, you query the writer.

A simple paraphrase of this problem statement might be this:

> Researchers have been limited in their attempts to breed superior bulb onions by an inferior method of marking the genes. In this method, the markers were for morphological characters. Because these characters could be influenced by the environment, the data they yielded were unreliable. A new method, electrophoresis, allows the identification of molecular markers (isozymes). In the proposed research, electrophoresis will be used to analyze isozyme variability in onions.

Figure 16.4 shows the problem statement reorganized to present conceptual information early and explicitly and to form paragraphs by topic. The sentences in boldface have been moved, and the two phrases in boldface italics (paragraphs 2 and 4) have been added to clarify information. No other editing has been done. The terms remain undefined, the dangling modifier remains in paragraph 4, and the writing style is the same. But the document is now more comprehensible and persuasive for readers who are not genetics experts. The terms become more clear once the concept is more clear. Given the hasty way in which the proposal will be read, it will be sufficient for readers to know that electrophoresis is a method and that an isozyme is a molecular marker, without knowing the specifics. Depending on how much time is available, the editor may work on sentences for clarity and persuasiveness. But even if he or she must stop now, the document will be more effective than the original.

As an editor, you may have other ideas for the organization of this problem statement. The version in Figure 16.4 is not necessarily the "right" way. Because we are not applying rules, we cannot assert with certainty that one way is right and another is wrong. However, each emendation here can be explained in terms of

Isozyme Variability of Allium cepa Accessions

Statement of the Research Problem

The original first sentence remains to establish the concept of the research. The familiar "genetic variability" also prepares for "isozyme variability." (See guideline 3.)

This easily understood information announces the significance in familiar and persuasive terms. The paragraph also establishes the context for the proposed research—the ongoing research at the university. (See guideline 3.) The last sentence introduces the rest of the paper by stating the limitations of the old method of research.

The rearranged and new information states the specific limitation of the old method of research. By knowing the limits of the old, reviewers will see the importance of the proposed method.

This paragraph and the next one are rearranged to create separate paragraphs on electrophoresis and on isozyme markers, rather than mixing the subjects as the original paragraphs 2 and 3 do. (See guideline 5.)

The subject of isozymes could precede the subject of electrophoresis. However, the material on isozymes is mostly background material. The material on electrophoresis focuses attention on the proposed project. (See guideline 3: place important information early in the sequence.)

This background information can be skimmed; it basically demonstrates the proposer's knowledge rather than giving information on the proposed research.

The explicit purpose statement leads nicely to the next section of the proposal, a statement of specific objectives.

Genetic variability is essential in the improvement of crop plants. **Researchers in this project are particularly interested in analyzing isozyme variability within the bulb onion (<u>Allium cepa L.</u>). The bulb onion is a major horticultural crop in Texas with a cash value of $75 million in 1982 (<u>Tx. Veg. Stats. 1983</u>). As onions are a major crop, techniques are continually being explored to facilitate the breeding of superior varieties. However, a major factor limiting advances in breeding is the identification of selection criteria, i.e., genetic variation.**

Until recent years most of the single gene markers used in higher plant genetics were those affecting morphological characters, i.e., dwarfism, chlorophyll deficiency, or leaf characteristics. **However, these characters are easily influenced by the environment,** *so the data* **are** *unreliable.* Molecular markers offer new possibilities of identifying variations useful in basic and applied research.

Proteins are an easily utilized type of molecular marker. Protein markers code for proteins that can be separated by electrophoresis to determine the presence or absence of specific alleles. The most widely used protein markers in plant breeding and genetics are isozymes.

Electrophoresis <as discussed above> provides a tool for selection of genetic characteristics. Electrophoretic studies of isozyme variation within a plant population provide information on the genetic structure (Sibinsky et al. 1984) without depending on morphological characters. **Electrophoretic techniques for analyzing isozyme variability in onions have been established (Hadacova et al. 1983; Peffley et al. 1985).**

Isozyme markers have many applications including introgression of genes from wild species, identification of breeding stocks, measurement of genetic variability, determination of genetic purity of hybrid seed lots, and varietal planting and protection (Tanksley 1983). Genetic studies of isozymes have been conducted on more than 30 crop species (Tanksley and Orton 1983). Considerable variation can exist among plant populations as well as among individuals within a given population. Domestic and wild barley accessions were assayed and substantial differences within and between accessions were found (Kahler and Allard 1981). Significant variation in allelic frequency and polymorphism has been detected in lentil collections (Sibinsky et al. 1984) as well as in *Zea mays* (Stuber and Goodman 1983).

This research intends to explore isozyme variability of Allium cepa accessions with the intent of identifying molecular markers in onion breeding and genetics.

FIGURE 16.4   The Problem Statement in Figure 16.3 Reorganized.   Code: Boldface indicates rearranged material; boldface italic indicates inserted information; <> indicates material that should be deleted because of rearrangement.

how readers learn from a text. Nothing has been done arbitrarily. Furthermore, the editing is based on an analysis of the context of use, including the purpose of the document, the intended readers, and the conditions of reading. Thus, decisions have been made to accommodate those readers and purpose. Had the problem statement appeared in the research report for publication, the organization would have been different. The review of literature in paragraph 5, for example, would have moved forward because of the established pattern in research reports of setting the research context with a literature review.

You may be interested to know that the edited proposal succeeded in winning funding!

Summary

A well-organized document makes information easy to learn and easy to use. Poor organization, more than any other quality, destroys the effectiveness of a document for comprehension and use.

Good organization means matching the arrangement of the material to the readers' use of the material or to the inherent structure of the content. Organization by task reflects what readers need to do and is appropriate for instructions and procedures. Organization for comprehension helps a reader to form schemata, or patterns for making sense of and remembering information. Standard organizational patterns show relationships of data and tap into readers' existing schemata. They help readers interpret meaning by showing how the parts connect to form a whole. Structure reinforces meaning.

Editing for organization takes place early, ideally at the document planning stage. By reviewing the document outline, the editor can work with the writer to develop the document according to the most effective order. The review requires analysis of the context of document use and what readers know. By anticipating readers' use of the material and the questions they will ask, writer and editor can arrange the material to facilitate completion of tasks or learning or both. When the document is complete, it still will require review to determine whether the plan for organization works overall and within sections. Editing for organization is also a way to determine whether the content is complete.

Further Reading

Gregory G. Colomb and Joseph M. Williams. 1985. "Perceiving Structure in Professional Prose: A Multiply Determined Experience." In Lee Odell and Dixie Goswami, eds., *Writing in Nonacademic Settings.* New York: Guilford, 87–128.

T. N. Huckin. 1983. "A Cognitive Approach to Readability." In P. V. Anderson, R. J. Brockmann, and C. R. Miller, eds., *New Essays in*

Technical and Scientific Communication: Research, Theory, Practice. Farmingdale, NY: Baywood, 90–108.

D. E. Rumelhart and A. Ortony. 1977. "The Representation of Knowledge in Memory." In R. C. Anderson, R. J. Spiro, and W. E. Montague, eds., *Schooling and the Acquisition of Knowledge.* Hillsdale, NJ: Lawrence Erlbaum, 99–135.

Discussion and Application

1. Perform a second editorial pass on the example in Figure 16.2. Are the categories satisfactory? Should any items be combined? Prepare a list of questions for the writer about items you don't understand to make sure the wording is clear and the grouping accurate. Could the list incorporate issues of time—frequency with which the manager should perform the tasks? If the main structure were chronological, what would be gained in understanding, and what could be lost?

   Discuss possibilities for making this list more effective other than by regrouping. For example, should each item have a rationale or explanation, as do items 27, 32, and 42 in Figure 16.1? What would be the basis for making such a decision?

2. Determine conventional patterns of organization in your subject field by consulting three periodicals, including at least one research journal. In a class discussion, compare the patterns of organizing research articles in different subjects, such as psychology and chemistry.

3. Compare several user manuals for word processing programs or several manuals in a series for one computer system. Are there any consistent patterns in the manuals? What organizational strategies seem particularly effective? If there are inconsistent patterns, did the document designers have a good rationale for modifying the patterns?

4. The following paragraphs and tables are from the time and cost analysis section of a proposal for landscaping an office park in two phases. Analyze the order of ideas and grouping in the paragraphs and tables. For example, consider whether the Phase 1 and Phase 2 sections present information in parallel order. Also consider how the tables group and sequence information. What would be some options for spatial patterns for Table 1? Edit for organization according to the guidelines presented in this chapter.

   [1]The estimated cost of this project is based on a cost of $6 per square foot for the patio and $4 per square foot for the sidewalks. [2]The first phase includes the upper patio, lower walks, planters, and stairway. [3]The cost for the patio is higher because it is made of bricks. [4]There are 80 linear feet of planters at $10 per square foot and 4,600 square feet of lower sidewalk at $4 per square foot. (See Table 1.)
   [5]The estimated time of completion of phase one is four to six weeks depending on the weather.

   TABLE 1.   Phase One Cost Analysis

	Sq. Feet	Cost
Upper level	6,400	$38,000
Planters	80	$800
Lower walk	4,600	$18,400
Stairway	300	$3,000
Total	11,380	$60,200

   [6]The second phase of construction will consist of the installation of six concrete sidewalks and the reconstruction of a blacktop sidewalk into concrete. [7]The estimated time of completion of phase two is two to three weeks depending on the weather.

TABLE 2.   Phase Two Cost Analysis

	Sq. Feet	Cost
Sidewalk 1	1,926	$7,680
Sidewalk 2	2,400	$9,600
Sidewalk 3	1,280	$5,120
Sidewalk 4	1,280	$5,120
Sidewalk 5	1,760	$7,040
Sidewalk 6	640	$2,560
Sidewalk 7	2,000	$20,000
Total	11,286	$57,120

[8]The total minimum cost for the proposed changes is $117,720.

5. A six-page guide for users of a major city library is organized alphabetically. Topics are listed below. Each topic is followed by one to four sentences explaining location or use. Discuss the merits and limitations of alphabetical organization for this document if the readers are first-time users of the library. How well will the alphabetical organization work if the readers are experienced users? What other patterns of organization might work for this document? Arrange the topics in another order based on the needs and interests of first-time users. What are their questions? What information may they need to reference more than once? (If they have to refer to certain facts, one way to make that information easy to find is to place it early.) Does the reorganization suggest any revisions in content? That is, should any topics be added or deleted? For your information, the comment for the "personal property" topic advises patrons not to leave their property unattended.

acquisitions
arrangement of the
    library
bindery
book returns
borrowing books
call numbers
card catalog
cataloging
change machine
circulation
computers
computer assisted
    search
services
copy center
current periodicals
director's office
elevators

fines
food and drink
gifts/exchange
government documents
hours of the main library
information
interlibrary loan
internet access
library cards
loan periods
lost and found
lounge areas
magazines (see periodicals)
materials processing
meeting rooms
microforms
newspapers
online computer search (see
    computer assisted search)

oversize books
periodicals
periodicals list
personal property
personnel office
rare books
reserve materials
restrooms
smoking
sorting areas
stacks
stairs
study areas
technical processing
telephone directories
telephones
water fountains

6. Examine several paragraphs from your own writing for paragraph cohesion. Mark the verbal links from sentence to sentence. Underline repetitions or variations of key terms. How could you edit your own writing to increase cohesion?

7. Vocabulary: These terms and concepts should be familiar enough to you that you can use them in making decisions about organization: *chunking, cohesion, macrostructure, microstructure, schema theory, top-down learning.*

C h a p t e r *17*

Visual Design

When speakers prepare speeches, they plan and research, organize content, and construct sentences. But the speech is not complete until it is delivered, with gestures, tone of voice, facial expressions, and possibly visual displays. What the listeners see and hear affects their responses to the content—whether they understand, fall asleep, remember, or ask questions. Unless speakers connect with their audience through delivery, other preparations may not matter.

Likewise, an architect plans a space based on intended uses of a building, but the visual features, such as exterior materials (brick, wood, glass), overall shape (tower, one-story), and entries, become signs of those uses. These features help people to define the building's uses (a factory does not look like an office building). They also help people anticipate where to find particular rooms (the receptionist's office is near a main entry, identified by its size, signs, and landscaping). The visual design affects recognition and usability.

As in speech and architecture, the successful delivery of a print or online document requires a plan for its presentation. Options for visual design include use of paragraphs, graphics, or lists; use, placement, and type style of headings; placement of document components on the page or screen, such as the heading of a letter at the top of the letter or a navigation table at the top of a screen; margins and number of columns; features of type, such as double or single spacing, paragraph indentation, and type style (bold, italic, roman); document size, shape, and binding; color (paper or background, type, graphics); use of page numbers and running headers; and inclusion of front and back matter, such as a glossary and index. Effective design not only attracts the readers' attention but also helps them understand the information and find selections within the document. *Visual design* refers to the visual and physical features of the document as selected to facilitate comprehension and use. Visual design, like organization and style, is functional, not just aesthetic: it points readers to key information and helps them interpret the information.

This chapter defines terms used in discussing visual design—*document design, graphic design, layout, format,* and *genre*—and identifies visual design options. It then discusses functions of visual design as bases for editorial choices. It includes a section on headings that pertains to several functions of design. The chapter concludes with guidelines for editorial choices about design.

Terms Related to Visual Design

Visual design is a component of *document design,* a broad concept that refers to all the choices a designer makes about a document, including choices about content, organization, and style. Like a designer in any other field, a document designer (who may be the editor) considers uses of the item being designed and makes choices to enable those uses. For technical documents these uses include reading to get information, to understand, to learn, to complete tasks, or to make decisions. Design choices are good if they enable these functions. Visual design is an important component of document design because it indicates the structure of the information, provides guidelines for locating information, and invites or discourages reading. The appearance of the document and its components provides information.

Document design and visual design are related to *graphic design,* but whereas visual design is concerned with how the presentation enhances function, graphic design is concerned with specifics of production and with aesthetics. The document designer considering visual presentation decides to use running headers to help readers find particular sections in the document (a function) whereas the graphic designer determines the typeface, size, and space from the top margin for the running header, based in part on aesthetics (what looks good). The document designer plans two columns of text, but the graphic designer decides how long the lines of type are and how much space is between columns. With publishing at the desktop or on the screen, the document designer or editor may also do the graphic design.

Layout is the graphic designer's plan for arrangement of text and visuals on a page and includes specifications for type and spacing. The layout may include a grid to show line length, placement of headings and visuals, and justification. The grid includes a sample for any format choices, such as a list, that the writer or editor has made. It organizes space and creates visual consistency. Figure 17.1 shows a designer's grid for the layout of a page in this book.

Visual design is similar to *format,* a term which emphasizes the result of the choices about placement of information on the page or screen rather than the process of design. Because format is conventional for some types of documents (genres), such as letters and memos, format can visually identify the genre, and the terms are sometimes confused. But some formats apply across genres: a report may look like a proposal—the genres may share a format, though they have different functions. Furthermore, not all reports use the same format. The term *visual design* emphasizes the process and avoids the confusion of *format.*

FIGURE 17.1   Designer's Grid for This Book

The editor participates in the visual design and evaluates whether the choices as implemented work well. This chapter reviews principles of design so that editors can make informed choices.

Visual Design Options

Writers and editors have some choices about design, just as they have choices about writing style. Choices depend on the complexity of the reading material, the probable uses, the characteristics of readers, related documents, and methods of production. The following list introduces some bases for choosing among the various options and indicates some of the ways in which the intended use of a document affects design choices.

- **Prose or visual representations.** Paragraphs are best for explanation and narrative. When the subject is highly complex, or when the reading is selective, a graphic display of text may aid in comprehension and selective reading. For example, flowcharts visually show the sequence of steps, and outlines show the hierarchy of ideas. Tables are appropriate when the readers must locate specific items of information.

For an international audience or subject matter about two- or three-dimensional objects, line drawings or photographs may be preferable to words. Visuals lessen the need for translation, and they show spatial relationships more readily than words.

- **Number and length of paragraphs.** Short paragraphs are quicker to read, but longer paragraphs permit the development of complex thoughts and reveal the relationships between ideas.

- **Indentation.** Indentation is the conventional signal for a new paragraph in print. In outline form, indentation signals subordination with the indented item subordinate to the item it follows. A hanging indent or outdent, where a heading protrudes into the left margin, can distinguish a major heading from a minor heading. Indentation of a long quotation (block style) visually signals a quotation, just as quotation marks signal a shorter quotation within a paragraph.

- **Alignment.** Text may be aligned on the left margin (left justified), on the right margin (right justified), on both margins, or centered. Left alignment or justification provides a common point for the eye to return to in left-to-right reading. Centering and right justification can be used for headings but increases reading difficulty in large blocks of text.

- **Lists.** When there are parallel ideas or a clear sequence of items, a list will reveal the structure more quickly than will embedding the items within prose paragraphs. Lists in the midst of paragraphs also call attention to the listed material. But too many lists can create the sense of reading an outline rather than the finished text.

- **Headings.** Headings identify key points, serve as transitions, show the overall structure of the document, and identify specific sections for selective reading.

- **Menu bars.** In online presentation, the arrangement of the categories of information in a menu bar across the top of the screen encourages the interactivity that is such a strong feature of online documents. Navigation buttons or lists at the top, side, or bottom of the screen enable readers to find the information they want within the document or a related document. Navigation devices, however, reduce the space available for content.

- **Line length and number of columns.** Lines that are too long or too short slow down reading. The maximum line length for easy reading in print is 2 ½ alphabets or two times the point size. This means that 65 characters per line across the page is the limit, whatever your type size. It also means that a line 20 picas long is about right for 10-point type whereas a line of 24 picas is appropriate for 12-point type. ("Point" and "pica" are printers' measures, defined in more detail in Chapter 22.) A large page or wide screen may require large margins or multiple columns to keep the line length right. Newspapers and magazines conventionally use multiple narrow columns to facilitate quick reading down the page rather than line-by-line across the page.

Lines tend to be longer than readable on World Wide Web sites because monitor screens are proportionately wider than paper. The reader's browser and monitor may determine how long the lines will be unless the code includes absolute values for length.

- **Numbers, letters, bullets.** Numbers, letters, and bullets distinguish items in a list. Numbers and letters may be used in lists embedded in paragraphs or in offset lists. They imply that the order of items is significant. Numbers can show different levels in the structure, as in the decimal system of numbering where "2.3.3" would mean that the part is the third item within main part 2, subpart 3. Numbers and letters are both useful as cross-references, as when a step in instructions refers the reader back to "step 2." Bullets (dark or open circles or squares or other shapes preceding an item) are used with offset lists in which the sequence of information does not matter. If itemization alone is the goal, bullets are sufficient and do not suggest unintended meaning.

- **Page orientation.** Most books use portrait orientation; that is, they are taller than they are wide. Because these books fit on most bookshelves, they are easy to store. However, the wide or landscape orientation may better suit the text or its storage requirements. Many user manuals use landscape orientation. Pages with the landscape orientation have multiple columns or margins as wide as columns so that the lines will not be too long for easy reading. Shifts in orientation, such as printing tables landscape in a portrait text, distract from reading or performance because they require readers to flip the document around. Computer monitors are wider than tall, and though reorienting the screen is not an option, a designer works with the available space to display items for coherence and to minimize scrolling

- **Variation in type style.** Any change in type invites attention. Shifts from roman to italic type, from lowercase to capital letters, from the primary typeface to boldface, or from a serif to sans serif typeface all attract notice that can be useful to point to safety issues or key points. Too many shifts create a busy look that readers shun.

- **Underlining.** Like changes in type, underlining emphasizes the material, but it is usually restricted to typewritten documents because it reduces legibility. The line itself interferes with perception of the descenders on letters (the tails below the line), information a reader uses in recognizing the word. The computer offers more professional and readable options for emphasis, including boldface, italics, and variations in type size. Professional typesetters, set underlined words in italics unless instructed otherwise. Underlining in online documents not only reduces legibility but may also be confused with hypertext links.

- **Boxes, white space, shading, pointers, symbols.** These visual devices also call attention to a part of the text. A box around a warning, white space surrounding type, and shading all draw the reader's eye. Pointers are arrows or hands with pointed fingers that direct a reader's attention to specific information. Certain symbols, such as a skull and crossbones, have developed universal meanings as signals of danger and can be useful in calling attention to warnings or substituting for verbal warnings.

- **Binding, folding, size.** Printed documents may be bound or not and folded or not. The decision depends on use, storage, and economics. For example, a brochure that will be mailed may need to fit into an envelope. A spiral binding lets a manual rest open on a desk while it is used, but it is less durable than a hard cover with a library binding, and it does not permit a printed title on the spine. Perfect binding, the attachment of paper covers with adhesive, allows easy armchair reading and is inexpensive but is less durable than a stitched library binding.

- **Color.** Color choices affect readability and meaning. High contrast, as achieved with black type on a white or yellow background, makes type easy to read. Reversing the pattern (light type on a dark background) can attract attention but at some sacrifice of readability. Colors have specific cultural meanings. Nancy Hoft reports, for example, that green means environmentally sound in many countries but disease in countries with dense jungles (p. 267). The international warning symbols use color in consistent ways, with red meaning danger to life, orange warning of risk of serious injury, and yellow urging caution. Color adds to the cost of print documents and slows the display of documents on the World Wide Web, but these disadvantages may be outweighed by gains in comprehension, usability, and motivation.

Functions of Visual Design

Like designs for buildings, interiors, or cars, document design is functional. Design enables people to use spaces or equipment. An architect designs spaces so people have room to do their work, store their equipment, look out of windows, find the bathroom—a building accommodates specific human purposes. Document designers make choices with type, page layout, paper, and types of information so that people can use and understand a document. The four main functions of visual design—comprehension, usability, motivation, and ongoing use—establish some principles for editing.

Comprehension

As you learned in Chapter 16, comprehension depends in part on perception of the structure of the information. An organized document is easier to understand than a disorganized one, and the structure of the document must match the structure of the information. Page or screen design and headings aid in comprehension because they reveal visually the structure of the document and, in turn, of the information. Headings offer an outline of the information—the hierarchical, sequential, and other relationships that readers use to interpret what they read. Paragraphs signal new topics. These visual signals complement verbal ones, such as forecasting statements in introductions, that explain how a chapter or section will develop.

In addition, using design for emphasis indicates what ideas are most important. If you highlight information with boldface type, color, shading, or listing, you alert readers to pay attention. Readers will more likely remember and use the information that gets the special treatment. Emphasis should be thoughtful: not every series of three in a sentence should be offset in a list because the information may not be important enough to warrant the attention. Only the most significant warnings should be boxed.

Usability

As Chapter 2 notes, readers are likely to read technical documents selectively. They *use* documents more than they *read* them. They may look for specific points of information, or they may look away from the text to complete a task and then find the place where they were. A usable document enables readers and users to find the information they need for a specific purpose. Good design enhances usability.

Selective Reading: Finding Points of Interest
The table of contents and index point to topics, but visual signals on the page also help readers find information. Headings within the text and running heads (section or chapter titles—sometimes abbreviated—at the top or bottom of each page), page numbers, and chapter numbers as well as titles identify topics and location within the overall document. Also, boldface or italic type or shading direct readers to material of particular importance.

Navigation in online documents is analogous to access in print; a usable document requires easy access to specific parts. The table of contents or site map on the home page or root screen is as important (or more so) than a table of contents in a book or manual for revealing the structure and scope of a document that the reader cannot see completely. Links to related topics enable selective reading, but they must be meaningful, lead to where they promise, and offer readers a way back to where they were.

To ensure usability, you imagine the probable methods a reader will use to search a text and assess the adequacy of access devices for those methods. You also check the accuracy and frequency of the headings and the wording in the table of contents against the headings in the text. In an online document you check the links for accuracy and completeness.

Page and Screen Design: Finding Information in Predictable Places
Consistent placement on the page aids readers. For example, in North America business letters include the return address and date in the heading at the top of the letter, enabling readers to respond without looking up an address and to keep a sequential record of correspondence. Writers do have the option of aligning all the parts of the letter on the left margin (full-block format) or aligning the heading, closing, and signature on a second margin just right of center (semi-block format).

But both options are consistent in the placement of the major parts. The heading will always be at the top, and the signature will always be at the bottom. If these conventions are violated, readers are forced to hunt for the information they need to respond or they may not find it at all. In addition, a writer who doesn't know conventions appears unsophisticated. Business letter format varies in different countries, and good editors, always concerned about readers, learn the formats that will be familiar to them.

Because of conventions for types of documents, readers expect to find certain kinds of information on the page in predictable places. For example, software menu items conventionally run across the top of a computer screen, and online help usually displays on the right side of the screen. Arbitrary shifts in placement of types of information, such as illustrations and text, will be confusing. The manual or help screen design should also be consistent with that of related manuals in the document set or with help screens in a family of products.

Being able to find information could be critical in an emergency. Figure 17.2 depicts a 4 ¾ × 11-inch flier intended for people to tape to their medicine chests or keep in a car first-aid box. A larger version (13 × 24) appears as a poster in restaurants. These documents provide first-aid instructions to be used in case of emergency. However, the verbal instructions sometimes appear to the left of the illustrations and sometimes to the right. On the poster version, instructions in Spanish alternate right and left with the English instructions. Thus, readers who have completed one step will have to hunt for the instructions for the second step; the arrangement forces their eyes to scan the page rather than guiding them back to a fixed point. Perhaps that arrangement established aesthetic balance, but the inconsistency will slow the reader in an emergency, and critical time may be lost.

A more functional design for the two-language version would establish all the instructions in one language to the left of the illustrations and all the instructions in the second language to the right. In the single language version, the most efficient design would place the pictures consistently to the left where they would provide a quick concept of the step to be performed. The text would verify the concept and provide details. After reading the first example, the reader would expect the illustrations to fall on the left and to identify the beginning of a new step. Because of the visual consistency, a reader looking up from the task to review the next step in the process would return to the right place on the poster without searching.

Successful design choices on the flier are the use of space between steps and numbers to identify where one step ends and another begins. The space may prompt the reader to look away from the flier to complete a step, but it also provides a point to return to when the reader is ready for new information. The white space and short lines also facilitate reading.

Frequent changes of design place the visual identity of the document at risk, can disorient readers, and are expensive in terms of decision time and redefinition of publication templates. It is rarely appropriate to change a design of a technical document simply for variety. But if design is inadequate for comprehension, usability, motivation, or use, it should change. Design changes may be legitimate just

First Aid for Choking

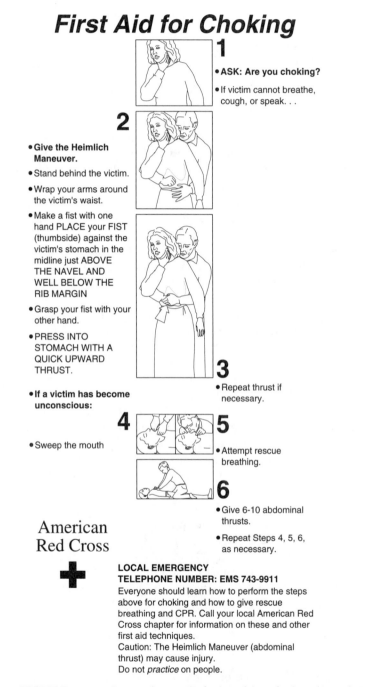

1

- **ASK: Are you choking?**
- If victim cannot breathe, cough, or speak. . .

2

- **Give the Heimlich Maneuver.**
- Stand behind the victim.
- Wrap your arms around the victim's waist.
- Make a fist with one hand PLACE your FIST (thumbside) against the victim's stomach in the midline just ABOVE THE NAVEL AND WELL BELOW THE RIB MARGIN
- Grasp your fist with your other hand.
- PRESS INTO STOMACH WITH A QUICK UPWARD THRUST.

- **If a victim has become unconscious:**

- Sweep the mouth

3

- Repeat thrust if necessary.

4   **5**

- Attempt rescue breathing.

6

- Give 6-10 abdominal thrusts.
- Repeat Steps 4, 5, 6, as necessary.

American
Red Cross

**LOCAL EMERGENCY
TELEPHONE NUMBER: EMS 743-9911**
Everyone should learn how to perform the steps above for choking and how to give rescue breathing and CPR. Call your local American Red Cross chapter for information on these and other first aid techniques.
Caution: The Heimlich Maneuver (abdominal thrust) may cause injury.
Do not *practice* on people.

FIGURE 17.2   Inconsistent Relationship of Visuals and Text

Source: Courtesy of the American Red Cross. All rights reserved in all countries.

to freshen the look of the document. Even when function overrides aesthetics, the document can still be attractive.

Motivation

Visual design creates the first impression of a document and invites or discourages reading. An attractive document with some white space on the page or open space on the screen is more inviting than a crowded page with solid lines of text running from margin to margin. But what is motivating in design depends on readers.

Some readers are highly motivated and will read no matter what the design (assuming that the content meets their expectations). Professional journals use limited formatting and small print because the readers must keep up with developments in their profession. (These journals also have small budgets because advertising revenues and circulation are limited.) Popular magazines, by contrast, compete with other magazines for a reader's discretionary time. Their elaborate and colorful designs attract attention and interest. Popular magazines also have a larger readership, greater advertising income, and bigger budgets than do professional journals. But even if the journals had larger budgets, they would not be likely to adopt the design of popular magazines. The professional nature of the journal is identified by its austere design, and a flashier design could trivialize the content. Web sites reflect the character of the organization or individual that they represent as well as purpose, such as marketing or information delivery.

A document's design may discourage readers. Users of equipment who can't find the instructions they need may abandon the manual and simply experiment with their machines or procedures. Managers may skim introductions and conclusions without reading the whole report. Such experimentation may be satisfactory and even desirable; an editor should not feel a moral obligation to make readers absorb every word. However, unmotivated readers may overlook necessary information. Users may discover just a small part of a machine's capacity or may fail at some procedure. And if they fail or get stuck, they may need a lot of costly time and effort from the customer support department, or they may take their business elsewhere.

Design can direct a reader's attention to vital information about health and safety. Highlighting with boldface type, boxes, space, or icons should encourage readers to read at least those sections. Design of a report can draw the attention of a manager to a warning and recommendation for a modification in equipment.

Readers without motivation need the encouragement of an attractive document. Budgets permitting, colors, graphics, large margins, illustrations, and the liberal use of formats other than paragraphs of text may encourage reluctant readers.

Ongoing Use

If a print document will be used more than once, the size, shape, material, and binding must permit easy storage and continued use. An editor makes design decisions by imagining the document in use or storage: fitting into an envelope, hanging on a wall, stored in a file folder, staying open on a desk, carried in a pocket, standing on a bookshelf, or attached to a piece of equipment, as with a safety warning on a portable electric tool.

If the document will be used outside or carried around, lamination can protect it from weather and from torn edges. If it will be used in low light, large type size and white space compensate for the low visibility. Spiral binding will be a good choice for a manual that remains open but unsatisfactory for a book stored on a shelf and identified by its spine.

Digital documents can be created in multiple separate files or in one long file. If readers are likely to print, they will appreciate a single document. If not, smaller files will be easier to scroll. Some organizations offer two versions—separate files as well as a complete one.

Headings

Headings can aid comprehension and usability. The headings must be effectively worded, the levels must accurately represent the hierarchy and relationship of ideas, and the frequency must match the readers' needs for signals.

Wording

Informative headings provide more information than structural headings. For example, the headings "Equipment Costs," "Installation Costs," and "Purchase Costs" tell more than do "Part 1," "Part 2," and "Part 3." Exceptions are "introduction" and "conclusion," which indicate both content and structure. One-word headings generally tell less than do headings containing several words; for example, "Costs" tells less than "Installation Costs," and "Discussion" tells less than "Analysis of Environmental Impact."

Some research results indicate that headings phrased as questions work better than headings phrased as statements because questions invite readers to examine what they are reading. Readers may remember more when the headings are worded as questions rather than statements. The heading "How Do I Apply?" may appeal more to readers than "Application Procedures." Either heading, however, identifies the two key concepts, "application" and "procedures." Questions would be less functional in a scholarly article than in a brochure because of reader expectations for each document type.

Parallel structure in headings provides consistency. Groups of headings within a section should all be phrases, questions, or complete sentences. Steps in a task should be identified in descriptive terms or as commands but not as a mixture of the two.

Levels of Headings

Different levels of headings reveal structural levels of document parts—main parts and subordinate parts. A short document, such as a brochure, will probably have only one heading level, but a longer, more complex document, such as this book, may have as many as four levels. Theoretically, it could have even more levels, but a complex structure can get confusing. A report of four to eight pages is not structurally complex enough to support more than one or two levels.

If the headings are to reveal the structure, readers must visually distinguish first-level (main) headings from second- and third-level headings. Centering or hanging indent, space before and after the heading, boldface type, type size, capitalization, color, and shading all create the visual power associated with main headings. Italics are less powerful visually than roman type and therefore used more for minor headings, such as paragraph headings, than for main headings. Paragraph headings run into a paragraph rather than being displayed on a separate line.

Boldface type is more readable than either capitalized or underlined type. Variations in type size can show differences in levels of headings, but readers probably won't recognize a difference of less than four points. (A point is a printer's measure; 72 points equal approximately one inch.) For example, if the second-level heading is set in 10-point type, the main heading will have to be at least 14-point type unless (or even if) there are other distinguishing features such as spacing.

Graphic designers advise left-justified headings rather than centered ones because of the left-to-right reading pattern. The left margin is a common point of orientation for readers from North American countries whether the text is display copy (headings and titles) or body copy. Centering may not be noticed on the headings for the narrow columns of a brochure. At least two lines of text should follow a heading on a page; an unusually large bottom margin is preferable to allowing a heading to appear alone at the bottom of the page.

The rules for outlining generally apply to headings. Just as a roman numeral I implies that at least a II will follow and an A implies at least a B, one heading at any level implies at least a second heading at the same level. Two headings of different levels should be separated by text.

An editor verifies that the headings accurately reflect the structure of the document and that the different levels, if any, can be distinguished. An editor may also advise on the style and placement of headings.

Frequency of Headings

If headings are good, are more headings better? Not necessarily. Headings are signals. They should not replace the text, nor should they interfere with the continuity of reading. Too many headings can distract from the content.

There are no rules for the amount of text per heading. Some types of documents, such as brochures, have proportionately more headings than do textbooks or reports. Each chunk of information may be identified by a heading in a brochure. If each paragraph of a book had a heading, however, it would be difficult to read long sections—the book would seem to hiccup.

To judge an appropriate frequency of headings, think about reading patterns and the structural complexity of the document at hand. For a brochure, with small chunks of information meant to be absorbed quickly, headings will predominate on the page. Long reports and textbooks need longer sections in order to develop complex ideas.

Application: Radar Target Classification Program

To illustrate the effect of visual design on comprehension, usability, and motivation, various designs for a single document are shown in Figures 17.3–17.7. The document was part of a program plan for research on the use of radar systems on aircraft or missiles to identify stationary military targets. Such a radar system has to distinguish targets from background clutter, that is, everything else in the scene (trees, grass, buildings).

Target identification begins with a detection algorithm, a computer program that uses a mathematical procedure to search the radar image for bright objects. Then a classification algorithm (the subject of this example) would examine each detected bright object and classify it as a tank, a truck, or neither. The classification algorithm would evaluate the object's size and characteristic features (for example, trucks have wheels whereas tanks have treads and gun barrels).

An algorithm is robust if it works well under a variety of conditions (wet, dry, snowy). Development of classification algorithms involves devising different approaches, testing how well they work, and then using the test results to plan improvements. A "seeker" is a missile programmed to search for a particular type of target. "MMW" means millimeter-wave, a particular band of radar wavelengths. "SAR" stands for synthetic aperture radar, a type of radar that produces detailed imagery. A "standoff sensor" is a radar or other system on an aircraft designed to view a battlefield from a long, relatively safe distance.

The purpose of the program plan is to persuade a manager that the research is well conceived and will lead to the development of effective algorithms. The manager is expected to have a good technical background but to want to read quickly.

The five versions of the document in Figures 17.3–17.7 illustrate how different format choices influence both the speed of reading and the accuracy of comprehension. Figure 17.3, the original version, uses minimal formatting. It is one long prose paragraph, with indentation to identify it as a paragraph, plus a heading with numeric label. The sentences in the original were not numbered, but numbers are included here for reference. Figures 17.3–17.5 show the information with increasing levels of formatting. Figures 17.5, 17.6, and 17.7 use a list format but with different emphases that will affect interpretation.

The writer indicates document structure verbally. In sentence 4, the transition "for example" invites the reader to look for examples of scenarios. In sentence 5, the transition "another scenario" signals that the list of scenarios continues. The words "intermediate complexity" recall "simple" in the previous sentence, thus signaling a simple to complex structure in the list. ("Simple" would have made more sense when readers first encountered it if a forecasting statement had been included to establish the order from simple to complex.) Sentence 6 affirms the structure of both the list and the order from simple to complex: "Third" signals that

4.3 CLASSIFICATION ALGORITHM PERFORMANCE AND UNDERSTANDING

[1]This section of the program plan describes the goals and objectives of the stationary target classification effort and outlines the proposed utilization of radar data to achieve these goals and objectives. [2]The most fundamental goal of the target classification effort is to develop and understand target classifiers which work reliably in a highly variable background clutter environment. [3]Different levels of algorithm complexity would be based upon various target classification scenarios. [4]For example, one simple scenario is the MMW seeker searching a small acquisition region in search of a priority (e.g., tank) target. [5]Another scenario of intermediate complexity might be the MMW seeker having the ability to gather multiple looks at the target or even a reduced resolution SAR. [6]A third, more complex scenario is the standoff sensor application which would have the capability of utilizing high range and cross-range SAR data with a 2-D classification algorithm. [7]Algorithms for these three typical applications might be considerably different in complexity and performance, but the goal is to develop algorithms, understand the underlying signal processing, and to characterize their robustness in highly variable clutter environments. [8]Finally, the promise of new algorithms to be developed in the future, both by universities and industry, will be evaluated as part of the effort in search of a promising solution to this difficult problem.

FIGURE 17.3   Original Version of a Section from a Program Plan

this is the third example, and "more complex" defines this scenario in relation to the first two examples. These verbal signals increase the chances that the reader will share the writer's concept of the scenarios in terms of number and complexity.

In the final sentence, the word "finally" is somewhat misleading. Because of the strong verbal signals about the three scenarios, a reader might first assume that the word signals the fourth item on the list. However, because the sentence does not describe a fourth scenario, the reader can conclude either that "finally" merely signals the end of the paragraph (which we can see visually) or that it identifies the final goal or task in the research project. The misdirection of the signal may require readers to reread at least the final sentence, if not others, to determine how the sentence relates to preceding ones.

Figure 17.4 maintains the paragraph format but breaks the one paragraph into three. The three paragraphs are the introduction, the elaboration of the primary goal (to develop target classifiers), and the statement of other, less elaborate goals. In this version, the list items are numbered. The numbers are not necessarily superior to verbal signals, but they are more readily apparent on quick reading.

In Figure 17.5, the three scenarios are set off, or blocked out, from the paragraph. The visual signal now predominates over the verbal signals. The space between paragraphs indicates when a new paragraph begins, making indentation redundant.

Figure 17.6 differs from Figure 17.5 only in the use of subheadings in the list to indicate levels of complexity. A reader who is skimming will quickly see, because of the boldface, that the concept of complexity is important. The subheads thus elevate

4.3 CLASSIFICATION ALGORITHM PERFORMANCE AND UNDERSTANDING

This section of the program plan describes the goals and objectives of the stationary target classification effort and outlines the proposed utilization of radar data to achieve these goals and objectives.

The most fundamental goal of the target classification effort is to develop and understand target classifiers which work reliably in a highly variable background clutter environment. Different levels of algorithm complexity would be based upon target classification scenarios. (1) One simple scenario is the MMW seeker searching a small acquisition region in search of a priority (e.g., tank) target. (2) Another scenario of intermediate complexity might be the MMW seeker having the ability to gather multiple looks at the target or even a reduced resolution SAR. (3) A third, more complex scenario is the standoff sensor application which would have the capability of utilizing high range and cross-range SAR data with a 2-D classification algorithm.

Algorithms for these three typical applications might be considerably different in complexity and performance, but the goal is to develop algorithms, understand the underlying signal processing, and to characterize their robustness in highly variable clutter environments. Finally, the promise of new algorithms to be developed in the future, both by universities and industry, will be evaluated as part of the effort in search of a promising solution to this difficult problem.

FIGURE 17.4   Paragraph Format with Numeric Indicators

4.3 CLASSIFICATION ALGORITHM PERFORMANCE AND UNDERSTANDING

This section of the program plan describes the goals and objectives of the stationary target classification effort and outlines the proposed utilization of radar data to achieve these goals and objectives.

The most fundamental goal of the target classification effort is to develop and understand target classifiers which work reliably in a highly variable background clutter environment. Different levels of algorithm complexity would be based upon target classification scenarios:

1. One simple scenario is the MMW seeker searching a small acquisition region in search of a priority (e.g., tank) target.
2. Another scenario of intermediate complexity might be the MMW seeker having the ability to gather multiple looks at the target or even a reduced resolution SAR.
3. A third, more complex scenario is the standoff sensor application which would have the capability of utilizing high range and cross range SAR data with a 2-D classification algorithm.

Algorithms for these three typical applications might be considerably different in complexity and performance, but the goal is to develop algorithms, understand the underlying signal processing, and to characterize their robustness in highly variable clutter environments. Finally, the promise of new algorithms to be developed in the future, both by universities and industry, will be evaluated as part of the effort in search of a promising solution to this difficult problem.

FIGURE 17.5   Paragraph Reformatted to Display the List

4.3 CLASSIFICATION ALGORITHM PERFORMANCE AND UNDERSTANDING

This section of the program plan describes the goals and objectives of the stationary target classification effort and outlines the proposed utilization of radar data to achieve these goals and objectives.

The most fundamental goal of the target classification effort is to develop and understand target classifiers which work reliably in a highly variable background clutter environment. Different levels of algorithm complexity would be based upon three target classification scenarios:

1. **Simple.** The MMW seeker searches a small acquisition region in search of a priority (e.g., tank) target.
2. **Intermediate complexity.** The MMW seeker has the ability to gather multiple looks at the target or even a reduced resolution SAR.
3. **Most complex.** The standoff sensor application would have the capability of utilizing high range and cross-range SAR data with a 2-D classification algorithm.

Algorithms for these three typical applications might be considerably different in complexity and performance, but the goal is to develop algorithms, understand the underlying signal processing, and to characterize their robustness in highly variable clutter environments. Finally, the promise of new algorithms to be developed in the future, both by universities and industry, will be evaluated as part of the effort in search of a promising solution to this difficult problem.

FIGURE 17.6   List Format with Subheadings

the concept of complexity to a higher level in the document structure and make readers more aware of this concept. To justify such an emphasis, the concept must be important—a matter of judgment based on knowledge of the subject.

The version of the document in Figure 17.7 tries to solve the problem of the troublesome "finally" in the last sentence. It uses a list format, but the two levels of lists show a hierarchy of ideas. Subheadings and boldface are omitted from the "scenarios" list because they increase the clutter of the page. Verbal information added to the list introduction establishes the simple-to-complex organization.

This design elevates the importance of project objectives above the three scenarios. This version of the document will produce a different concept by the reader than the versions in 17.5 and 17.6, even though all three use lists and all contain the same information. The headings give readers an initial concept of the information; that is, they help readers form a schema for the data into which the verbal details will fit. If this emphasis is wrong—if the main structure is the three scenarios rather than the three goals—the format will mislead readers. Changing their concept later will be difficult because the concept outlined by the visual display will be established. The paragraph version would be preferable to a highly formatted version if the formatting would mislead. Readers of the paragraphs might still comprehend incorrectly, but the format would not contribute to their misunderstanding.

Is it more important for readers of the radar document to understand the three scenarios or to understand the three goals? The answer will tell you whether the version in Figure 17.7 is preferable to the versions in Figures 17.5 and 17.6. The section

4.3 CLASSIFICATION ALGORITHM PERFORMANCE AND UNDERSTANDING

This section of the program plan describes the goals and objectives of the stationary target classification effort and outlines the proposed utilization of radar data to achieve these goals and objectives. There are three main goals:

1. **To develop target classifiers which work reliably in a highly variable background clutter environment.** This is the most fundamental goal of the target classification effort. Different levels of algorithm complexity would be based upon three target classification scenarios of increasing complexity:

   a. The MMW seeker searches a small acquisition region in search of a priority (e.g., tank) target.
   b. The MMW seeker has the ability to gather multiple looks at the target or even a reduced resolution SAR.
   c. The standoff sensor application would have the capability of utilizing high range and cross-range SAR data with a 2-D classification algorithm.

2. **To develop and understand algorithms based on target classifications.** Algorithms for these three typical applications might be considerably different in complexity and performance. Subgoals for the understanding of algorithms will be:

   a. To understand the underlying signal processing.
   b. To characterize their robustness in highly variable clutter environments.

3. **To evaluate the promise of new algorithms.** These new algorithms would be developed in the future, both by universities and industry, as part of the effort in search of a promising solution to this difficult problem.

FIGURE 17.7   List Format with Two Levels; Emphasis on the Goals Rather Than on the Scenarios

of the document you have before you is too small a part of the whole for you to answer the question with confidence, though the introductory statement suggests that the section is about goals, and the "finally" in the last sentence suggests that the dominant structure is the goals. On the other hand, there is so little elaboration of the goals after the first one that an editor may rightly assume that the three scenarios are dominant or that there is one goal with subgoals. Reading closely is the first task of the editor. Your editing will not improve the document if it distorts the meaning through powerful format signals.

In addition to close reading, you have the option of consulting other similar documents. Is the pattern of comparable sections in other program plans to list goals and objectives? Such a pattern would reinforce the decision to list goals. Even more important, you can query the writer about the appropriate emphasis, perhaps illustrating two or more possibilities to clarify your query.

Which format—paragraph, internal numbering, listing—is best? There is no easy answer. List format is not always preferable to paragraph format or vice versa, nor is the two-level list necessarily preferable to the single-level list. Formats are not good or bad in any absolute sense. Design must enable readers to comprehend both accurately and quickly. Headings are good if they enable readers to comprehend the text accurately, and they are bad if they distract or mislead. A list may help with complex material because it reveals the structure of the information. But it would become tiresome and counterproductive if used to the exclusion of paragraphs in a long book meant to be read with concentration.

Your efforts at increasing comprehensibility through design may reveal gaps in the information. In the sample passage, the writer may wish to add details to some of the goals. Alternatively, the writer may collapse two or more goals into one—maybe there aren't really three goals after all. Some questions about word choice also emerge from the close reading that reformatting requires. For example, in the last sentence of Figures 17.3–17.6, is it accurate to say that the "promise" of new algorithms will be evaluated? Is the title of the section clear enough?

In Figure 17.8, the document has been edited for style, organization, and completeness of information as well as for visual design. Some of the content changes were based on inquiries to the researcher who wrote the original paragraph.

One point should be clear: if you as editor have questions about the structure of the information—about the number of items or about their relative importance—the reader probably will, too. The reader is less likely than you are to probe for the correct meaning and may "learn" the information incorrectly—unless you help by providing the correct visual or verbal signals.

4.3 CLASSIFICATION ALGORITHM DEVELOPMENT

Introduction   This section describes classification algorithm development as well as the proposed utilization of radar data in this development effort.

The basic purpose of the radar target classification effort is to develop target classification algorithms that work reliably in highly variable clutter environments, and to understand how they work. Different levels of algorithm complexity will be based on different target classification scenarios:

A. One relatively simple scenario would be a MMW seeker searching a small acquisition region for a priority target (e.g., a tank).
B. A scenario of intermediate complexity would involve a MMW seeker that could gather multiple looks at the target and then process them by (1) noncoherently integrating them to reduce radar scintillation or (2) coherently processing them to construct a reduced-cross-range-resolution SAR image.
C. A relatively complex scenario would involve a standoff sensor capable of utilizing high-resolution range and cross-range SAR data in a 2-D classification algorithm.

Algorithms for these three typical applications might be considerably different in complexity and performance; in each case, however, the goal would be to develop the algorithm, understand the underlying signal processing, and characterize the algorithm's robustness in highly variable clutter environments. An additional part of classification algorithm development will be evaluation of new algorithms that will be developed in the future by both universities and industry.

FIGURE 17.8   Paragraph Edited for Format and for Style, Organization, and Completeness of Information

Guidelines for Editing for Visual Design

Document design, like writing, is an inexact task, especially as an editor moves beyond conventions and uses design to aid comprehension and usability. Your judgments will be sound if you understand the function of each choice you make and if you can imagine the document in use, not just as type on a page or screen. The choices should relate to the complexity of the content and its demands on the reader as well as to production and budgets. The bases for decisions may conflict with one another. For example, the need to make a document inviting to read will suggest a liberal use of white space, but budgets may require the editor to conserve space. Some specific suggestions for editing for visual design follow.

1. **Know the conventions of design.** You can find universal conventions, such as letter format, in textbooks and in style manuals such as *The Chicago Manual of Style*. Style manuals for disciplines (see Chapter 6) also provide information on conventions within the disciplines. Your own publishing organization may have a style manual with a section on visual design, or it may use templates in the publishing software that ensure consistency of design in related documents. Even if it does not, you can check previously published copies of documents of the type you are editing. If the document is a response to a request for manuscripts, such as a proposal in response to an RFP or a journal article in response to a call for papers, consult the instructions for preparing manuscripts. If you are editing for an international audience, learn the conventions for countries where the document will be read. At the least, as editor, you must ensure that the document conforms to the conventions. Conventions for online documents are being established by practice and research and are increasingly being collected in style manuals.
2. **Make design decisions early in document planning.** If the writer knows the design specifications and options from the start, he or she can prepare the first draft of the document according to these decisions. That means less work for you as editor and less chance for conflict if you make significant changes. Use templates in your word processing or publication software to establish type and spacing conventions for different types of documents. (See Chapter 4.)
3. **Know production options and budget.** Production methods and economics may limit design options. Inhouse reports do not justify elaborate and expensive design, but support materials for an expensive product and a large number of readers do. If you are editing online documents, consider that different browsers and platforms may result in different displays. Check the display using several browsers and platforms. Keep the design simple to accommodate different browsers and the comparative difficulty of reading from the screen.
4. **Read for meaning before emending format.** Visual design is such a powerful signal to content because it quickly identifies the structure of the information. If you give the wrong visual signal, it will be difficult, if not impossible, for the words to correct that impression. Be sure that your visual signals accurately reflect the information.

5. **Use design to enhance content, not distract from it.** Visual design should be subtle. If it calls attention to itself and away from the meaning, then the document suffers. Readers are distracted by anything unconventional. Writers and editors can use that knowledge to the document's advantage, as when they increase the type size and use boldface for warnings. But extensive shifts at best annoy and at worst interfere with comprehension.

   If the reader first sees a visual hodgepodge rather than a document, the design needs to be simplified. Generally, the best design is the simplest one that will achieve the goals of aiding comprehension and usability, motivating readers, and facilitating ongoing use. One good policy is to force yourself to articulate the reason for any emendation. For example, you might insert a heading saying to yourself, "The reader needs to know here that the discussion is complete and the conclusion begins." If you have changed the display merely because "it looks good" or "just for variety" or because you recognized an embedded list in a paragraph, your reasons are fuzzy and your priorities in editing are skewed.

6. **Match the level of visual design to the demands of the text.** For short and simple documents and documents that will likely be read straight through or at least in large sections, the paragraphs are probably best. Most books consist primarily of paragraphs because the books are meant to be read in sections. Imagine trying to read a book of several hundred pages that was formatted as extensively as the document in Figure 17.7. You would tire quickly because of the excessive visual signals.

   For documents longer than one or two pages or documents that readers may read selectively, headings can be useful for purposes of both usability and comprehension. Documents with complex technical information may benefit from more elaborate visual design, such as lists and numbering to indicate sections. Documents that are meant to be read quickly and selectively, such as resumes, are also highly formatted.

   Instructions require particular attention to visual design. Steps in the task need to be clearly differentiated, perhaps separated by white space or rules (lines). Numbering of steps helps readers find the right place on the page when glancing at the text, and numbers are useful in cross-references, as when you need to refer to a previous step.

7. **Check headings.** Confirm that levels of headings match the information—that is, that level-one headings indicate the main points. Check for parallel structure. Evaluate their frequency: Are there enough that readers can identify main points? Are there so many that they interrupt reading?

Summary

Visual design aids comprehension and usability, motivates reading, and facilitates ongoing use. Design is always subordinate to meaning and use; that is, editors do not use design to decorate a page but to enable a reader to understand and use the document. Desktop and online publishing offers multiple design options but also requires responsible and informed choices. More is not necessarily better, and simpler is usually safer. Editors must make judgments about design based on their knowledge of the document as it will be used.

Further Reading

Thomas M. Duffy and Robert Waller, eds. 1985. *Designing Usable Texts.* Orlando: Academic Press.

James Hartley. 1985. *Designing Instructional Text.* 2nd ed. London: Kogan Page Ltd.

Nancy L. Hoft. 1995. *International Technical Communication.* New York: Wiley.

Robert Kramer and Stephen A. Bernhardt. "Teaching Text Design." *Technical Communication Quarterly* 5.1 (Winter 1996): 35–60, Useful for designers as well as for teachers. Maps principles of visual design, including seeing the page as a grid and using active white space.

Karen Schriver. 1997. *Dynamics in Document Design: Creating Texts for Readers.* New York: Wiley.

Discussion and Application

1. In the library, locate a professional journal and a trade journal, preferably on the same subject. For example, consult the *Journal of Clinical Psychology* and *Psychology Today* or *Mechanical Engineering* and *Popular Mechanics* or the *Journal of Finance* and *Money.* Identify the visual features of both journals, including the document as a whole (size, shape, binding, and paper) and page format (number of columns and line length, use of headings, white space, illustrations, and color). Using the evidence of content, format, and journal preface (if any), describe the readers for the two journals, considering their interest in the subject, probable reading habits, and backgrounds. Then explain how the design decisions relate to the assumptions about readers. Compare and contrast the two journals, and evaluate the design for both on the criteria of usability, comprehension, and motivation. If you were editor of either journal, would you recommend changes? If so, what would be your reasons?

2. Continue the analysis and evaluation of the document in Figure 17.2. What design options do you identify? What other choices might have been better ones?

3. The subject of the following two paragraphs is radar. Both paragraphs, like the document in Figure 17.3, have embedded lists. Format the two paragraphs to display the lists. Edit for punctuation, spelling, and completeness of information as well as for format. Then, with other class members or in writing, discuss the bases for making a decision about whether the original paragraph format, the list format, or perhaps a third alternative is the best choice for these paragraphs. If you have questions about meaning, formulate queries for the writer.

Example 1

Test planning—Mission planning defines the number and type of missions together with the actual flight plans for the mission to efficiently gather the data.

The needed data includes distributed clutter, discretes, and targets in clutter under varying conditions such as: clutter types: meadows, trees, tree lines, desert; target types: both civilian and military, in these clutter environments and in various target configurations; and environmental conditions: wet, dry, snow. The mission planning will reflect inputs solicited from the MMW government and industry communities to make the data base widely useful.

Example 2

Three difficulties exist with 2-D images from a data analysis point of view. 2-D images are typically processed over a narrow angle of rotation (1 degree is typical). 2-D images in radar coordinates are also difficult to compare at different aspect angles because the target orientation is different in each image. The third difficulty is the elevation ambiguity inherent in 2-D radar imagery.

4. In your own words, explain how format affects comprehension. Give some specific examples of how an editor can use format to influence what a reader learns from a document.

5. Find three different documents, such as a letter, announcement, and manual. Identify the different visual design options used for each. Explain what functions each of the options fulfills. Identify any options that seem superfluous or that seem to interfere with the document purpose. As editor, suggest alternatives and explain why they might be preferable.

6. Find three different web sites and evaluate the design options as you did for the print documents in #5. As your instructor guides, choose three sites representing the same kind of purpose or three sites with contrasting purposes.

7. Compare the documents in Figures 17.3 and 17.8. Identify the editorial changes in content, organization, and style. Evaluate the editing: What is better about the version in 17.8? What information is new? Imagine the editorial procedure: How did the editor begin? What questions did he or she ask the writer?

8. Find a document with at least two levels of headings. Identify the type of wording: Informative or structural? One word or several? Phrases or questions? Identify the ways in which the levels are distinguished visually. Do the headings provide the information a reader needs? Do they enable selective reading? Can a reader distinguish the levels visually? Do they distract?

9. Vocabulary: These terms and concepts should be familiar enough to you that you can use them to make editorial decisions about visual design: *hanging indent, outdent, left justification, informative heading, structural heading, offset list.*

Chapter 18

Illustrations

When technical communication emerged as a career specialization during World War II, the typical technical document was mostly words. Furthermore, the page was packed with words. Narrow margins, scant interlinear spacing, and the lack of headings let the words crowd the page. Technical documents have changed dramatically since then as writers have discovered the power of graphics and format. Writers use typography and space to enhance comprehension and access, and documents without visuals are rare. Some documents are wholly visual, with illustrations substituting for text. The term *illustrations* is comprehensive: it refers to tables, graphs, flowcharts, diagrams, line drawings, and photographs. The term *graphics* is sometimes used as a comprehensive term, especially in identifying a type of computer program, but it is used in this chapter to mean a specific type of illustration, the graph.

A parallel move from words to visual information is evident in online instructions for using computers. When computers became available to general users, commands changed from programming codes to words such as "delete" and "move." Now instructions increasingly rely on icons, or visual representations of procedures, rather than on words. Users and readers, too, have discovered the power of illustrations. The increasing prominence of the World Wide Web, a graphical system of information distribution, suggests that technical communicators will work even more with visual information than in the past.

Illustrations, just like text, require comprehensive editing. Copyeditors check accuracy, correctness, consistency, and completeness. Editors with comprehensive responsibilities advise in the planning and construction of illustrations and confirm in completed illustrations the match of form and content, the logical arrangement of elements within the illustration, and the integration of illustrations with the text. The same principles of good communication that editors apply to text also apply to illustrations. Knowing those principles prepares editors for working with illustrations. Just as with text, readers seek information in illustrations and ease of access to the information. Just as with text, visual information must be organized in a way that shows the overall concept as well as the relationship of parts to the whole. The presentation must be free of the distraction of messiness or undue clutter.

This chapter summarizes reasons to use illustrations, reviews types of illustrations, and discusses the work an editor performs in comprehensive editing. It also examines the process of preparing illustrations for print and the use of computers in editing illustrations. Chapter 5 covers copyediting of illustrations, and Chapter 10 discusses tables.

Reasons to Use Illustrations

In technical communication, the reasons to use illustrations parallel the reasons for making any other choice about a document. These reasons derive from what we know about readers of technical documents (see Chapter 2). Readers seek information, read selectively, and expect the document to follow conventions. They often read in order to act. They create meaning based on previous learning. They learn organized information more readily than random or chaotic information.

Illustrations may be the main source of information in a document. They may negate the need for words in international technical communication. They may support verbal text. Thus, illustrations in technical documents are functional rather than decorative. Editors evaluate how well illustrations achieve their function. Aesthetic issues are secondary to function.

These reasons for using illustrations form a basis for evaluating them. All the illustrations in a document should achieve one or more of these objectives.

Convey Information

Illustrations in technical documents may convey information more readily than text would convey the same information. A line drawing or photograph shows readers the whole object as well as the shapes, proportions, and relationship of parts more readily than could words. Some instructions are mostly or even wholly visual because users will work with objects. Illustrations help users relate the information in the document to the object in their hands. Illustrations also to some extent bypass words in instructions. Whereas verbal instructions must be translated for international users, visual instructions may save the costs of translation and printings in multiple languages. And they may enable visual learners to complete a task.

Illustrations also represent quantitative and structural data efficiently. For example, lists of numbers get buried in sentences, but they are more accessible in tables or graphs. A flowchart readily reveals the structure of an organization or process, whereas sentences that attempted to present the same information would be tedious.

Illustrations aid with interpretation of information because they show relationships and enable comparisons. Facts do not tell much in isolation. The fact that a company's earnings in the third quarter of the fiscal year were $2.3 million means less than a comparison of those earnings to the same period of the previous year or to the earnings of a rival company. A line graph would show those earnings as they relate to the earnings of the previous quarters, and a bar graph could compare third-quarter earnings in several years. These graphs facilitate the formation of schemata or concepts in the reader's mind. They provide a structure for organizing information. Assuming the structure works for the content, it will increase the comprehensibility of the information.

Support the Text

Illustrations may stand alone in instructions for readers who learn better from visual instruction or for international users. When they support text, they reinforce the meaning in the words by presenting it in a different mode. Most people learn better if they can check the verbal information with the visual, and vice versa. One document is likely to contain some information that is better represented visually and some that is better represented verbally. The two sources of information complement each other.

Illustrations can support the text by motivating readers. For example, a cartoon included with intimidating material can relax readers. An illustration can provide relief from the steady columns of print. A photograph of a person can rouse sympathies, and photographs can dramatize as well as record damage to property and equipment. All illustrations require valid reasons for inclusion. Readers of technical documents expect information from the illustrations. Merely decorative illustrations (those without information) distract from the text rather than enhancing it. Instructions for hanging wallpaper may be enhanced by a line drawing of a person pressing a strip of paper against the wall, but a drawing of a person engaged in no particular action would simply distract. Efforts to be cute—for example, adding hands and a face to a piece of electronic equipment—may insult readers. As with all editing, you must know your readers, and your choices about illustrations must relate to the overall purpose of and design for the document.

In order to support the text, illustrations must be integrated with it. Integration means, in part, that readers can tell when illustrations and text are connected. Placement of the illustration near the text that discusses it helps readers connect related parts of the document. Other devices of integration include verbal references to the illustration and consistent use of terms in the illustration and the text. Integration of text and illustrations also means that both forms communicate the same message. The message would be mixed if the text gave instructions on a serious safety procedure while the illustration illustrated a smiling employee in a casual pose.

Enable Action

Action for a user of illustrations may consist simply of finding a number in a table. The table itself makes the numbers more accessible than would sentences. In complex tables, devices such as shading and spacing increase the ease with which a user can find the right number. Shading blocks of data and inserting extra space between groups of lines propel the eye across the page and help readers stay on the right line.

Readers also depend on illustrations in learning to perform a task. Although writers and editors often assume readers will use both the text and illustrations, readers may try to take shortcuts and read only the illustrations. They will appreciate illustrations that enable them to perform the task quickly, without extensive reading. Knowing that readers may pay attention only to the illustrations makes editing for completeness and coherence a priority.

The purpose of enabling action dictates a consistent placement of illustrations and corresponding text on the page. Readers want to find the text and illustrations in the same place each time they look up from the task rather than finding illustrations sometimes to the right of the text, sometimes to the left, sometimes on top, sometimes beneath.

Types of Illustrations

Different types of illustrations serve different functions and convey different information. For example, a table provides access to specific data while graphs provide a concept of how pieces of quantitative data relate to one another. Table 18.1 summarizes types of illustrations and their uses. As editor, you will need to know what illustrations achieve what purposes in order to edit for form, organization, and content. One of your tasks will be to check the match of form and content. You also need to use the correct terms in referring to types of illustrations. A correct technical vocabulary enables accurate communication and helps to establish your professional expertise.

Illustrations may be classified according to four broad categories: tabular, graphic, structural, and representational. Tables place verbal or quantitative items in rows and columns for easy access to particular pieces of information and for comparison. Graphs display quantitative information for purposes of comparison. Structural illustrations—flowcharts, schematic diagrams, and maps—emphasize the structure of the data rather than comparisons or representation. Representational visuals—drawings and photographs—depict two- and three-dimensional items and show spatial relationships. Examples of these categories of illustrations are displayed in Table 18.1.

Comprehensive Editing of Illustrations

Comprehensive editing of illustrations means evaluating their content, organization, form, and style in terms of overall document purpose and readers' needs. The illustration must make sense. The parts must cohere, and readers must be able to identify and interpret what they see. Editing may result in restructuring of the illustrations or in additions or deletions. Illustrations may be rescaled or resized depending on the desired emphasis. Extra lines as well as unnecessary information may need to be deleted. As with all comprehensive editing decisions, guidelines, judgment, and the ability to imagine readers interacting with a document help editors improve and not just change the illustrations.

TABLE 18.1   Types of Illustrations

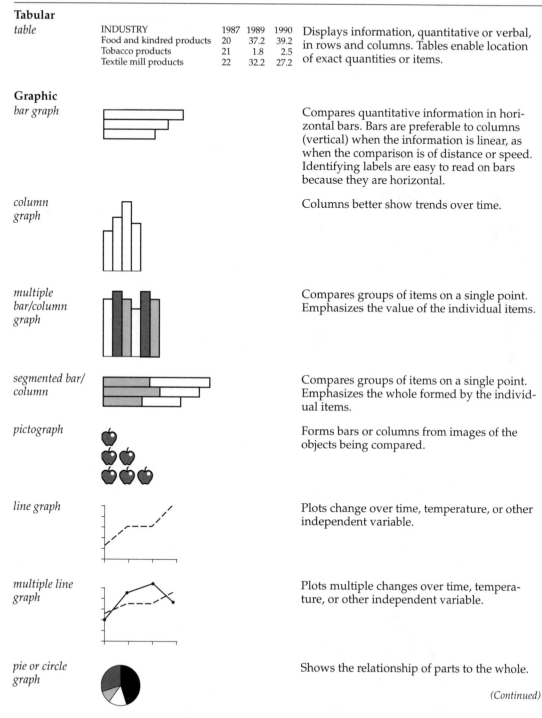

Tabular

table

INDUSTRY	1987	1989	1990
Food and kindred products	20	37.2	39.2
Tobacco products	21	1.8	2.5
Textile mill products	22	32.2	27.2

Displays information, quantitative or verbal, in rows and columns. Tables enable location of exact quantities or items.

Graphic

bar graph

Compares quantitative information in horizontal bars. Bars are preferable to columns (vertical) when the information is linear, as when the comparison is of distance or speed. Identifying labels are easy to read on bars because they are horizontal.

column graph

Columns better show trends over time.

multiple bar/column graph

Compares groups of items on a single point. Emphasizes the value of the individual items.

segmented bar/ column

Compares groups of items on a single point. Emphasizes the whole formed by the individual items.

pictograph

Forms bars or columns from images of the objects being compared.

line graph

Plots change over time, temperature, or other independent variable.

multiple line graph

Plots multiple changes over time, temperature, or other independent variable.

pie or circle graph

Shows the relationship of parts to the whole.

(Continued)

TABLE 18.1 *Continued*

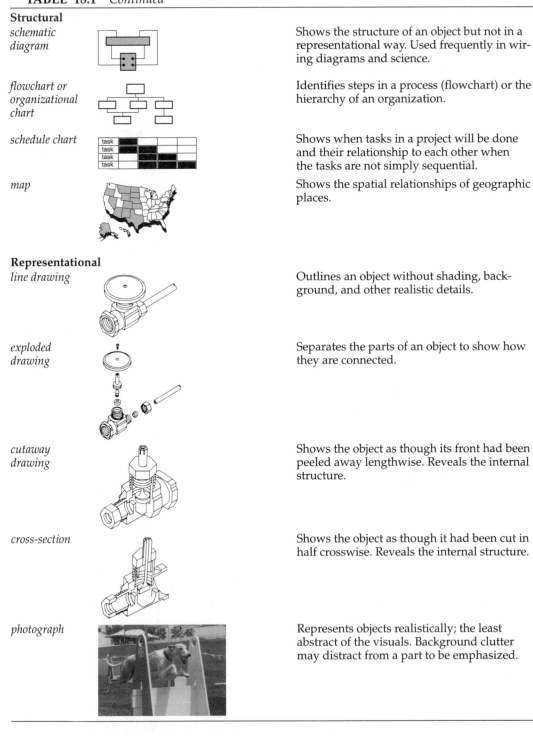

Structural

schematic diagram — Shows the structure of an object but not in a representational way. Used frequently in wiring diagrams and science.

flowchart or organizational chart — Identifies steps in a process (flowchart) or the hierarchy of an organization.

schedule chart — Shows when tasks in a project will be done and their relationship to each other when the tasks are not simply sequential.

map — Shows the spatial relationships of geographic places.

Representational

line drawing — Outlines an object without shading, background, and other realistic details.

exploded drawing — Separates the parts of an object to show how they are connected.

cutaway drawing — Shows the object as though its front had been peeled away lengthwise. Reveals the internal structure.

cross-section — Shows the object as though it had been cut in half crosswise. Reveals the internal structure.

photograph — Represents objects realistically; the least abstract of the visuals. Background clutter may distract from a part to be emphasized.

Appropriateness and Number

Decisions about whether to include illustrations and how many to include are based in part on budget considerations—visuals cost money. They may also relate to the document set. If other manuals in a series or other newsletter issues are illustrated in one way, the set has established conventions that subsequent documents will respect. Preferably, some policies about illustrations will be established before a draft is prepared. For example, a writer should find out before having 30 photographs made that only three line drawings will be acceptable. Editors should participate in the making of these policies.

Beyond these constraints of circumstance, editors consider whether visual or verbal presentation will better convey information and enable action and how illustrations may support the document purposes and reader needs.

Match of Form, Content, and Purpose

The form of the illustration (for example, line graph, table, photograph) should reinforce the content. It should also enable readers to use the information in it. This principle relates to the selection of types of illustrations.

A graph shows almost instantly the shape of the data, so a reader's initial interpretation is very fast. The graph thus helps in concept formation. The graph in Figure 18.1, however, confuses readers by conveying a misleading message. Its purpose is a comparison of three people's scores on a test. However, the line graph conveys the message of movement over time rather than a comparison of values. It also suggests visually that subject A progresses toward subjects B and C.

If the purpose were to compare one person's score at three different periods, the line graph would work fine to show the change. A column graph would work better than a bar graph if the intent were to compare the three scores of three sep-

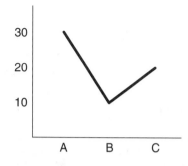

FIGURE 18.1 Mismatch of Form and Purpose.
This line graph misrepresents the data by suggesting some progression from point A to point C. All three scores are one-time scores by different people. A column graph would better compare absolute variables.

arate subjects. Measuring scores implies "high" and "low" scores. The vertical presentation would suggest that vertical concept.

Arrangement: Sequential and Spatial

Illustrations must make sense in relationship to one another and to the text. The arrangement of parts within an illustration shows their relationship spatially.

In a sequence of illustrations, the whole should come before the part—a general-to-specific arrangement. Readers need to comprehend the whole before the parts make sense. Readers might not recognize a part of a mechanism if they have not seen the context in which the part exists.

Ordering of details within the illustration shows the relationship of parts and helps readers form concepts. Some standard arrangements are more important-to-less important, larger-to-smaller, and first-to-last. Thus, the bars in a bar graph might be ordered from largest to smallest unless some other information (such as chronology) dictates a different arrangement.

Illustrations linked with text in instructions should generally appear on the left with the text to their right. This arrangement corresponds to the left-to-right reading pattern in Western countries. Readers form the concept of the step from the illustration and then consult the text for the details. The arrangement can also be vertical. If users are reading to perform a task, the illustration may be at the top and the text below. A step number or heading should introduce the illustration to show that it belongs with the text that follows rather than the text that precedes. If users are reading to get information, illustrations usually follow the text referring to them.

If the illustration illustrates component parts, the arrangement of details should reflect the spatial relationships of the item represented. A computer's mouse, for example, would be inappropriately displayed above the monitor. Instead, it should appear to the right of the monitor where it is used by right-handed people. This functional arrangement overrides aesthetic considerations, such as balance.

Should graphs and tables be placed in the text or in an appendix? The answer depends on whether most readers need the illustrations at the point where they are mentioned or whether the illustrations present supplementary information that only some readers may check. Supplementary illustrations belong in an appendix. If they provide important concepts that all readers must know, place them in the text.

Emphasis and Detail

Because readers tend to pay more attention to visuals than to text, the inclusion of a visual announces that this information is important. Emphasis helps readers interpret information by showing the hierarchical structure.

Emphasis of details within an illustration draws readers' attention. Details may be highlighted in these ways:

- **Size:** large attracts more attention than does small.
- **Color or shading:** bright colors and dark shading attract the most attention.

- **Labels:** important parts are labeled, others are not.
- **Foregrounding:** important parts appear in the front.
- **White space:** details surrounded by space generally attract more attention than do details crowded by others.
- **Shape:** in a matrix of circles, one notices the lone triangle; in a segmented circle, one notices the segment separated from the others.
- **Sequence:** first and last seem more important than middle.
- **Typographic cues:** boxes and arrows focus attention.

The graph in Figure 18.2 appeared in the brochure of a charitable organization seeking donations. Many donors want to know that their gifts will be used for charitable purposes rather than for administration and base their decisions about giving in part on how efficient the organization is. Anticipating that some donors would question expenditures and being proud of its relatively low administrative expenses (13%), the organization prepared this pie graph—an appropriate form for illustrating how income is distributed.

Inadvertently, instead of emphasizing its efficiency, the graph emphasizes administrative costs through the placement of this segment at the front and through the three-dimensional display. These methods make the 13% seem visually a larger part of the expenses than the numbers suggest. In fact, the campaign and administration expenses segment seems almost as large as the segment reflecting the amount spent on youth, though the numbers verify that almost three times as much is spent on youth as on campaign and administration. Because the visual representation is more powerful than the numbers, most readers will not use the numbers to reinterpret what the graph has interpreted for them. The graph may interfere with the purpose of convincing potential donors that their dollars will be well spent in this organization.

The document purposes and readers' needs determine which details should be emphasized. Readers must be able to identify in the illustration any details they

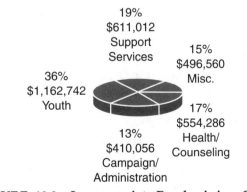

FIGURE 18.2   Inappropriate Emphasis in a Graph

will use, such as parts they need to assemble or numbers they need to interpret. Details must be accurate and complete. By contrast, details unrelated to the concept or task diminish the power of visuals. The form can be buried in the clutter, so that readers spend as much time plowing through excessive details as they would reading the information in paragraph form. Thus, as editor, you should simplify or use multiple visuals if the clutter hides the important information. Remember from Chapter 16 that seven items is generally the maximum a person can remember. Because the visual power of a pie graph or a multiple line graph diminishes once the number of segments or lines exceeds seven, a graph may not be the right form for such data. Likewise, the detail of a photograph may make it difficult for readers to locate specific parts they need to connect, in which case several photographs, or perhaps a line drawing, should be used.

In evaluating the level of detail, ask what readers need to know and do from the illustration. If it presents more detail than necessary, delete what is not needed. If readers need to know all the information but the details intimidate or inhibit quick reading, consider using more than one illustration.

Perspective, Size, and Scale

Perspective, size, and scale give readers useful information and influence the interpretation of the information. Perspective shows depth: the place of an object in a three-dimensional field from the front of the illustration to the back. Whether an object is standing on end or on a horizontal plane is a matter of perspective. Because objects appear smaller in proportion to their distance from the viewer, objects that seem farther away also seem proportionately small. The inappropriate emphasis in Figure 18.2 results partly from choosing a three-dimensional graph, which places the front segment closer to the reader.

Size is a factor in identification. Illustrations should be big enough that they can be read, but rarely so large that they dominate the text and never so large that they seem to shout on the page. Size can also indicate importance. Illustrations should be sized proportionately within a document so that shifts in size indicate shifts in importance. Assuming comparable complexity and importance, one line graph will have essentially the same size boundaries as another line graph, but if one line graph is more important or complex than another, it could be larger. Parts of objects may be drawn proportionately larger than the whole to show detail.

The scale of graphs should reflect numeric values. Distortion results from expansion or condensation of the x or y axis or from improper measurement. Editors determine whether the graph conveys its information accurately. If the graph gives the wrong first impression, it will be difficult to correct it later with words or other details because the schema will already be formed in short-term memory.

Bar and column graphs and pie graphs can be scaled quantitatively. Assuming that the bar or column begins at zero, each bar can be measured so that its length equates to its numeric value. The segments on a pie graph can be calculated with a protractor. Computer graphics programs will calculate the right size from data you provide. By contrast, scale on line graphs is more difficult to establish because

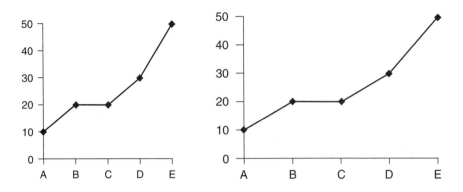

FIGURE 18.3 Scale in Line Graphs. Contraction of the horizontal scale in the graph on the left exaggerates the incline.

two dimensions can be manipulated. Distortion is easy with computer adjustments as well. The correct scale is a matter of judgment rather than measurement. The two line graphs in Figure 18.3 present the same data. However, the graph on the left gives the impression of a much greater incline than the graph on the right. The contraction of the x axis increases the incline. Contraction of the y axis can also steepen an incline.

Psychological factors and conventions will contribute to the judgment about whether a scale is right. For example, if a 20 percent increase in sales is an astonishing gain for a company, a fairly steep incline will be appropriate. But if a company gains 20 percent every year, the increase will be correctly represented in a flatter line. The line graph that appears in the business section of the newspaper each week exaggerates the movement of the Dow Jones average of stock prices, but convention has established the scale, and readers interpret accurately on the basis of prior experience.

In technical documents, even persuasive ones, the goal is always to provide accurate information. Purposeful distortion of data may be accepted in sales documents, whose readers expect to be sold, but it is not acceptable in technical documents, whose readers expect to be informed.

Relationship of Text and Illustrations

Illustrations convey information, help readers form concepts, and enable action. Often, though, text is needed for explanations and interpretation of significance. Even in documents that are mostly visual, text is the glue that unifies the document and creates coherence. Editors must consider whether the visual and verbal information together convey the whole message. Often the illustration, whether tabular, graphic, structural, or representational, provides the initial concept, and the text elaborates.

Problems with interpretation may occur when the creators of documents try to bypass words altogether in order to reach an international audience. Emergency instructions in airplanes, for example, are frequently visual only, using no words. Some office equipment includes only visual instructions for installation and use. The measure of effectiveness for such documents is the same as the measure for verbal documents: Can readers interpret and use the information? Because visual information, like verbal, has limitations, the illustrations alone may not provide enough information.

Some concepts and procedures are easier to represent visually than others. The goal of using nonverbal instructions only may not be possible for some topics. The illustration in Figure 18.4 appeared in the instructions for a machine that makes transparencies for overhead projectors. It is recognizable as a dial in context, but the illustration by itself does not convey the entire instruction; a user has to determine, by trial and error, the best setting of the dial for a given purpose and mark the illustration so that subsequent users will know where to set the dial. Because this illustration is the first of four in a sequence, users might interpret that the first step is to set the dial, but then they will be frustrated trying to determine how and where. It is unlikely that they will use the illustration to mark the setting unless they have a verbal prompt. If an arrow is added to the illustration to indicate visually a direction to move or place something, readers will know immediately that they are to perform some action. The illustration in Figure 18.4 does more than identify a thing (a knob or dial)—it tells the user to turn it clockwise. Words will still be necessary to recommend marking the dial on the instruction sheet to indicate the optimal setting.

In editing illustrations, use words as necessary for identifying and explanatory information. If your company cannot afford translators and printings in different languages, take the chance that at least some users will speak English. Remember, too, that the visual information must complement the verbal information. The text and illustration should use the same terms. For example, if the text refers to a piece of data or a part of a mechanism, that information should be labeled on the illustration. Likewise, if the illustration names a part, the text should probably discuss

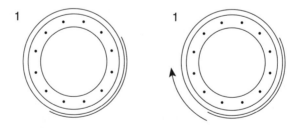

FIGURE 18.4   Visual and Verbal Information.
The visual information in the dial on the left is insufficient to tell readers to mark the appropriate setting on the instruction sheet. The arrow in the dial on the right substitutes for a verb phrase: "turn the dial clockwise"

it. Placement of the illustrations in relation to the text can also reinforce the parallels between verbal and visual information.

Discriminatory Language and Good Taste

The guidelines for avoiding discriminatory language and maintaining good taste in prose (see Chapter 15) apply equally to illustrations. Illustrations can discriminate whether they use words or not. For example, a document illustrating white males in professional positions and females in menial positions shows bias whether the accompanying text is biased or not. Representations of people in line drawings and pictographs should suggest racial and sexual equality.

Good taste involves respect for adult readers and the seriousness of document purpose. For example, comics can motivate and reinforce ideas, but lewdness insults. Corny cartoon characters are risky because they can equate readers with children or with incompetent or subliterate adults. The formality of illustrations should match the formality of the style and page design to maintain a consistent persona and tone. Although cartoons may work for casual brochures, they would be detrimental to a bank's annual report printed on heavy, cream-colored paper.

Maximizing Data Ink

Edward Tufte theorizes that most of a graphic's ink should represent the interesting variations of data (see Further Reading). But many visuals are cluttered with "chartjunk." Excessive details and callouts, three-dimensional portrayals of lines and bars in graphs, unnecessary shadows and shading, grids that are too much foregrounded, elaborate legends typed in all caps, lines between rows and columns in tables when space suffices to separate them, and decorations all interfere with the reader's ability to interpret what is important. Illustrations should provide readers with meaning, and readers don't appreciate the distraction of sorting through material they do not need. Generally, clean and simple visual representations are preferable to cluttered ones.

Tufte advises maximizing data ink by erasing non-data ink and redundant data ink. He also observes,

> Just as a good editor of prose ruthlessly prunes out unnecessary words, so a designer of statistical graphics should prune out ink that fails to present fresh data-information. Although nothing can replace a good graphical idea applied to an interesting set of numbers, editing and revision are as essential to sound graphical design work as they are to writing (*The Visual Display...*, p. 100).

Tufte focused on printed visuals, but plenty of comparable junk accompanies graphics and graphic representation on the World Wide Web and other online documents. For example, graphic backgrounds for screens can overpower the information that is on the screens, and animations or flashing banners for the sake only of drawing attention can become irritating if a reader has to stay on a page very long or has to return to it. Editors can advise during planning and revision knowing the principle that "more" does not usually mean "better."

Application: Cassette Instructions

Figure 18.5 shows a portion of the owner's manual for a cassette player. The manual as a whole is an 8½ × 11-inch booklet, with pages stapled down the center.

Safeguard Against Accidental Erasing

Every time a recording is made, the sound previously recorded is erased. The cassette and the recorder are equipped with a special device to safeguard valuable recordings from being erased accidentally. On the back of the cassette on either side are two lugs. If you want to be sure that a recording cannot be accidentally erased, break out these lugs with a knife. If only one track is to be protected, break out the lug to the left when the tape is in position for using that track. When the lug is broken out, the RECORD switch cannot be depressed.

To record on a cartridge in which the lug has been broken, place a piece of tape over that area and proceed as in the instructions for Recorder.

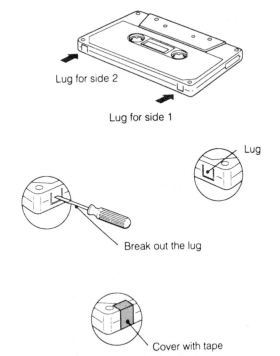

FIGURE 18.5   **Part of an Owner's Manual for Using a Cassette Player.**   The form suits the content and purpose. Objects are easily identified, sequence clarifies the part-whole relationship, and circling identifies parts and a shift in scale. Editorial analysis, however, reveals some potential improvements.

Because readers look for shortcuts, they may try first to make sense of it just by looking at the pictures. Thus, the visual will have to be self-contained and self-explanatory. These particular instructions tell how to safeguard the cassette against accidental erasure. Editing follows the familiar procedure of analyzing the readers and communication situation, evaluating the illustration, and establishing editing goals.

An analysis of the drawings and establishment of editing goals may begin with an understanding of readers' use of manuals and the goals of illustrations. Readers of these particular illustrations will perform an unfamiliar task using familiar objects, a screwdriver and cassette tape. Visual representations, in this case line drawings, make sense because readers must recognize the objects and the location of parts on the objects in order to complete the task. This illustration meets the criteria of appropriateness and match of form, content, and purpose.

Readers must be able to identify the objects and procedures in order to complete the task. The drawings and the type are large enough to read, and the white space around the separate parts both encourages attention to the illustrations and establishes the boundaries of the separate details, the way space between words establishes the boundaries of the words. The whole mechanism appears first, establishing the context for the specific parts that the cassette owner will manipulate. The cassette and screwdriver are easy to identify. Perspective is clear—the cassette does not stand on its end but rather lies on a horizontal plane, much the way a reader will hold it in order to complete the task. The device of circling indicates parts of the whole. Circling also clarifies that the scale for the illustration of the part is bigger than the scale for the whole. The callouts clearly attach to the part or process they identify.

Chances are your first impression is positive because of quality reproduction and professional use of type and space. However, the positive features of the illustration—appropriateness of the concept, easy recognition of items, use of space, overall attractiveness—may actually interfere with an evaluation of whether the readers will get accurate, usable information. Comprehensive evaluation will need to consider other features, such as the relationship of text and illustrations, the arrangement of information, and the emphasis and details.

- **Relationship of text and illustrations.** As editor, you will make sure the text and the illustration convey the same message, that they use the same terms, and that location on the page indicates the relationship of verbal to visual information.

  The text and illustration do not correspond in some terminology. The text directs the owner to use a knife to break out the lugs, but the visual illustrates a screwdriver. The text refers to protecting "tracks," but the illustration refers to "sides." The device is called both a "cassette" and a "cartridge." The reference to "recorder" in the last line may misdirect readers. You cannot tell from the selection illustrated here, but the heading that follows the illustration is "Cassette Tape Player/Recorder." Thus, readers may assume that the next section provides the instructions for recording. In fact, the instructions appear three columns back, identified by a buried (indented) heading "Recorder." The potential for confusion could be eliminated by retitling the appropriate section "How to Record."

The terms in the drawing must also parallel terms in the unwritten text, the terms readers use in referring to the illustrated items. "Lug" may be technically correct, but more readers will recognize "tab." Similarly, blank tapes typically identify the sides by letters rather than by numbers. Readers, relating the drawing to the objects in their hands, may more readily recognize sides A and B than sides 1 and 2. Terms should be edited to match in the text and the drawing.

- **Arrangement.** The structure of the illustration is to move from definition of parts to steps for completing the process. The two instructions "Break out the lug" and "Cover with tape" suggest a two-part procedure. However, readers who follow the steps literally will defeat their purpose because the tape simply replaces the tab. Additional words are necessary in the callouts or as headings to indicate that the second step reverses the first rather than completing it.

To prevent erasure
Break out the tab

To record after the tab has been removed
Cover the opening with tape

Another possibility to show the relationships of steps would be to place the paragraph about recording on a cartridge in which the lug has been broken between the "break out" and "cover" steps. Figure 18.6 depicts this arrangement.

The check for arrangement continues with evaluation of the orientation of the illustrations in relationship to the way the readers will use the objects. A good feature of the illustration is the orientation for the "break out the lug" step. The screwdriver and cassette are positioned in the illustration to emulate the actual position if a right-handed owner performs this step. Thus, the orientation is accurate for about 85 percent of owners. The illustration of the cassette, however, shows the opposite orientation. If readers begin literally or mentally with the orientation they see in the first illustration, they will have to reorient the cassette. Perhaps the orientation shift will not confuse readers for this simple procedure, but readers working with complex objects and unfamiliar procedures, such as assembling complex equipment, may be confused by arbitrary shifts. The cassette should be rotated in the first illustration to emulate a reader holding it in his or her left hand.

- **Emphasis and details.** A check for emphasis and details reveals that the two procedures receive equal emphasis, yet the first procedure is the more essential one. Separating these steps visually, as suggested previously, will also solve the emphasis problem by clarifying the separate aims, one of which is more important than the other.

The illustration is confusing because owners who wish to safeguard just one side of the tape will have to work to interpret which lug to break out. If side 1 faces

Safeguard Against Accidental Erasing

Every time a recording is made, the sound previously recorded is erased. The cassette and the recorder are equipped with a special device to safeguard valuable recordings from being erased accidentally. On the back of the cassette on either side are two lugs. If you want to be sure that a recording cannot be accidentally erased, break out these lugs with a knife. If only one track is to be protected, break out the lug to the left when the tape is in position for using that track. When the lug is broken out, the RECORD switch cannot be depressed.

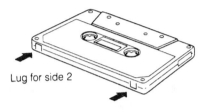

Lug for side 2

Lug for side 1

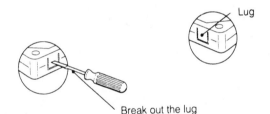

Lug

Break out the lug

To record on a cartridge in which the lug has been broken, place a piece of tape over that area and proceed as in the instructions for Recorder.

Cover with tape

FIGURE 18.6 Rearrangement of Text and Illustrations to Reflect the User Point of View. Separation of the two procedures with text helps to clarify that the procedures are not steps in sequence.

up in the illustration, the designation of side 1 and side 2 lugs is correct, but if side 2 faces up, they are backwards. The words in the text help, but they may not be read: "Break out the lug to the left when the tape is in position for using that track." Not all users know when the tape is in position for using the track, especially for machines in which insertion is horizontal rather than vertical. Furthermore, the information about playing position is unnecessary (and therefore gives too much detail).

Marking the cassette in the illustration "side A" will simplify the interpretation of which tab protects which side. The instructions could read: "To protect side A, break out the tab to the left as side A faces you with the label at the top. Break out the opposite tab for side B."

In the illustration, the tabs are inaccurately attached to the cassette. The U shape indicates three open sides. On three major brands of cassette tapes, the tabs are attached on the opposite side from those in the illustration. Top and bottom may matter little in this situation, but they can matter quite a lot in other illustrations. Imagine an illustration showing the "on" and "off" positions for a switch in reverse position.

Figure 18.7 shows the edited illustration and text. The text has been simplified to follow principles of good style and to work better with the drawing. The reference to the "special device" is deleted because it is unnecessary and prompts readers to locate something that is never referred to again.

The original illustration has a number of positive features, and because it looks good, editors may assume that it is good. Because of the easy identification of items and the attractiveness, it would be difficult to anticipate, on first observation, any problems. The discovery of the problems is a chainlike process that begins with imagining a reader using the illustration. When some obvious problems emerge, an editor looks further, guided by principles of good communication, such as accuracy, consistency, and patterning instructions according to patterns of learning and use, as well as by the specific checks in the editing of illustrations. The check of the mechanism itself results from the confusion over which tab to break out for a given side of the cassette.

Does the editing really matter? Owners can probably determine by trial and error which tab to break to protect a given side, and common sense may tell them that taping a hole just created is counterproductive. However, good instructions should prevent the need for trials by ensuring that owners do the job right on the first attempt. Otherwise, we'd just leave owners to their own devices to begin with and skip the manual. A few editorial emendations can save interpretation time and prevent errors. The gains from emendations such as those made here would be proportionately greater with a more complex procedure. In more complex procedures, a reader does not so readily correct a simple misstep or make the adjustments in orientation required to do the job correctly.

Preparing Illustrations for Print

The task of editing illustrations may include selecting photographs and noting instructions for printing as well as deciding about form and content. The instructions relate to reductions and enlargements, cropping, redrawing, and halftones for photographs.

Safeguard Against Accidental Erasing

Recording erases the sound previously recorded. To prevent accidental erasure by a new recording, you can break out the tabs on the back of the cassette with a screwdriver. To protect the recording on only one side of the tape, break out the tab to the left when the label for that side faces you and the label is at the top. When the tab is broken out, the RECORD switch cannot be depressed.

To record on a cassette after the tab has been broken, place a piece of tape over that area and proceed as in the instructions in the section "How To Record."

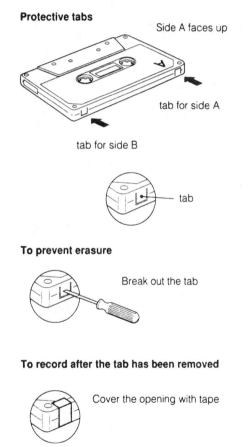

Protective tabs

Side A faces up

tab for side A

tab for side B

tab

To prevent erasure

Break out the tab

To record after the tab has been removed

Cover the opening with tape

FIGURE 18.7   Instructions after Comprehensive Editing.   Text and illustrations are parallel. The cassette is rotated to correspond to the mechanism in use and to subsequent illustrations. Words clarify which side of the cassette faces up so that the owner can select the correct tab. Headings identify the two procedures as separate steps.

You may recommend reductions and enlargements to fit the illustration in the available space, to indicate emphasis, and to ensure readability. Reduction may improve the quality of photographs and line drawings by de-emphasizing irregular or broken details in the original. Be sure, however, to consider the legibility of the type as well as of the illustration itself. The amount of reduction or enlargement may be expressed as a percentage, but it is easier to communicate the desired dimensions by specifying one of the measurements, for example, "reduce to 18 picas wide." The reduction in height will be proportional. (A pica is a printer's measure, equivalent to about one-sixth of an inch.)

Cropping means cutting off a portion of an illustration. The effect will be a reduction, but instead of an overall reduction, part of the length or width or both is removed. A cropped illustration can subsequently be enlarged or reduced. Cropping serves both to center and to emphasize the focal point of the illustration. Cropping removes extraneous borders or details so that the illustration focuses on the important information. Figure 18.8 illustrates reduction and cropping. Both procedures reduce the size, but they do so in different ways and with different objectives.

Some illustrations will be submitted in rough sketch form and will need to be drawn by an artist or computer specialist. These illustrations will require marking for copy clarification (labels and callouts) and for size and type of drawing.

Photographs and shaded drawings, unlike drawings made with black lines, must be converted to halftones before printing in fullscale printing. That is because ink prints in one color rather than in shades. A halftone, made by lasers or by photographing the original through a screen, converts the shadings into dots of various sizes and density. (See an example on page 380.) Large dots densely grouped create the image of a dark shade while small dots well spaced create a light shade. When the figure is printed most viewers will not recognize the dot composition.

Illustrations are handled in printing separately from text. Thus, they should be separated from the typescript and marked with the instructions. Because some illustrations require special treatment, unless they are submitted in camera ready form, they should be handled early in the production process so that they will not hold up the text.

Editing Illustrations with Computers

Computers and graphics programs have enabled writers and editors to prepare high-quality illustrations that previously would have been drawn by an artist. Graphics programs allow users to draw freehand or to create geometric shapes and graphs. Some programs permit rotating and shading objects, and most allow users to expand, shrink, and crop graphics.

A second type of useful program is an electronic clip art, or click art, program. It includes ready-made illustrations of a variety of subjects that can be pasted electronically into your document. These programs vary in quality as well as in subject matter, so check the examples before you choose one.

If you prepare graphics for desktop publishing, you will need a high-quality laser printer and scanner. The scanner will allow you to create a halftone-like print

FIGURE 18.8   Cropping and Reducing Illustrations.   Top, the original; lower left, cropped and enlarged; lower right, reduced.

of a photograph and to convert line drawings to an electronic file for modification and insertion into text files. If you are preparing graphics for the World Wide Web, you will need software to convert the files to GIF or JPEG format, or whatever may supercede these formats in the future.

Summary

Illustrations communicate according to the same principles by which words communicate. In technical documents, they convey information, support the text, and enable action. Editors consider the appropriateness of the type of illustration for its content and purpose; arrangement of visuals in the text; emphasis and details; perspective, size, and scale; and relationship to the text. Editors apply knowledge of organization, style, and design to enhance visual communication.

Further Reading

Judith Butcher. 1992. *Copy-Editing: The Cambridge Handbook*, 3rd ed. Cambridge: Cambridge University Press. See especially Chapter 4, "Illustrations."

Sam Dragga and Gwendolyn Gong. 1989. *Editing: The Design of Rhetoric*. Farmingdale, NY: Baywood.

Charles Kostelnick and David D. Roberts. 1998. *Designing Visual Language: Strategies for Professional Communicators*. Boston: Allyn & Bacon.

A. J. MacGregor. 1979. *Graphics Simplified: How to Plan and Prepare Effective Charts, Graphs, Illustrations, and Other Visual Aids*. Toronto: University of Toronto Press.

Frank R. Smith. 1973. "Editing Technical Illustrations." *Journal of Technical Writing and Communication* 3: 177–204.

Edward R. Tufte. 1983. *The Visual Display of Quantitative Information*. Cheshire, CN: Graphics Press.

Edward R. Tufte. 1997. *Visual Explanations: Images and Quantities, Evidence and Narrative*. Cheshire, CN: Graphics Press.

Discussion and Application

1. Vocabulary: To test your understanding of key concepts,

   a. Distinguish these four types of visuals and give an example of each: *tabular, graphic, structural,* and *representational.*
   b. Distinguish *figures* from *tables.*
   c. Distinguish *illustrations* from *graphics* as category words.
   d. Explain *sequential* and *spatial* arrangement.
   e. Define *emphasis, perspective, halftone,* and *cropping.*

2. The tax table in Figure 18.9 comes from the IRS manual for completing an income tax return. After a person determines taxable income, he or she locates the tax due according to filing status. Identify ways in which the table conveys information, supports the text (the instructions in the manual), and enables action.

1996 Tax Table—*Continued*

23,000

At least	But less than	Single	Married filing jointly *	Married filing separately	Head of a household
23,000	23,050	3,454	3,454	3,841	3,454
23,050	23,100	3,461	3,461	3,855	3,461
23,100	23,150	3,469	3,469	3,869	3,469
23,150	23,200	3,476	3,476	3,883	3,476
23,200	23,250	3,484	3,484	3,897	3,484
23,250	23,300	3,491	3,491	3,911	3,491
23,300	23,350	3,499	3,499	3,925	3,499
23,350	23,400	3,506	3,506	3,939	3,506
23,400	23,450	3,514	3,514	3,953	3,514
23,450	23,500	3,521	3,521	3,967	3,521
23,500	23,550	3,529	3,529	3,981	3,529
23,550	23,600	3,536	3,536	3,995	3,536
23,600	23,650	3,544	3,544	4,009	3,544
23,650	23,700	3,551	3,551	4,023	3,551
23,700	23,750	3,559	3,559	4,037	3,559
23,750	23,800	3,566	3,566	4,051	3,566
23,800	23,850	3,574	3,574	4,065	3,574
23,850	23,900	3,581	3,581	4,079	3,581
23,900	23,950	3,589	3,589	4,093	3,589
23,950	24,000	3,596	3,596	4,107	3,596

24,000

At least	But less than	Single	Married filing jointly *	Married filing separately	Head of a household
24,000	24,050	3,607	3,604	4,121	3,604
24,050	24,100	3,621	3,611	4,135	3,611
24,100	24,150	3,635	3,619	4,149	3,619
24,150	24,200	3,649	3,626	4,163	3,626
24,200	24,250	3,663	3,634	4,177	3,634
24,250	24,300	3,677	3,641	4,191	3,641
24,300	24,350	3,691	3,649	4,205	3,649
24,350	24,400	3,705	3,656	4,219	3,656
24,400	24,450	3,719	3,664	4,233	3,664
24,450	24,500	3,733	3,671	4,247	3,671
24,500	24,550	3,747	3,679	4,261	3,679
24,550	24,600	3,761	3,686	4,275	3,686
24,600	24,650	3,775	3,694	4,289	3,694
24,650	24,700	3,789	3,701	4,303	3,701
24,700	24,750	3,803	3,709	4,317	3,709
24,750	24,800	3,817	3,716	4,331	3,716
24,800	24,850	3,831	3,724	4,345	3,724
24,850	24,900	3,845	3,731	4,359	3,731
24,900	24,950	3,859	3,739	4,373	3,739
24,950	25,000	3,873	3,746	4,387	3,746

25,000

At least	But less than	Single	Married filing jointly *	Married filing separately	Head of a household
25,000	25,050	3,887	3,754	4,401	3,754
25,050	25,100	3,901	3,761	4,415	3,761
25,100	25,150	3,915	3,769	4,429	3,769
25,150	25,200	3,929	3,776	4,443	3,776
25,200	25,250	3,943	3,784	4,457	3,784
25,250	25,300	3,957	3,791	4,471	3,791
25,300	25,350	3,971	3,799	4,485	3,799
25,350	25,400	3,985	3,806	4,499	3,806
25,400	25,450	3,999	3,814	4,513	3,814
25,450	25,500	4,013	3,821	4,527	3,821
25,500	25,550	4,027	3,829	4,541	3,829
25,550	25,600	4,041	3,836	4,555	3,836
25,600	25,650	4,055	3,844	4,569	3,844
25,650	25,700	4,069	3,851	4,583	3,851
25,700	25,750	4,083	3,859	4,597	3,859
25,750	25,800	4,097	3,866	4,611	3,866
25,800	25,850	4,111	3,874	4,625	3,874
25,850	25,900	4,125	3,881	4,639	3,881
25,900	25,950	4,139	3,889	4,653	3,889
25,950	26,000	4,153	3,896	4,667	3,896

26,000

At least	But less than	Single	Married filing jointly *	Married filing separately	Head of a household
26,000	26,050	4,167	3,904	4,681	3,904
26,050	26,100	4,181	3,911	4,695	3,911
26,100	26,150	4,195	3,919	4,709	3,919
26,150	26,200	4,209	3,926	4,723	3,926
26,200	26,250	4,223	3,934	4,737	3,934
26,250	26,300	4,237	3,941	4,751	3,941
26,300	26,350	4,251	3,949	4,765	3,949
26,350	26,400	4,265	3,956	4,779	3,956
26,400	26,450	4,279	3,964	4,793	3,964
26,450	26,500	4,293	3,971	4,807	3,971
26,500	26,550	4,307	3,979	4,821	3,979
26,550	26,600	4,321	3,986	4,835	3,986
26,600	26,650	4,335	3,994	4,849	3,994
26,650	26,700	4,349	4,001	4,863	4,001
26,700	26,750	4,363	4,009	4,877	4,009
26,750	26,800	4,377	4,016	4,891	4,016
26,800	26,850	4,391	4,024	4,905	4,024
26,850	26,900	4,405	4,031	4,919	4,031
26,900	26,950	4,419	4,039	4,933	4,039
26,950	27,000	4,433	4,046	4,947	4,046

27,000

At least	But less than	Single	Married filing jointly *	Married filing separately	Head of a household
27,000	27,050	4,447	4,054	4,961	4,054
27,050	27,100	4,461	4,061	4,975	4,061
27,100	27,150	4,475	4,069	4,989	4,069
27,150	27,200	4,489	4,076	5,003	4,076
27,200	27,250	4,503	4,084	5,017	4,084
27,250	27,300	4,517	4,091	5,031	4,091
27,300	27,350	4,531	4,099	5,045	4,099
27,350	27,400	4,545	4,106	5,059	4,106
27,400	27,450	4,559	4,114	5,073	4,114
27,450	27,500	4,573	4,121	5,087	4,121
27,500	27,550	4,587	4,129	5,101	4,129
27,550	27,600	4,601	4,136	5,115	4,136
27,600	27,650	4,615	4,144	5,129	4,144
27,650	27,700	4,629	4,151	5,143	4,151
27,700	27,750	4,643	4,159	5,157	4,159
27,750	27,800	4,657	4,166	5,171	4,166
27,800	27,850	4,671	4,174	5,185	4,174
27,850	27,900	4,685	4,181	5,199	4,181
27,900	27,950	4,699	4,189	5,213	4,189
27,950	28,000	4,713	4,196	5,227	4,196

28,000

At least	But less than	Single	Married filing jointly *	Married filing separately	Head of a household
28,000	28,050	4,727	4,204	5,241	4,204
28,050	28,100	4,741	4,211	5,255	4,211
28,100	28,150	4,755	4,219	5,269	4,219
28,150	28,200	4,769	4,226	5,283	4,226
28,200	28,250	4,783	4,234	5,297	4,234
28,250	28,300	4,797	4,241	5,311	4,241
28,300	28,350	4,811	4,249	5,325	4,249
28,350	28,400	4,825	4,256	5,339	4,256
28,400	28,450	4,839	4,264	5,353	4,264
28,450	28,500	4,853	4,271	5,367	4,271
28,500	28,550	4,867	4,279	5,381	4,279
28,550	28,600	4,881	4,286	5,395	4,286
28,600	28,650	4,895	4,294	5,409	4,294
28,650	28,700	4,909	4,301	5,423	4,301
28,700	28,750	4,923	4,309	5,437	4,309
28,750	28,800	4,937	4,316	5,451	4,316
28,800	28,850	4,951	4,324	5,465	4,324
28,850	28,900	4,965	4,331	5,479	4,331
28,900	28,950	4,979	4,339	5,493	4,339
28,950	29,000	4,993	4,346	5,507	4,346

29,000

At least	But less than	Single	Married filing jointly *	Married filing separately	Head of a household
29,000	29,050	5,007	4,354	5,521	4,354
29,050	29,100	5,021	4,361	5,535	4,361
29,100	29,150	5,035	4,369	5,549	4,369
29,150	29,200	5,049	4,376	5,563	4,376
29,200	29,250	5,063	4,384	5,577	4,384
29,250	29,300	5,077	4,391	5,591	4,391
29,300	29,350	5,091	4,399	5,605	4,399
29,350	29,400	5,105	4,406	5,619	4,406
29,400	29,450	5,119	4,414	5,633	4,414
29,450	29,500	5,133	4,421	5,647	4,421
29,500	29,550	5,147	4,429	5,661	4,429
29,550	29,600	5,161	4,436	5,675	4,436
29,600	29,650	5,175	4,444	5,689	4,444
29,650	29,700	5,189	4,451	5,703	4,451
29,700	29,750	5,203	4,459	5,717	4,459
29,750	29,800	5,217	4,466	5,731	4,466
29,800	29,850	5,231	4,474	5,745	4,474
29,850	29,900	5,245	4,481	5,759	4,481
29,900	29,950	5,259	4,489	5,773	4,489
29,950	30,000	5,273	4,496	5,787	4,496

30,000

At least	But less than	Single	Married filing jointly *	Married filing separately	Head of a household
30,000	30,050	5,287	4,504	5,801	4,504
30,050	30,100	5,301	4,511	5,815	4,511
30,100	30,150	5,315	4,519	5,829	4,519
30,150	30,200	5,329	4,526	5,843	4,526
30,200	30,250	5,343	4,534	5,857	4,534
30,250	30,300	5,357	4,541	5,871	4,541
30,300	30,350	5,371	4,549	5,885	4,549
30,350	30,400	5,385	4,556	5,899	4,556
30,400	30,450	5,399	4,564	5,913	4,564
30,450	30,500	5,413	4,571	5,927	4,571
30,500	30,550	5,427	4,579	5,941	4,579
30,550	30,600	5,441	4,586	5,955	4,586
30,600	30,650	5,455	4,594	5,969	4,594
30,650	30,700	5,469	4,601	5,983	4,601
30,700	30,750	5,483	4,609	5,997	4,609
30,750	30,800	5,497	4,616	6,011	4,616
30,800	30,850	5,511	4,624	6,025	4,624
30,850	30,900	5,525	4,631	6,039	4,631
30,900	30,950	5,539	4,639	6,053	4,639
30,950	31,000	5,553	4,646	6,067	4,646

31,000

At least	But less than	Single	Married filing jointly *	Married filing separately	Head of a household
31,000	31,050	5,567	4,654	6,081	4,654
31,050	31,100	5,581	4,661	6,095	4,661
31,100	31,150	5,595	4,669	6,109	4,669
31,150	31,200	5,609	4,676	6,123	4,676
31,200	31,250	5,623	4,684	6,137	4,684
31,250	31,300	5,637	4,691	6,151	4,691
31,300	31,350	5,651	4,699	6,165	4,699
31,350	31,400	5,665	4,706	6,179	4,706
31,400	31,450	5,679	4,714	6,193	4,714
31,450	31,500	5,693	4,721	6,207	4,721
31,500	31,550	5,707	4,729	6,221	4,729
31,550	31,600	5,721	4,736	6,235	4,736
31,600	31,650	5,735	4,744	6,249	4,744
31,650	31,700	5,749	4,751	6,263	4,751
31,700	31,750	5,763	4,759	6,277	4,759
31,750	31,800	5,777	4,766	6,291	4,766
31,800	31,850	5,791	4,774	6,305	4,774
31,850	31,900	5,805	4,781	6,319	4,781
31,900	31,950	5,819	4,789	6,333	4,789
31,950	32,000	5,833	4,796	6,347	4,796

(If line 37 (taxable income) is— / And you are— / Your tax is—)

* This column must also be used by a qualifying widow(er).

Continued on next page

FIGURE 18.9   Tax Table, Internal Revenue Service

At the supermarket

> As of 1989, 62% of all supermarkets—grocers with at least $2 million in annual sales—used computerized scanners.
> The machines can be programmed to automatically reduce the price of sale items, eliminating the need for price tags.
> Some even have been connected to computerized voice synthesizers that recite the price of each item as it is being scanned.
> Scanners allow merchants to keep precise track of what has been sold and what remains in inventory.

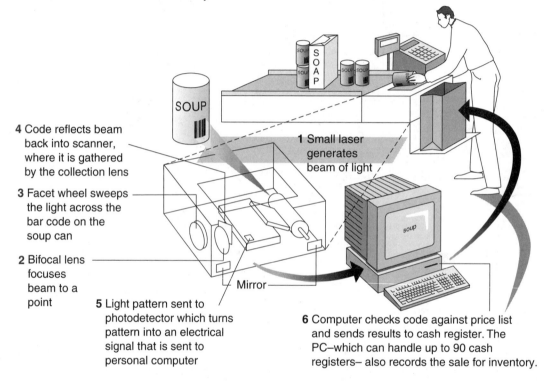

4 Code reflects beam back into scanner, where it is gathered by the collection lens

1 Small laser generates beam of light

3 Facet wheel sweeps the light across the bar code on the soup can

2 Bifocal lens focuses beam to a point

Mirror

5 Light pattern sent to photodetector which turns pattern into an electrical signal that is sent to personal computer

6 Computer checks code against price list and sends results to cash register. The PC–which can handle up to 90 cash registers– also records the sale for inventory.

FIGURE 18.10 **Line Drawing to Illustrate How a Bar Code Reader at a Grocery Store Checkout Counter Works.**

Copyright 1991, *USA Today.* Reprinted with permission.

3. The line drawing in Figure 18.10 illustrates the how a bar code reader works at a grocery store checkout counter. The drawing appeared in a newspaper for a general reading public. Discuss this visual in terms of the following:

information conveyed	perspective
appropriateness of the visual information	relationship of text and visuals; labeling
match of form, content, and purpose	proportion of data ink to non-data ink
arrangement: sequential and spatial	
emphasis and detail—are the important facts illustrated?	

OPERATIONS

The Society promotes a wide variety of activities under the auspices of some 67 committees. The work of these committees ranges from developing professional standards and Society goals to planning for future directions of the Society, conference planning, membership analysis and needs, and publications. Each of these activities requires financial support, some of which is donated by individual members or corporate sponsors. The bulk of the costs, however, is supported by the Society. The chart below shows how our funds are spent. Other expenses—particularly dues, rebates to chapters, and office operations—account for the balance. Note that our costs exceed the income received from membership dues; dues received pay for 87.4 percent of our expenses.

EXPENSES

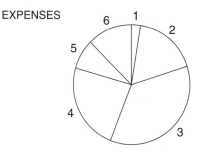

1	Miscellaneous	2.6%
2	Member Dues Refund to Chapters	17.6%
3	Office	35.7%
4	Publications	23.6%
5	Committees	8%
6	Awards & Grants	12.5%
	Total	100%

Income to the Society comes primarily from four sources: dues, conferences, sale of publications and interest from investments. Income varies somewhat from year to year depending on the number of members, the success of the publications sales program, and the financial success of our annual conference. The Society has been fortunate to have had two well-managed and financially successful conferences back to back—Pittsburgh and Boston. The success of these conferences has significantly improved the net worth of the Society.

INCOME

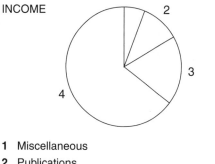

1	Miscellaneous	5.9%
2	Publications	11%
3	Conferences	18.4%
4	Dues	64.7%
	Total	100%

A widely-accepted standard within the professional society community calls for reserves (net worth) of about three years operating expenses. During the past five years our net worth has im-proved from $163,600 to $399,071, an increase of $235,471. Since 1980, the increase has been $183,171. The last two conferences, with other fac-tors, have significantly contributed to our present position of 1.3 years of reserve for our more than $300,000 annual budget. We cannot rely on the continual financial success of the conferences. Therefore, the Board has taken under consideration several options to improve our financial base. Increased membership is one of these options as is an increased technical exhibit at the conference. We have made significant strides and are taking the necessary actions to assure our continued financial stability.

FIGURE 18.11   Part of a Brochure for Members of a Professional Society

4. The two columns of information in Figure 18.11 are part of a brochure for members of a professional society. It is a simple kind of annual report. Evaluate the text and graphs. Begin by analyzing the purpose of the information: What should readers know, and what should their attitude toward the organization be after reading this information? Evaluate the graphs according to the criteria identified in this chapter. Also consider the guidelines for copyediting visuals in Chapter 5. Suggest editing goals.

5. The three circle graphs in the following figure illustrate the proportion of women in the workforce who achieve managerial positions. The purpose of the graphs is to reinforce the text, which claims that women do not rise above the "glass ceiling" in proportion to their numbers in the workforce. They should enable readers to interpret the text or even to substitute for it. Evaluate the graphs according to match of form, content, and purpose. Recommend revisions.

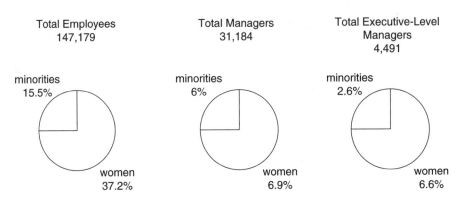

Chapter *19*

Editing Online Documents

Online publication makes information available electronically, through the Internet, CD-ROM, floppy disks, or software installed on a hard drive or network server. Instead of pulling a book or manual from a shelf to search for information or help with procedures, a user of an online publication opens a computer file. An online document can be as long as a book or as short as one computer screen.

Online or electronic publication, often taking advantage of hypertext, makes a lot of sense when

- the material must be updated frequently
- the volume of material in print would require extensive storage space
- users have easy access to computers and the software to use the online documents
- users need quick answers to questions while they work and are impatient with searching printed manuals
- multimedia presentation (graphics, animation, video, sound) can increase comprehension and motivation

Online help and tutorials are common ways to document software, enabling users to learn how to use the software to complete their tasks and to search for answers to their questions without much interruption in their work. Companies promote their products on the Internet and use internal networks for training and for sharing materials. CD-ROM publication allows the equivalent of multivolume information in a tiny bit of space.

As a technical editor, you almost certainly will work with online documents. Editors can contribute to the design and redesign of online documents because they think about readers and purposes and because they know about organization, style, and visual information. Editors bring design literacy to the task of creating and improving online documents. This knowledge complements subject-matter and technical expertise and results in a design that enables users to locate, understand, and use information.

Much of what you know about hard-copy editing will apply to online documents, but because the medium and structure of these documents differ from those of hard copy, it is limiting to think of online documents as electronic versions of paper documents. To edit online documents effectively, you need to consider its special features and the ways people will use it. Ideally you design from the ground up, not merely adapt an existing document to a new medium, but even in redesign, you can think about readers, uses, and purposes, not just about the coding that enables electronic display.

This chapter applies the editing principles you already know to online documents but also distinguishes the characteristics and uses of print and online documents to show where new principles and processes apply. The chapter begins with an overview of hypertext, common in online technical communication. It continues with discussion of issues in editing online documents—content, organization and navigation, style, page design, multimedia, grammar, and consistency. The chapter then suggests a process for editing that leads to functional and effective online documents. The term *hypertext* is used here in its broad sense to refer to any online information with links, including multimedia, not just to verbal information. The emphasis in this chapter is on hypertext for the World Wide Web.

Understanding Hypertext

Hypertext on the World Wide Web is coded with HTML (Hypertext Markup Language—see Chapter 4) so that it can be stored and displayed electronically and linked electronically to other hypertext documents. The links—connections between hypertext documents or parts of the same document—enable users to locate documents or sections within them by a simple click on a word, phrase, or image identified on the screen as a link (usually by a color and underlining or as a box around an image). The links, in a way that is invisible to users, identify the addresses, or URLs (uniform resource locators), of related documents or sections. When the user clicks on a link, the click instructs the computer to find the document or section in the address. As readers have a lot of choice about which links to pursue, different readers may locate and read different information in different sequences.

Documents on the Internet and much online help use hypertext. When the document contains not just words but also sound or animation and links among these media, it is called *hypermedia*. The hypertext may reside on your own computer as part of installed software, on a CD-ROM, or on another computer that you contact through the Internet.

The relationship of hypertext to print is somewhat like the relationship of computers to typewriters: in each pair the items have similar purposes, but the differences are just as important. Using print and typewriters as metaphors for hypertext and computers can help with comprehension of online presentation, but because hypertext and computers offer options for presentation of information that print and typewriters lack, understanding of how they function must go beyond these metaphors and even require unlearning.

Part of the task of editing hypertext is to consider what is to be linked and how to keep readers oriented and in control when they jump via links to another document or another section of the document.

Hypertext Links

Electronic links in hypertext can directly join parts of the document that are not joined sequentially or link the document to other related documents. These links, coded in HTML or another language, enable users to select sections of documents or to pursue themes of interest. The code, represented as tags for different types of text, tells the computer what to find and where to find it. For example, in the source code for the STC salary survey (Figure 4.2, page 56), you will see this notation at the top:

```
<A NAME="top"> <H1> 1995<BR> Technical Communicator<BR> Salary Survey</H1> </A>
```

The tags are distinguished from the text by the angle brackets. The code *A NAME=* identifies an "anchor" named "top." At the end of the text (not shown in Figure 4.2), readers will be given the option to return directly to the top of the document without scrolling back. The code that enables them to jump directly to the top is this:

```
<A HREF="top"> Return to the top.</A>
```

"A HREF" indicates a **h**ypertext **ref**erence to a previously named anchor. The code within the brackets does not display on the screen. The reader sees only the words, "Return to the top." These words are blue and underlined on the screen to signal that the link is "hot," meaning that by clicking on the words, the reader directs the computer to display the top of the document. The tag </A> means that the anchor has ended. The slash distinguishes the end of the anchor from its beginning.

The same type of tags can link different documents, including documents that reside on computers other than the host computer that act as file servers. The reference names are the file names or URLs. For example, the people reading the salary survey may wish to go to the STC home page. The link refers to this URL:

```
http://www.stc-va.com
```

Links can also be part of graphics or embedded in image maps. Users can click on the image in the same way that they click on words.

This linking affects the reading process. Although nobody has to read a printed book in a linear way, from beginning to end, the book's binding imposes a linear structure on it. Although one can read a hypertext document in a linear way, the links can encourage skipping and selecting.

Links function especially well when the user's task is searching for specific information rather than comprehension of the whole document. A user of a software application with a specific question searches the help materials by selecting keywords in the help menu that are linked to their explanations. A consumer seeking medical information uses the link in one document to find another document created by a different individual or organization. A researcher reads selectively in several articles, looking just for the information on a particular topic by following the links identified in the text of one article. A shopper selects a price list and skips the history of the store. A trainee completes one tutorial but not the others. Readers impose their own structure on the text according to their interests—and the options for linking that the text provides.

The editor's comprehensive task regarding links is to anticipate how readers may want to link and what associations they may benefit from making. A copyediting task is to check that the links work—that is, that they take readers to the specified places.

Navigation

The process of moving among documents or among sections of a document by using links is called *navigation*. You can navigate by direction (forward, back) or by topic. When you open a hypertext document (such as one on the Web), you will probably see navigation buttons at the top of the screen. On the Web interface (browser), the buttons may say *forward, back,* and *home.* They take you from one site to another: back to where you have been, forward to a site where you have been before, and home to your own home page. The browser also provides a history trail that reveals your route through multiple sites. Buttons within a document take you back or forward in the document or to its home page (main menu). In addition to these directional options, the document may provide a list of the main divisions, much like a table of contents, to let you search topics. Links to topics may be available at the tops, bottoms, or sides of pages or embedded in text.

Figure 19.1 illustrates one way a reader might navigate through the various divisions of a hypertext document. Another reader might follow an entirely different path. Figure 19.1 also illustrates how a reader can get "lost in cyberspace." A reader who reaches a level-three section without a sense of how that section connects to a higher-level section or even to parallel sections may become disoriented and have to work backwards to retrace steps or move blindly forward. Both procedures are inefficient and frustrating. On the other hand, a reader with a good sense of the way the information is organized and with links that anticipate where readers may wish to go next can select links according to goals.

Even though the reader's path of navigating a hypertext document may be unpredictable, the structure of the document must be predictable and apparent. Making the structure apparent and the document easy to navigate depends on these qualities:

- organization according to a meaningful and simple pattern
- links embedded in the text that anticipate the readers' needs

- links at the top and bottom or side of the screen that take readers to other parts of the document or site
- generic navigation devices, such as backward, forward, and home buttons
- consistent use of color and icons to identify sections of the document

An editor of hypertext considers likely paths that readers will take through the document and whether the navigational aids are adequate to enable them to go where they want.

Principles of Design for Online Documents

This section considers the design features of online documents and offers some principles for editing. Like print documents, online documents consist of information arranged and displayed in a way that should suit the purpose and needs of readers. Likewise, they include devices to aid navigation. Information is visual as well as verbal. Because most computers use color monitors, these documents probably use color, and they may use other media besides print and graphics. They demonstrate style through word choices and sentence patterns as well as through visual information. As with print documents, quality and clarity require good grammar, spelling, and consistency.

Content and Information Design

Probably anything that can be published in print can also be published online: instructions and tutorials, company information, style manuals, catalogs, reports, articles, visuals, policies and procedures, games, and more. Much that cannot be published in print can be used in online documents, including sound and animation. The range of possibilities for content and types of documents is wide.

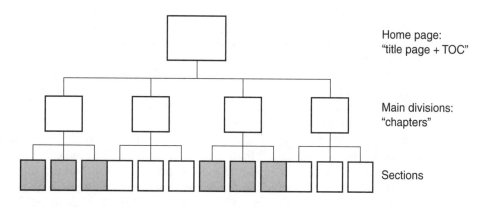

FIGURE 19.1   **Hypertext Document Structure**

Uses and purposes determine what content is appropriate. If the goal is performance, the content will focus on tasks. If the goal is comprehension of an organization or system or theory, content will focus on analysis of issues or divisions and the connections between them. Selection of content begins with analysis of uses and purposes. Information design means that someone with visual and verbal design skills as well as a sense of purpose and content thinks about both what should be included and how it should be organized and displayed.

Some designers recommend including the URL on each World Wide Web site along with a signature, address, update information, and copyright notice. This information enables users to go back to a site from a printed version and is valuable as well in citation of sources. The update information is analogous to a publication date in a book and lets a user determine how current the information is. The signature and address are analogous to the author's name or publisher information in a book and provide a contact as well as identification. Because copyright law protects online as well as print documents, the copyright notice is not necessary for that protection, but it does caution users about appropriate uses.

Organization and Navigation

Hypertext encourages readers to structure the text in the way that is most meaningful to them. But it would be naive to assume that as a result writers and editors bear no responsibility for organization. To allow readers freedom to select, the internal structure of hypertext must be carefully designed. Hypertext, like paper text with its sections connected in a linear way, must accommodate the fact that meaning derives from a reader's sense of connections. To comprehend, to interpret, and to draw conclusions, people look for relationships between general and specific, whole and part, causes and effects, and relationships of time, space, or topics. Readers also orient themselves to their place in the document, knowing what comes before and after and what other information is available.

The inherently linear structure of a book may interfere with the perceptions of relationships that are something other than linear, but the parts stay fixed, and a good index, table of contents, section headings, running headers, and page numbers, as well as a reader's ability to flip through pages, can keep a reader oriented to a particular topic as it relates to others. The size and feel of a book tell a reader something about its scope.

The connections are more transparent in hypertext, and because a reader sees the text one screen at a time, it is harder to get a sense of the document as a whole. The limited size of a computer screen, slowness of scrolling in comparison with flipping pages, and even the lower resolution of text on the screen as compared with the printed page mean that a reader of hypertext will depend on explicit structural information rather than the look and feel of the whole.

Figure 19.2 shows the hierarchical structure of a well-organized hypertext document. Although represented spatially, the structure is analogous to the structure of a print document, which might be represented in a linear outline form. The top level (the home page of a Web site or root screen in online documentation) is analogous

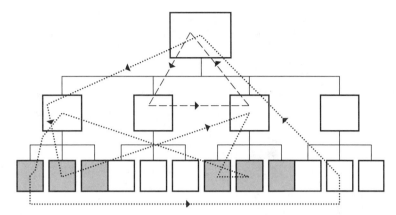

FIGURE 19.2   Possible Reading Sequences for Hypertext.
Dotted lines indicate the links between sections that the reader
chooses. This diagram illustrates two opposite patterns depending
on the first move from the home page.

to the cover or title page of a book or manual. The home page or root screen may
also identify the main structural divisions, much as a table of contents identifies
the divisions of a print document. The overview gives a sense of the whole that will
help readers perceive the parts. The main divisions (level two) enable readers to select the category of information or function that meets their interest and need.
From this level of detail, readers develop a sense of how the parts relate to the
whole. And, if the opening screen establishes a consistent page design, it teaches
readers where to look on the screen for different kinds of information, including
the navigation buttons.

Figure 19.3 shows the root screen of an online help system for Microsoft Word.
It illustrates that just like a book divided into parts or chapters an online document
has divisions that group related information or functions. The Microsoft Word help
is divided into different types of help, including a tutorial, keyword search, and
reference information. Within the main divisions, information is further divided
into sections, just as a chapter in a book is divided into sections.

A well structured online document is easier to use than fragments of information connected according to no apparent pattern. Users who perceive the pattern can
navigate both systematically and intuitively. Their sense of the structure will help
them navigate the document even without the explicit navigation information—
just as shoppers become familiar with the layout of a department store or grocery
store and can go to its various sections without checking the signs.

Editors can advise about matching structure with purpose and content and
about grouping within categories of information, using the principles of organization explained in Chapter 16. The best time for an editor to advise about structure
of an online document is during planning, before the sections are developed. Publications teams often begin with a flowchart or storyboard, displaying graphically

Word Help

| Contents | Search | Back | History | Index |

| On Top |

Word Help Contents
To learn how to use Help, press the HELP key or ⌘+/.

Using Word

Step-by-step instructions to
help you complete your tasks

Examples and Demos

Visual examples and
demonstrations to help you
learn Word
(Available only with
Complete/Custom Setup)

Reference Information

Answers to common questions;
tips; and guides to terminology,
commands, and the keyboard

**Programming with
Microsoft Word**

Complete reference information
about the WordBasic macro
language

Technical Support

Available support options so that
you can get the most from your
Microsoft product

FIGURE 19.3 Root Screen of Microsoft Word® Online Help. The root screen shows the main divisions of help according to function. Underlined words are links to the sections they name. Navigation buttons at the top of the screen also allow users to link to particular kinds of information. Each type of help has a corresponding icon to aid in recognition.

Source: Copyright © 1983–1995. Portions reprinted with permission from Microsoft Corporation.

what information will appear in what sections and at what level. (Refer to Scenario Two in Chapter One.) The flowchart as a planning tool invites developers to consider content and arrangement. The main divisions may represent major types of questions that readers will have, major tasks they need to accomplish, or the goals or projects of an organization. The divisions could even be a series of steps to be read sequentially, as in a series of lessons for training.

Style and Ease of Reading

Style refers comprehensively to the impression and coherence created by words and sentence structures. The same principles of good style that apply to print also apply to online documents, but the requirements for conciseness are greater because of limitations of screen and window size and difficulty of reading from screens. Fitting the text on the screen or in a help window mean that each word must be necessary and appropriate. A concise style can also minimize the length of the whole document, an important goal for reducing the fatigue that results from relatively poor legibility of characters and contrast of print and background. Yet conciseness should not preempt coherence, nor should abruptness substitute for thoughtful selection of words. A good editor balances the needs for information with the needs for conciseness and the limits of screen size.

Style refers not just to words but also to the image on the screen—whether a no-nonsense, serious image, a sophisticated image, or an image of fun, or something else. This image is analogous to the persona of the writer created by a text-only document. The image should support the document purpose.

Screen Design and Color

Screen design refers to the way the information will be positioned on the screen and what combinations of type and graphics will represent the information. Online documents have different looks and styles depending on purposes just as print documents do. Informative documents tend to be text and minimally formatted, whereas marketing sites often take advantage of graphics, color, sound, and animation. Effective screen design creates a page that supports the purpose of the document, is easy to read, minimizes scrolling, and allows readers to navigate easily. Meeting these goals requires awareness of the users' hardware and software, creating legible type, and following the goals of consistency and simplicity.

Effects of Hardware and Software on Screen Design Choices

Designers must consider that readers and their equipment have a lot of control over how screens will display. The software that enables users of the World Wide Web to find and display documents that are stored on other computers is called a *browser*. The reader's browser, not the writer's, determines what the reader sees from what is coded. It interprets the HTML code. Most browsers are graphical, but some show only text, and some do not see tables. Different browsers for the Internet display the same information with different fonts and sizes. The length of the displayed text line adjusts to fit the available space of the monitor or window.

The operating speed of the computer and modem determine how quickly information will load. Low-end equipment will load the information slowly, and the wait may frustrate readers. Color, graphics, patterned backgrounds, sound, and animation all slow the loading time.

Legibility of Type

Another screen design challenge is that type on screens is generally harder to read than type in quality print reproduction. The wide screen of a typical computer monitor invites a long line length that violates easy reading, and HTML codes do not encourage the range of display options, such as multiple columns and side headings, that word processors and page layout programs permit to achieve an optimum line length in print. Italics are hard or impossible to read on many monitors, and screen resolution is probably worse than resolution in print. These variations of print and screen explain, in part, why paper documents coded for the World Wide Web without redesign for the medium may succeed in making information available but rarely succeed in making information readable.

To solve the problem of line length, some designers use tables and frames to section the screen. These sections increase their control over placing text and graphics for maximum readability. The sections can be used to create columns, side headings, side-by-side text and graphics, and vertical navigation bars.

The difficulty of reading from screens argues for simple backgrounds as opposed to elaborate patterns that minimize the contrast between characters and background. To maintain contrast, designers also choose colors of different intensities for text and background. Dark characters on a light background or light characters on a dark background (perhaps yellow on dark blue) will be easier to read than background and text created with two colors of the same intensity, such as red and black.

Font choice also affects legibility. Simple shapes of characters formed with broad strokes offer advantages. Arial and Times New Roman are some choices preferred by designers. Size of text should be at least 12 points except for notes as for copyright, update, and page signature information.

Visual Consistency and Simplicity

Given the challenge of creating readable text and graphics for computer screens, design principles such as consistency and simplicity can be even more important for online design than for print. Some conventions for selection and placement of different types of information are familiar to users from their use of word processing programs. Functional information (menus, toolbars) typically spans the top of the screen. Structural and content information (topics) may go across the top or down one side. These functional and structural divisions help users to keep their place in a long document and to search it. Variations interrupt users from their main task of locating and understanding the content.

Consistency in the use of type, icons, and color to signal types of information and structural divisions also increases the usability of a document. A document style sheet and templates for design encourage consistent choices just as they do for print. Variations in different parts of text, such as headings and paragraphs, should be apparent from typeface, type size, and spacing, but each of these parts should be displayed consistently.

Color can enhance comprehension and motivation, or it can contribute to a confusing and busy look. Arbitrary shifts of background color or other inconsistent uses of color will confuse because readers expect shifts to mean something. On the other hand, using different colors consistently to distinguish types of information could enhance comprehension. Perhaps, for example, all of an organization's research activities appear on pages with blue banners at the top and its advocacy activities appear on pages with green banners.

Simplicity follows from determining what is essential in contrast to what may be only decorative or novel and focusing on usefulness to readers rather than on cleverness of the designer. A simple design that is easy for users to comprehend functions better than an elaborate design that may be breathtaking on first viewing but clumsy to use.

Grammar and Verbal Consistency

Online documents, just like print, require copyediting for details. Errors suggest carelessness at best and risk misunderstanding at worst. Typos may be more apparent on the screen than on the page if only because the screen is less familiar but also because a reader sees the words straight on rather than at an angle looking down. Errors seem especially out of place in documents with creative graphic displays and complex structures. Inconsistencies raise questions about whether the variations are meaningful or mistakes.

Graphics and Other Media

Online publication readily offers options for information design beyond print, including color, graphics, sound, video, and animation. All of these options have the potential to enhance comprehension, performance, motivation, attention, and interest. In functional documents, as opposed to entertaining ones, each design choice should have a function (which might be entertainment as a hook into the document). All these media choices are expensive to produce in terms of time, and they exclude users who lack the equipment to hear and see them. They also require significant storage space and load slowly. In time these limitations will diminish. For the present a designer who can include these media effectively in a document can attract users who might find plain text tedious. These options may also improve comprehension and retention.

Political, Ethical, and Legal Issues

The explosion of online information, especially on the World Wide Web, raises some political and ethical questions regarding access to and use of information. The value of sharing information freely via the Internet competes with the value of controlling and protecting information and the right to profit from it. Technology enables the restriction of information to certain people. The selection of users and restriction of access raise political and ethical questions about privilege and freedom. Technology also enables the collection of information about users that is considered private in other circumstances. At the same time, organizations have a responsibility for protecting the security of personal information they request, such as credit card numbers. These unsettled values and issues will, as they develop, determine what limits on information sharing and information protection people are willing to accept. Editors should keep informed about these issues and articulate them in planning and review of online documents.

Other issues of possible misrepresentation and quality concern editors as well. Because anyone with the necessary equipment and software can publish on the Web, there is no consistent quality control of what is published. The same statement could be made about print, but readers of materials from major publishing houses can count on a review process that aims to maintain high standards of accuracy, significance, and fair use. Editors can encourage responsible self-monitoring of material that an organization publishes and also be cautious about using information retrieved from the Web, checking for accuracy of facts and respecting the rights of people who created the work.

Guidelines for fair use of materials described by existing copyright laws apply to online documents as well as to print. (See Chapter 21.) All expression—online and print, visual and verbal—is protected by these laws and cannot be used by another person except with permission of the individual or organization that created the work. These laws prohibit replication of someone else's materials even though it is technically easy to do so.

All documents exist in a context that is not limited to the goals of the organization and the clients who use the organization's materials. Political and ethical impacts of online documents affect their value just as do textual features such as organization.

Process of Editing Online Documents

Although good management strategy encourages editors to participate in design of the document from its concept stage, editing online documents almost requires early participation. Editors bring a special expertise—the ability to think with the user's perspective and a knowledge of organization, style, and design as tools of usability—to creating documents that work. Online materials can be updated and revised, but reconceptualization and redesign are time consuming and inefficient.

Without participating in upfront planning and design, editors are almost certainly relegated to cleanup work—fixing typos and inconsistencies.

Editing scenario two in Chapter 1 describes the editor's role in the whole process of planning, developing, and reviewing a hypertext tutorial. The editor served as a coordinator and advisor throughout the development process. Such an arrangement takes best advantage of an editor's expertise with information design, not just with sentence structures.

The first three steps that the following paragraphs describe—defining readers and purpose, establishing document structure and links, and establishing page design and style—are part of planning. Team review and editing for content, organization, style, grammar, and consistency should be ongoing as the document develops. The document's links should be tested when a complete draft is ready. All but the simplest documents require a usability test, which the editor or a testing specialist may conduct. Managing and updating the information after the document is first published are necessary to keep it functioning as intended.

1. **Define readers and purpose.** Two or three groups of people have an interest in the content and design of hypertext: the organization that develops it, the readers who use it, and possibly third-party advertisers who hope to attract the people who read your site to their own products. Effective hypertext design must enable users to find what they want if the other two groups can achieve theirs.

   In defining purpose, the questions and possible answers in Table 19.1 may help your organization focus its plans for hypertext. Other questions to stimulate document planning appear in Chapter 2.

   If an organization wants readers to visit and revisit the document, it will have to meet readers' needs, not just the organization's. For example, a company that wants an advertising presence on the Web will need to provide information as well as a "billboard" to give readers a reason to come back to the site more than once. Online help must answer a reader's question without requiring an extensive search. Editors contribute to the document planning precisely because they think in terms of readers, users, browsers, and problem solvers whereas managers may think more about the organization, and technical experts think about coding.

   Thinking about the readers' use of the document will lead to design decisions. For example, you will have to decide whether to present the information in a series of linked documents (segmented) or in one long document with links to topics within it (connected), whether to design a site that encourages exploration or one that allows a quick exit, and whether to include graphics and color or to depend on text. Different purposes, such as marketing and instruction, point to different design choices. Table 19.2 shows some design implications of different purposes.

2. **Define document structure and links.** The principle by which the document will be divided into parts and how the parts will connect serves as a kind of

TABLE 19.1   Questions for Analyzing Readers and Purposes of Online
             Documents

Designer's Questions	Possible Answers

What do we hope to accomplish with this online document?	Sell products or services.
	Instruct readers so that they can use our product.
	    Provide a tutorial.
	    Provide a quick answer to a specific question.
	Provide information.
	Persuade: convey a message.

Why choose online publication? Will it supplement print, be another version of print, or stand alone? What will it accomplish that print will not?	The text needs to be updated frequently. Production and distribution of revisions are quicker with hypertext. Readers will use the text primarily for reference. Readers want to work with minimal interruption. We can reach readers that we would not reach in print. Readers have access to the equipment and software for reading the online document.
What will readers do or know as a result of using this document?	Buy a product or service; enroll; donate. Complete a task. Accept our point of view. Browse selectively; read the entire document. Print the document. Look for other related documents: ours, those of other organizations. Return to the document at a later time.
What do readers hope to accomplish with it?	Buy products or services. Accomplish a task; learn a procedure. Get information; find sources of related information. Be entertained. Finish the task efficiently and quickly. Print. Return to the document later for similar goals.
What kind of equipment and software do readers have?	Powerful equipment (fast clock time, large screens, graphical browser, sound and video cards): documents load quickly, reflect our color, design, and graphics. Low-end equipment (slow load time, small screens, nongraphical; not equipped for multimedia): Slow speed and display may discourage use.
Who are these readers, and what do they know about our subject and hypertext?	Novice users; timid. Expert users; risk takers.

blueprint for document development. A flowchart may represent the structure as well as possible links among parts. Among other functions, the chart defines tasks for collaborative groups dividing the responsibilities of document development. If regular review meetings are scheduled, the flowchart may be readily

TABLE 19.2   **Design Implications of Document Purposes**

Purpose	Design Implications: Content, Style Etc.
Marketing	Company information that defines products and services and creates a positive image of quality and service
	Product information: model numbers, price, warranty, materials, cost, technical specifications
	Ordering and shipping information
	Security to prevent misuse of credit card numbers
	Forms that allow a reader to respond with an order
	Segmented structure: enable users to bypass information that doesn't interest them
	Attractive graphics, screen design, and use of color; perhaps emphasis on "fun"
Entertainment	Attractive graphics, screen design, and use of color
	Animation, sound
	Emphasis on fun: games, quizzes, surprises
Instruction and reference	Good table of contents with links to each section
	Good keywords and cross-references that will enable users to find what they are looking for.
	Structure by tasks
	Position on the screen that won't interfere with the user's main task
	For online help, concise style to keep the help window small
Information	Connected rather than segmented structure: readers may save or print a report or other informative document

modified according to new information that emerges as the document develops, but the flowchart at least provides a starting point.

3. **Define the style and design the screens.** Guidelines for style and templates for design will save production time later. Being able to see what a user will see can focus attention on readers and perhaps provide some criteria for evaluation.

Before you commit too much to a particular design for a Web site, test it with several browsers that readers are likely to use. You can get test versions of several browsers from the Web. Also test the design on the type of equipment that users are likely to own, not necessarily what is in your office.

4. **Review and edit for content, structure, and style.** Opportunities for review and editing should be scheduled into the planning. A plan may be fine conceptually but fail for reasons the team cannot anticipate. Testing it requires some review, before too much effort goes into developing something that may later be scratched.

5. **Copyedit.** Online documents merit the same reviews that you use for print documents—for spelling, grammar, punctuation, consistency of capitalization and other mechanics, accuracy of data, and labeling of visuals. A check of hypertext links can be part of copyediting.

6. **Conduct a usability test.** Tests focusing on tasks that intended users will perform (including reading for comprehension) extend the contributions of editors to evaluation. Tests provide the opportunity to observe whether representative users can navigate and whether they find the information they need. They also measure whether time requirements and accuracy of performance meet standards.

7. **Maintain the site.** Because an online document is easy to update, it requires ongoing editing. New information may require new sections, which in turn must fit into the overall scheme. Changes of content may introduce inconsistencies and errors. The editor may not bear primary responsibility for site maintenance but will probably participate, especially if there are content changes.

Summary

Online publication offers a number of advantages over print and will undoubtedly increase in the future. The effectiveness of online documents depends less on the technical skill used to create them than on skill in information design. Editors contribute design literacy as well as language expertise to the task of creating effective online documents.

Further Reading

Scott Boggan, David Farkas, and Joe Welinske. 1996. *Designing Online Help for Windows 95*. International Thompson.

William Horton et al. 1996. *The Web Page Design Cookbook*. New York: Wiley.

World Wide Web Resources

Web specifications: World Wide Web Consortium
    http://www.w3.org
HTML primer and tags: National Center for Supercomputing Applications
    http://www.ncsa.uiuc.edu
Style guides:
    http://info.med.yale.edu/caim/StyleManual_Top.HTML
    http://www.w3.org/hypertext/WWW/Provider/Style/Overview.html

Discussion and Application

1. Examine examples of sites suggested below on the World Wide Web to determine
   - probable target readers
   - likely purposes of the readers and goals of the site sponsor

- organization (main divisions, parts—connected or segmented, sequential or by categories)
- navigation devices: what, where located, types of information
- screen layout
- visual design (text or tables; paragraph length; headings)
- use of color or icons to identify the section
- style (sentence style and illustrations)

Then evaluate each site to determine whether the choices have been effective. Summarize your findings overall by noting differences in choices and their match to intended uses.

copyright page: http://lcweb.loc.gov/copyright

cyberspace law page: http://www.ssrn.com/cyberlaw/

state government site for your state

site for your college or university

commercial site—advertising or sales

2. Examine the online documentation in your word processing program. What types of help are available? For example, does the help include a searchable index, balloon help, wizards, reference materials, a tutorial? How does a user locate these types of help? What differences do you note in content, style, organization, and screen design?

Part 4

Management and Production

Chapter 20
Collaborating with Writers

Chapter 21
Legal and Ethical Issues in Editing

Chapter 22
Type and Production

Chapter 23
Management

Glossary

$$C \quad h \quad a \quad p \quad t \quad e \quad r \quad 20$$

Collaborating with Writers

A topic that is frequently discussed in articles on technical editing is how to establish a productive working relationship between editor and writer. The literature points to potential for conflict between editor and writer. The proposed solutions usually relate to interpersonal skills.

The frequency of this topic is surprising in a sense. Perhaps the issue of the editor–writer relationship would not be so pervasive if the literature included more about editing effectively. Conflicts should diminish if the editing is well done. But in another sense, the frequency with which this issue is raised is not surprising. Editing requires contact with other people. When people collaborate, they must place the demands of the task above their personal whims and their egos.

Effective editing requires the editor to win the trust and cooperation of the writer. This chapter gives some suggestions for how to do that. It begins with a description of the relationship between editors and writers and explains why these relationships work or fail. It then identifies strategies of editing, management, and interpersonal skills to make the relationships work. It also offers suggestions for how and when to phrase queries to writers, how to conduct editor–writer conferences, and how to compose letters of transmittal to writers.

The Editor–Writer Relationship

Like other interpersonal relationships, the relationship between editor and writer may be positive and marked by cooperation, or it may be negative and marked by conflict and stress. Why should editors and writers cooperate? They both have something to gain from such a relationship. They may gain personally from having a positive work experience. But they also gain professionally: working as partners—

341

collaborating—they can produce a better document than either could alone. Both editor and writer work on behalf of readers.

Some writers can be described most accurately as researchers or subject matter experts. They may have no special training in language or document design. Editors provide that expertise. Even when writers are expert writers, editors provide a different, perhaps more objective, point of view about a document. Editors benefit from the writer's subject matter expertise and from the writer's research to generate the text. The collaborative ideal results from the realities of different specializations.

Relationships fail for three reasons: poor editing, poor management, and oversized egos. Writers have no reason to respect incompetent editors, and they justifiably resent unnecessary or incorrect intervention in a text. Stress in editor–writer relationships will be caused by unnecessary delays in work by either writer or editor or by poor communication among all the members of the production team. Writers and editors may place their egos above the more important goal of making the document work for readers. A writer may regard every editorial comment as personal criticism and become defensive or even reject the editing. An editor may regard the writer as an inferior and develop a contemptuous attitude that only encourages defensiveness.

Ernest Mazzatenta, former president of the Society for Technical Communication, has collected statements from writer-researchers about what they like and dislike about editors. They are representative of statements offered by hundreds of scientists and engineers who have participated in his technical writing courses. These statements are paraphrased in the following lists.

What Writers Like Most About Editors

- Restructures the report so that the train of thought is smooth and logical.
- Points out ideas and explanations in the report that are not clear to the reader and then rewrites them.
- Catches misspelled words.
- Generally improves readability.
- Usually returns the paper within five working days.
- Approaches the writer considerately concerning any changes.
- Edits fairly promptly. Does it without malice.
- Shows patience.

What Writers Dislike Most About Editors

- Asks the writer to rewrite a section without giving any indication of what's wrong with it or any direction to take.
- Makes changes only to incorporate the editor's style of writing.
- Is somewhat conservative in that the editor suggests qualifiers and disclaimers to analyses that, in the writer's professional judgment, are excessive.
- Uses words that are not acceptable to the writer or others and won't change them.
- Replaces words with synonyms.

- Requires too many iterations (for example, ten).
- Makes comments that are inconsistent with the department head's comments.

Relationships between editors and writers are potentially good—and potentially stressful. They can be characterized by collaboration or by competition. Editors and writers have choices about what type of relationship they will have, and editors especially can do much to create a productive working relationship.

Strategies for Working with Writers

One place to begin in developing strategies for working with writers is to analyze the statements in likes/dislikes list. Most of the likes and dislikes concern editing itself (the editor's work with organization, spelling, and style). Others concern good management (promptness, coordination with a department head or other reviewer). A few imply personal qualities, such as consideration or stubbornness. The strategies discussed here parallel this classification of likes and dislikes.

Edit Effectively

Editing effectively is the most important thing you can do to make your relationship with a writer productive and cooperative. Effective editing involves collaborating with the writer to design the document for readers even before the first drafts are written. During development and when the draft is complete, you approach editing as collaboration with the writer to make the document work for readers. If you focus on readers rather than on errors, writers will appreciate your rescuing them from writing clumsy, incoherent, or inaccurate documents.

As this textbook has emphasized throughout, effective editing also requires you to be correct, objective, and sensitive to the needs of readers and context of use. You must base your emendations on known principles of effective communication rather than on your taste. You must restrict your changes to those that will make the document more comprehensible and usable. You must resist making changes simply to satisfy your own style unless you can articulate reasons why those changes will make the document better for readers. And when you do make changes, you must be sure you are right, and not make changes out of ignorance.

Writers are justifiably impatient, and even angry, when editors change their meaning. They are frustrated when editors simply change things to no apparent benefit. One editor who edited instructions on Basic Life Support was disconcerted by the faulty parallelism in the ABCs of cardiopulmonary resuscitation (CPR). The list read: "Airway clear of obstructions, Breathing or ventilation, and Circulation by cardiac compression." So she changed the phrasing and gained parallelism—but lost the ABCs. The problem is that ABC is a mnemonic device taught in CPR classes nationwide. The instructions had to be re-edited (or unedited), and the editor had to rebuild her relationship with that writer. An effective editor knows the subject matter well enough to avoid introducing errors and knows the resources to check when content questions arise. These resources include subject matter dictionaries and handbooks as well as the writer.

Manage Efficiently

Efficient management saves time because the job gets done right the first time and does not have to be redone. Efficient management also respects production schedules and other people's deadlines. As editor, you can take a number of steps to manage the job efficiently.

1. **Participate early.** If you can participate in planning the project, you and the writer and team manager can agree on overall project goals from the start. This strategy helps to eliminate the conflict that arises when the editor enters the project at the end and has different ideas about the way it should have developed. Editing is conceptual and therefore part of the early stages of project development. The idea that editors are only fixers of errors at the end of development invites conflict.

2. **Clarify your expectations.** Guidelines that cover usage, punctuation, spelling of technical terms, documentation style, and visual design (headings, margins, spacing) should be available to writers before they write. These guidelines won't be as complete as the style sheet you prepare while editing because you won't be able to anticipate all the possible variations in style until you edit, but they can cover general issues. Guidelines save editing time because writers prepare the documents correctly. Figure 20.1 shows the guidelines for authors published in the front of every issue of *Technical Communication*. This page establishes standards for punctuation, style, manuscript preparation, and documentation. If your organization has a house style guide (see Chapter 6), make sure writers know about it and use it.

   Editors and writers should agree in advance about readers and purpose, illustrations (size, number, whether rough or ready for printing), and length. Write down these assumptions as document specifications.

   You can sometimes clarify expectations for content. If you know that a document will have to cover certain topics in a certain order, provide the outline. If some parts of the document will be boilerplate, that is, repetitions of existing text, include those sections with the outline. These aids can also prompt a writer to get started. The project may look manageable because something is already written.

   A schedule is also necessary from the start, whether it is prepared by the editor or team leader or collaboratively. If you will need a document by a certain date in order for production to proceed on schedule, clarify that from the start. You may be justifiably irritated if you receive documents so late that you cannot do your job well, but you can also help to prevent that situation by establishing when you will need the various drafts. Estimate your time realistically, but be aware that people who are not editors underestimate how long it

CONTENT

Technical Communication publishes articles of professional interest to technical communicators—including writers, editors, artists, teachers, managers, consultants, and others involved in preparing technical documents. Before writing an article, look over several recent issues of the journal to make sure your topic and approach are appropriate for our readers.

Also, please look over the literature in the field and cite any relevant publications, so that your article builds on and extends previous work, if there is any. For example, see previous issues of the journal, the *Proceedings* of the Society's annual conferences, the IEEE's *Transactions on Professional Communication*, the *Journal of Technical Writing and Communication*, and appropriate textbooks.

Because the primary audience of *Technical Communication* is enlightened practitioners, manuscripts reporting the results of research or proposing theories about topics in our field should ordinarily include descriptions of or suggestions for practical application of the research or theory. In some instances it may be more appropriate to discuss practical applications of theory or research in one or more additional articles (for example, manuscripts intended for a special section or special issue on a topic) by the same author(s) or one or more collaborators. In such cases, the author(s) should discuss this approach with the editor at the time the article is originally submitted.

Please do not submit a manuscript that is under consideration by another publisher.

STYLE

The purpose of *Technical Communication* is to inform, not impress. You should therefore write in a clear, informal style, avoiding jargon and acronyms. All decisions on style and usage should be guided by common sense: What is the most common, most clearly understandable way to present the information? Our authority on spelling and usage is *The American Heritage Dictionary*; on punctuation and format, the *Chicago Manual of Style*.

Avoid language that might be construed as sexist. The most common problem is use of gender pronouns in a way that implies sexual stereotyping, such as, "A writer delivers *his* completed manuscript to a secretary so *she* can type it." Easy ways to avoid this problem include using plurals ("Writers take their completed manuscripts to secretaries for typing"), using second person if appropriate ("Take your completed manuscript to the secretary for typing"), and recasting the sentence ("A secretary types each manuscript as soon as the writer completes it").

MANUSCRIPT PREPARATION

Articles should usually not exceed 5000 words (about 20 double-spaced pages). They should be typed or printed, either double- or triple-spaced, including references, on one side only of good-quality paper.

Please indent all paragraphs, and underline any words you want *italicized*. Number the pages, put your name *only* on a cover sheet to permit copies to be sent anonymously to referees, and submit three complete copies.

Journal format allows three levels of headings for paragraphs or sections of your article. Please indicate them as follows:

FIRST-LEVEL HEADING	(all caps, on a line by itself)
Second-Level Heading	(initial caps, on a line by itself)
Third-level heading	(first word only capitalized, in italics, as part of the first line of the paragraph)

Illustrations should be camera ready (just the way you want to see them in the journal). Submit one reproducible, plus two photocopies if the art is not reproduced in the text. Most illustrations will be printed 20.5 picas wide (approximately 3-3/8 inches), although the full page width of 42.5 picas (7-1/16 inches) is used on occasion. If your art is prepared oversize, allow for reduction in choosing type size and line widths.

Number figures and tables in separate sequences, using Arabic numerals. *List figure captions on a separate page.* In the text, be sure to refer to all figures and tables by number, and discuss the important features or summarize the message of each. (But don't, for example, simply repeat the numbers from a table.)

Provide a descriptive abstract (150 word maximum) of your manuscript with the first-level heading SUMMARY following the article's title.

Also include a biographical sketch (approximately 100 words) of yourself and similar sketches of any co-authors. If your article is selected for publication, we will request a glossy black-and-white photo of you and any co-authors. However, photos are optional.

Whenever possible, prepare your manuscript with word-processing software. If you have done so, you will be asked to provide an electronic file of your final draft along with the paper copy when your article has been accepted for publication. Articles may be submitted on 3.5 inch diskettes in ASCII or in one of the commonly used word-processing formats.

REFERENCES AND BIBLIOGRAPHY

Identify all sources of information in the text in the style specified in *The Chicago Manual of Style* for the natural sciences (chapter 16): the author's last name and the date of publication, enclosed in parentheses and inserted at the appropriate place in the text, thus: (Jones 1990); if a specific statement or quotation is being cited: (Jones 1990, p. 20); if text contains author's name: Jones (1990) reported that ". . ." (p. 20). At the end, list all references alphabetically by the last name of the author in *Chicago* style, using hanging indentation:

Book

Author, Last Name First, First Name Author. and F. N. Author. 19XX. *Full book title: Subtitle* [note u & lc]. 2d ed. [as appropriate]. Place of publication: publisher.

Journal Article

Author, L.N.F. 19XX. "Full article title." *Journal Title* [note no comma] 00 [volume number], no. 00 [omit issue number if journal is paged consecutively]: 00-00 [inclusive pages].

Anthology or Proceedings Paper

Author, L.N.F. 19XX. "Paper title." In *Proceedings* [or *Anthology*] *Title*, ed. John Smith [if known], pp. 00-00 [inclusive pages]. Place: publisher.

Thesis or Dissertation

Author, L.N.F. 19XX. "Title." Ph.D. dissertation, School.

Other comments

Unsigned periodical articles are listed periodical title first; do not use Anonymous: *Periodical*. Date. "Article," pp. 00-00.

A bibliography of other sources may be listed separately in the same style.

Please spell out abbreviations: for example, I[nstitute of] E[lectrical and] E[lectronics] E[ngineers] Press.

The entries in the list of references or bibliography may be annotated at the end of each listing if you wish to comment about them.

Avoid explanatory notes—either work the material into the text (perhaps in parentheses) or work it out. Similarly, identify an unpublished source in the text but do not list it in the references.

COPYRIGHTS

The Society for Technical Communication holds the copyright on all material published in *Technical Communication*. (The Society grants republication rights to authors on request.) If your article has been previously published or presented, please tell us when you submit your manuscript.

ADDRESS FOR SUBMITTING ARTICLES

All manuscripts should be sent to:

Dr. George F. Hayhoe
George Hayhoe Associates
194 Aberdeen Drive
Aiken, SC 29803-7100

Correspondence may be sent to the postal address above or by electronic mail to:

GFHayhoe@aol.com

or by fax to:

803-642-9325

FIGURE 20.1 Guidelines for Authors Submitting Articles to *Technical Communication*

Source: Reprinted by permission of the Society for Technical Communication

will take to edit. You have to educate them or they may abuse you innocently. (Scheduling and estimation of time are discussed in more detail in Chapter 23.)

3. **Work with the writer throughout development.** Ongoing collaboration follows from advance planning. The initial plan for document development should establish some points of review after sections are complete but before the whole is finished. If plans need to be adjusted for any reason, it's better to do so midway than to wait until the end. This plan keeps the writer on schedule, but it also allows rethinking of the plan before too much writing is complete.

4. **Don't surprise.** Share your plans for editing with the writer. If it is understood that you will edit only for spelling, consistency, grammar, and punctuation, don't surprise the writer with elaborate revisions of style and format. If you change your mind about your editing plans, have good reasons for doing so, and discuss your thinking with the writer before you show him or her a document smeared with red pencil marks.

   If you plan extensive revision, discuss the plans before you do revise or ask the writer to revise. You can also prepare a writer for bad news. In a phone call you might say, "I had to reorganize extensively." Then, when the document arrives, the writer expects reorganization, which may not look so ominous as he or she feared. Give the writer a chance to review the editing early enough in production to make changes. It's not fair or efficient to edit heavily and then show the writer the edited version only after it has been set in page proofs and cannot be changed.

5. **Be prompt.** Keep your expectations for yourself as high as for others. Get your work done on time. To avoid misunderstanding, let others know in advance when you will complete the work, and keep them informed about necessary schedule revisions.

Develop an Attitude of Professionalism

Even if the writer is not good at writing, you can respect the person and his or her expertise in other areas. You can focus on the task rather than on the person. Assume that your goals and the writer's are mutual: to make the document work for its readers and purpose. Assume that you are players on the same team, not that you are an expert correcting an incompetent neophyte.

You can encourage professionalism in writers by setting an example. Complete your work on time and be prompt for meetings. Even language, dress, and posture can communicate your attitude toward your job and to your colleagues. Flexibility and good humor can help as well. If the writer's choices are as good as yours, or if they are relatively unimportant, you can yield. That doesn't mean you have to compromise on principles of good writing, but you can recognize that to risk a relationship or an entire project for a comma or variant spelling is to misorder your priorities.

Sometimes you will be wrong. You will misinterpret information or emphasis, and you will miss some errors. The review of edited copy by the writer or other technical experts acknowledges the possibility of editorial error. (In fact, the review should be seen in that light rather than as a time to show the writer all the failings in the document.) The way you respond to someone who discovers imperfect editing will model the way you want writers to respond to you. If you are defensive and aggressive or try to blame someone else, you will encourage writers to respond in the same way to edited copy. You can be gracious. You can thank the writer for rescuing you from mistakes. You can say, "Now I understand," rather than accusing the writer of being unclear to begin with.

Clear, neat, and accurate marks show respect for the writer as reader of your marks and indicate your professional commitment to your task. You may gain some psychological advantage by using a black or green pencil rather than a red one. Red reminds some writers of critical schoolteachers and blood. Be cautious, however, about choosing blue if you will need to photocopy the edited typescript as blue does not photocopy well.

The Editor–Writer Conference

A good working relationship requires frequent face-to-face meetings or, in the case of long-distance editing, contact by phone, letter, or email. The way you manage yourself in these meetings will set the tone for your relationship. Conferences may occur for planning, progress reports, and review. The review conference, in which the edited document is examined or plans for further editing are made, is the most sensitive because of the potential for implied or real criticism. Documents may be passed back and forth between writer and editor without a conference, and it saves time to do so. However, if the editing will be extensive, it may be better to plan a conference to talk about the changes than to surprise a writer with a heavily marked document.

Some companies with product teams use a variation of the conference called an inspection meeting. All the people working on the product gather to review the edited version of the document and to direct the next stages of development. The inspection meeting combines technical review with editorial review. One advantage of such a meeting is that it reinforces the team concept of development instead of the one-to-one relationship between writer and editor. And, it is easier to resolve disagreements when all the people who might have suggestions are in one place at one time instead of depending on sequential and sometimes contradictory comments. Even with inspection meetings, though, it is likely that editor and writer will meet face-to-face or correspond one-to-one.

Even if a meeting has the potential to generate tension, it is preferable to exchanging heavily edited documents without discussion. It allows for give and take—the opportunity to question and clarify and to make nonverbal gestures of accommodation, such as nodding your head and smiling. You can make these conferences go well by organizing your comments, by using tactful and goal-oriented language, and even by arranging the furniture to suggest cooperation.

As you plan for the conference and its organization, think of the overall goals. The purpose of the conference is to clarify the next steps in document development. All conferences should result in some agreement and understanding among collaborators, arrived at through information sharing and negotiation. Planning and review conferences should end with understanding and clarification of goals, tasks, responsibilities, and schedule.

Using the review conference for instruction distracts from the main conference purpose and from the focus on the document at hand. Instruction also demotes the writer to the role of student rather than colleague and collaborator. Save instruction for another setting, or limit it to responding to questions the writer asks. If you repeatedly edit one writer for persistent, correctable problems, it's quite reasonable to suggest a time for instruction. But don't mix instruction with the development and production of a specific document.

Conference Organization

A conference is a form of communication, just as a printed, graphic, or electronic document is. Thus, the conference should adhere to principles for effective communication, such as organization. Before you meet with the writer, develop a plan for what you want to discuss and the order in which you will approach topics. Notes identifying topics and pages to which they apply will help the conference proceed efficiently and with a minimum of fumbling. An organized conference should increase the writer's confidence in you as editor and manager. It will reflect your comprehension of the whole project as well as your respect for the writer's time.

An overview statement, like the introduction to a document, identifies the topic and goals of the conference and the order in which you will proceed. You could organize by topic, perhaps asking questions about content and then offering comments about style. Working top down, from content and organization to the details, will keep the conference focused on substantial issues that require conversation. Or you could proceed through the document from beginning to end, chronologically (though not line by line). The conference could reasonably end with suggestions and a schedule for revision.

If you meet with several reviewers in a document inspection meeting, the importance of organizing the conference increases. Use their time efficiently by defining in advance the issues about which the reviewers may disagree or about which you need help. Work toward consensus on major issues and leave the minor issues alone.

Review of the Edited Typescript or Hypertext

In a review conference, do not feel that you have to call attention to every emendation in the document. You do not have to initiate a conversation about all the punctuation changes, for example, but you should be prepared to explain them if the writer should ask. Your goals are to verify that your editing is correct and consistent with the overall document goals and, working with the writer, to establish the next steps in project development, whether revision or approval for continuing production. It is inefficient to plan a line-by-line review of the document in conference. A writer who is determined to consider each emendation may appreciate a copy of the edited document before the conference. The writer can then review it line by line on his or her own time.

A heavily marked typescript may intimidate or alienate a writer and create defensiveness that can spoil the conference. If company policy gives you the privilege of editing electronically, you can easily prepare clean copy incorporating the editing. But keep records of the editing for style and organization so that you will be able to query about the accuracy and seek approval for substantive changes. Also keep the original version so that you will be able to restore it if necessary.

If policy and the writer's trust permit you to share clean rather than marked copy, three positive results can be achieved. First, a clean copy is less threatening than a marked one, so the egos of the writer and editor are less likely to interfere with the work of producing an effective document. Second, it will be easier to spot additional editing needs in a clean copy than in one in which the marks distract from reading. Finally, a clean copy will show the results of editing and demonstrate its value.

The Language of Good Relationships

The words you choose, your nonverbal expressions, and the way you listen all affect how well a writer receives your messages. Your language can communicate a sense of collaboration, or it can communicate power and arrogance.

Active Listening

One purpose of a review conference is information gathering. Thus, an important method of communication is listening. Active listening means drawing out the writer and working to understand his or her point of view. It contrasts with the pretense of listening while mentally formulating your own statements. If writer and editor listen to each other only superficially, the chances of real communication taking place are not good.

One strategy of active listening is to repeat or paraphrase something a writer says.

Are you saying, then, that...?

The echo of the writer's statement guards against misinterpretation. If your paraphrase is inaccurate, the writer can correct you. This strategy also forces you to listen carefully enough to repeat. An active listener also probes for more information when a writer expresses ideas incompletely. Probing statements encourage the writer to elaborate.

Please go on.

What do you mean by...?

How will readers use this information?

How does this point relate to...?

Please explain how...

These are not statements to challenge the writer's competence but to seek the information you need to complete the editing.

With active listening, you also encourage the writer to share and cooperate. You can show your interest with nonverbal signals (nodding your head, smiling, leaning forward) and with verbal signals ("I see," "uh-huh").

Positive Language

The words you choose in the conference can encourage partnership or they can provoke defensiveness. Writers will respond to goal-oriented language more positively than to criticism.

Critical language	Goal-oriented language
wordy	Condensing the text will increase the chance that a person in an emergency will find the necessary information quickly.
poorly organized	The tasks are rearranged chronologically so that the crew will know when to do each one.

The goals (finding information in an emergency, knowing when to do the tasks) are explanations that show the editing to be purposeful rather than arbitrary. Critical language may derive from modeling inappropriately after composition teachers. An editor, like a teacher, must evaluate, but the evaluation is not an end in itself. It should suggest goals for revision based on reader needs.

Another way to emphasize document goals rather than resort to criticism is to use "I" statements when you may seem critical, but "you" statements when you are praising. "I" statements do not suggest that you are stupid but rather that you are assuming the role of readers who may not have enough information.

I don't understand how this example explains the thesis. [not "You haven't been clear here"]

I can't tell whether I should turn off the hard drive or the monitor first. [not "Your instructions are incomplete"]

You did a good job of interpreting the table. [good news]

Try to focus on neutral subjects—the document and the reader—rather than on the writer and editor.

One final caution: avoid words that suggest inappropriate editorial intervention, especially *change*. No writer wants to hear that you have "changed" his or her text. In fact, writers fear that editors will change the meaning. While polysyllabic euphemisms are generally undesirable stylistically, it may be better when dealing with a sensitive writer to substitute *emendation* for *change*.

Confidence

Your own confidence in your editing will encourage the writer's confidence. Wishy-washy comments can turn off writers as much as aggressive comments can. You should not apologize for your good suggestions and honest explorations, nor should you avoid making recommendations for fear of alienating a writer. Keep in mind that you and the writer are partners working on behalf of the readers.

Furniture Arrangement

Psychologists talk about nonverbal communication, such as the information we give by our posture and facial expressions. Similarly, arrangement of furniture in a room suggests the relationship between the people in the room. In Figure 20.2, the executive arrangement, with an executive protected behind an imposing desk, suggests a relationship of superior to inferior. Such a relationship is inconsistent with the partnership of editor with writer, as well as being inefficient when writer and editor look at a document together. In this arrangement, editor and writer look at each other, perhaps confrontationally.

The collaborators arrangements, by contrast, place editor and writer in more equal positions, as partners. These arrangements allow editor and writer to look together at a document and therefore encourage the focus of the conversation on the document rather than on personalities. (See Rosemarie Arbur, cited in Further Reading.) A conference room may be a better place to meet than the office of writer or editor. In the words of psychologists, it is more neutral territory. Likewise, a table may be more appropriate than a desk that belongs to either the writer or the editor.

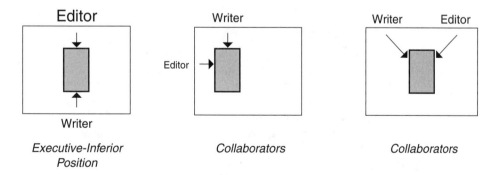

FIGURE 20.2 Chair Arrangement for the Editor–Writer Conference. The Executive-Inferior arrangement makes it difficult to look at the document and invites confrontation between writer and editor.

Correspondence with Writers

Some of your comments to writers will be on paper or online rather than face to face. Especially when the editing requires no major changes but rather simple revisions and clarification of points of information, you can pass the edited document back to the writer without having to schedule a conference. Correspondence takes the form of queries written on or in the document and letters or memos of transmittal or email.

Whether you are working long distance with the writer or face to face, you will probably write queries to the writer on the document itself, in the margin if you are editing on hard copy or annotations in the text if you are editing electronically. Queries may be questions about content, such as "What is CLHM?", they may ask for more information or for verification that an emendation is correct, or they may explain a substantive emendation, such as reorganization. If you place the queries right where the questions arise rather than keeping a separate list, the writer won't have to match and hunt. If you keep a list, though, include page numbers. Even if you use the queries only in the review conference, they are helpful reminders of points you wish to discuss with the writer. (See Chapter 3 for a discussion of the mechanics of using query slips.)

The same principles of positive language that apply to conferences apply to queries. Evaluative comments, especially negative evaluations, are inappropriate because they do not direct the writers in revision. An occasional positive evaluation ("interesting" or "good") may be encouraging, but such comments—whether positive or negative—may imply that your goal has been evaluation of the writer's work rather than preparation of the document for readers and publication.

Don't query every little change or go running to writers when the answers are fairly obvious. You not only interrupt the writer's work with constant queries, you convey lack of confidence. You will be able to answer some questions by careful reading of the text or by checking other sources. Save the queries for points of information that you cannot determine on your own.

Letters of Transmittal

A letter or memo of transmittal with an edited draft, like any letter of transmittal, tells what the document is, what the writer should do with it, and when he or she should return it to the editor. A secondary purpose may be to motivate. The letter of transmittal is used for long-distance editing or for inhouse editing when the writer will study the emendations before meeting with you. The letter or memo may be transmitted on paper or online. The tone of the letter should be formal (though some situations may allow for some informality), and the language should be in standard, high-quality English. Use the conventional letter or memo format, as the situation requires, and be sure to edit your own writing. You gain credibility as editor if your own writing meets the same standards you expect from writers.

Like all letters, a formal letter of transmittal has a three-part structure parallel-ing the three purposes, and the email version, though less formal, also includes these parts. The introduction is a statement of transmittal. The body of the letter explains the goal of editing and the goal of revision. The conclusion asks for a re-sponse. Figure 20.3 illustrates how the body of a letter of transmittal might read.

The letter focuses in a businesslike way on the task at hand, not on evaluation. The statement of transmittal quickly orients the writer to this project, and the letter clarifies what he or she is to do. The emendations are explained in terms of purpos-es that the writer and editor share. The letter does not substitute for queries at the point of question but it can call attention to important needs.

Writers should always have final review privileges. Do not send an edited copy directly to production without inviting the writer to check for content and ac-curacy. By all means, alert the writer to substantial or questionable changes.

Email Correspondence

Email offers a convenient way to stay in touch with the writer or other members of the product team throughout the course of editing. Because it is less formal than the letter of transmittal, it may seem more spontaneous and less intimidating. An email message is less intrusive than an interruption with a visit to someone's office. It is easy to correspond by email fairly frequently, which may help to keep the project on schedule.

You can attach documents to email messages so that all team members can see exactly what the issue is. If your group is networked, you may be saving updated versions of the document on a server that various people can access.

statement of transmittal	The edited version of the first draft of your proposal for the pulse welding research project is enclosed.
explanation of editing and request for revision	I have checked for accuracy of information and correctness of usage and punctuation. I have also verified that the format con-forms to the RFP. Please check the editing, especially the reword-ing on pages 2 and 3. The numbers in the table on page 3 don't add up to the total shown, so please check to see whether the costs for each item are accurate or whether the addition is incorrect.
request for response	I will need the revised copy for final proofreading by July 2 if the proposal is to be mailed to meet the deadline. Call at extension 292 if you have questions.

FIGURE 20.3 Letter of Transmittal to a Writer

Summary

Good relationships with writers follow from good editing and good management as well as from good interpersonal skills. Good editing makes a document better, not just different and certainly not wrong. Good management means establishing expectations early, communicating with all people involved with the project, and completing work promptly. Interpersonal skills and tactful communication encourage cooperation. The focus should be on the document, the task, and the reader rather than on the personalities of the writer or editor. If editor and writer collaborate, they create more effective documents than either could alone.

Further Reading

Rosemarie Arbur. 1977. "Student-Teacher Conference," *College Composition and Communication* 28: 338–342.

Ernest Mazzatenta. 1975. "GM Research Improves Chemistry Between Science Writers, Editors." *Proceedings, 22nd International Technical Communication Conference.* Washington, DC: Society for Technical Communication.

Lola M. Zook, ed. 1975. *Technical Editing: Principles and Practices.* Washington, DC: Society for Technical Communication. See the chapters by Nelson A. Briggs, "Editing by Dialogue," and Eva P. Dukes, "The Art of Editing."

Discussion and Application

1. Consider the following case studies as topics for class discussion or journal entries.

    a. You are editing an anthology of ten articles written by various subject matter experts. Part of your punctuation style is to use a comma before the final item in a series. When you return the edited copy to one writer who has omitted commas before the final item, he protests aggressively, challenges your knowledge, and threatens to withdraw the article from the collection unless you remove the commas you have inserted. What is your response? Before you decide finally how you will respond, think about your goals for the anthology.

    b. You are a recent college graduate editing a research proposal for a senior staff member with a Ph.D. The proposal is full of academic jargon that you think may invite the proposal reviewers to smirk at the researcher rather than to respect her. You simplify some of the sentences and vocabulary. You return the proposal to the writer, who, in turn, delivers it to the typist. Later you learn from the typist that the writer has written on the copy, "Ignore the editorial comments. Type as originally written." You're furious at the put-down and the waste of your time; furthermore, you are convinced that the pretentious style will jeopardize funding. What do you do?

    c. The writer whose work you are editing is a very nice person, but his writing is terrible. You can't understand many of the sentences, not because the subject matter is unfamiliar but because the construction is so bad. Frankly, you're appalled at the lack of writing skill, and you know your attitude is getting in the way of your work. What can you do?

d. For cases a, b, and c, identify specific ways in which efficient management might prevent the conflict.

2. Conduct editor–writer conferences in class and then answer the accompanying questions. Each student should bring a document he or she has written to a classmate for editing. The document should be in a technical rather than a creative genre. In a brief planning session, the writer explains to the editor the purpose and readers of the document and any particular editing needs that he or she has identified. The editor, in turn, may query the writer. If the document is substantial, the editing will take place out of class. After editing is complete, each student editor meets the writer for a review conference. After the conference, the writer evaluates the editing and the conference and makes suggestions for future conferences. These questions can be used for evaluation:

   a. Consider the editing. What needs did the editor discuss? Did the editor overlook any editing needs? Comment on accuracy, completeness, and objectivity. Was the editor too aggressive or too passive on some issues? Did the editor seem to be a collaborator? Comment.

   b. Consider the conference. How did the editor organize it? Did the editor clarify his or her plan for the conference at the beginning? Was the language goal oriented? Was the editor confident? How did you feel about being edited by this editor? Comment.

   The exchange may work best if the writer is not also the editor of that writer's work.

3. Describe a specific example of each of the three strategies discussed here for productive working relationships: good editing, good management, interpersonal skills.

4. If you have had experience—good or bad—working as an editor with a writer or as a writer with an editor, analyze where the relationships succeeded or failed by considering editing skill and procedures, management, and interpersonal skills.

5. Think of ways to express these critical statements in a way to encourage productive revision. You will have to invent specific details.

   a. These sentences are awkward.
   b. The whole section of the report is unclear.
   c. The brochure is poorly organized.
   d. It's impossible to tell why the project you propose is important.
   e. The instructions are confusing.

Chapter 21

Legal and Ethical Issues in Editing

Laws and codes of ethics aim to protect the good of a society as well as to protect individual rights. In technical publication, intellectual property laws govern copyright, trade secrets, and trademarks. Laws also require organizations to accept responsibility for product safety. Editors work with other members of product development teams and legal experts to verify adherence to these laws and the representation of them in the text through warnings and notices of copyright, trademarks, or patents.

Codes of ethics are less formally encoded than laws, as are the sanctions for violating them, but, like laws, they aim to protect individuals and groups from harm. They may be encoded as statements of professional responsibility, such as the "Ethical Guidelines" by the Society for Technical Communication, or derive from cultural values of right and wrong, such as the values that individuals should not harm other persons and that they have responsibility for maintaining the quality of the environment.

Legal and ethical issues pertain both to individuals and to organizations. Editors can be most effective as individuals if corporate policies establish commitment to legal and ethical behavior and if corporate procedures allow for review of products and documents by a variety of knowledgeable people, not just the editor or even just the legal department.

This chapter reviews legal and ethical issues that pertain to technical publication and suggests corporate policies to encourage ethical behavior.

Legal Issues in Editing

Editors share responsibility with writers, researchers, and managers for protecting documents legally and for ensuring that they do not violate intellectual property, product safety, and libel laws. Editors verify that permissions to reprint portions of other people's publications are obtained and review instructions and warnings for potentially dangerous products. Using the symbols ©, ® and ™ indicates copyrights, registered patents, or trademarks.

The editor can fulfill legal responsibilities more readily and thoroughly if the organization's policy manual and style manual include policies and guidelines about intellectual property, product safety, and misrepresentation.

Intellectual Property: Copyright, Trademarks, Patents, Trade Secrets

The law of many countries, the United States among them, recognizes that one can own intellectual property just as one can own land or a business. These protections are intended as incentives to develop ideas or products that may improve the quality of life for the community. Intellectual property includes original work of fiction or nonfiction, artwork and photographs, recordings, computer programs, and any other expression that is fixed in some form—printed, recorded, or posted on the World Wide Web. These expressions are protected by copyright law. Other intellectual property includes work protected as trademarks, patents, and trade secrets. The owner of intellectual property has the legal right to determine if and where the work may be reproduced or used.

Copyright

The United States Copyright Act of 1978 protects authors of "original works of authorship," whether or not the works are published. The Copyright Act gives the owner of a copyright the right to reproduce and distribute the work and to prepare derivative works based on the copyrighted work. No one else has these rights, and reproducing work copyrighted by someone else violates the copyright law. As editor, you verify permission to use copyrighted material and take steps to protect documents published by your company.

Ownership

Copyrights belong to the author who created the work unless the author wrote the work to meet responsibilities of employment. In the case of "work for hire," the employer owns the copyright. Most technical writers and editors work for hire, and the manuals and reports they write are the intellectual property of their employers. Some publishers require authors to surrender the copyright to them. Works by the U.S. government are not eligible for copyright protection; they are in the "public domain" and can be used by other people without permission. Collections with contributions by multiple authors are generally protected by a single copyright, but the sections may also be copyrighted individually.

The copyright extends 50 years beyond the copyright owner's death. Works written for hire are protected 75 years beyond publication. When the copyright expires, works are in the public domain and may be reproduced and distributed by others unless a copyright is renewed.

Copyright Notice, Registration, and Deposit

Copyright is automatic in the United States as soon as a work exists in fixed form, and protection does not require a notice or registration. However, registration with the Copyright Office gives maximum legal protection. Registration requires sending an application form (Form TX for most technical documents), a fee ($20 in 1997), and two copies of the document to the Register of Copyrights in the Copyright Office.

For the best protection, a published work should include a notice of copyright. In a book, the notice usually appears on the verso page of the spread that contains the title page and is sometimes called the copyright page. The notice includes the symbol ©, abbreviation "Copr.," or the word "Copyright"; the year of publication; and the owner's name. Here is an example: © 1998 Allyn & Bacon.

You can get more detailed information on U.S. copyrights from the Copyright Office, either in print or on the World Wide Web. Particularly useful publications are Circular l, "Copyright Basics," and Circular 3, "Copyright Notice." You can download application forms from the Web site. These and other helpful resources are listed at the end of this chapter.

Copyrightable work published in the United States is subject to mandatory deposit for use in the Library of Congress. Some publications are exempt from this requirement. Circular 7d gives more information on what must be deposited and what is exempt.

International Copyright Protection

Copyright in one country does not automatically extend to another. Use of works in a particular country depends on the laws of that country. Countries that have signed one or both of two multilateral treaties, the Universal Copyright Convention (UCC) and the Berne Convention, offer some protection to work published in member countries. Circular 38a lists countries that maintain copyright relations with the United States. Some countries offer little or no copyright protection for work published in other countries.

Permissions and "Fair Use"

Because work is protected by copyright, permission must be obtained to reproduce sections of someone else's work. Usually the writer requests permission from the copyright holder, but editors verify that permissions have been acquired before the document goes to print. The permission ought to exist in writing, whether in a letter or on a form. The correspondence will establish exactly what will be reprinted and where. The copyright owner may charge a fee for use of the material. If permission is denied, the material cannot be used. Permission must be acquired for each use. Permission to use copyrighted material in one publication does not give one the right to use it elsewhere.

A request for permission to reprint copyrighted material should include the following information:

- Title, author, and edition of the materials to be reprinted
- Exact material to be used: include page numbers and/or a photocopy
- How it will be used: nature of the document in which it will be reprinted, author, intended audience, publisher, where the material will appear (for example, quoted in a chapter, cited in a footnote)

Fair use allows some copying for educational or other noncommercial purposes. For example, you may photocopy an article in a journal to study for your research paper. But your professor cannot copy whole sections of a book for the class so that the class won't have to buy the book because doing so would deprive the book's publisher of sales of its intellectual property. In a work setting, it is best to be cautious about copying, especially if you will distribute the work widely and profit from this distribution.

Copyright prohibits duplication of software for multiple users unless an organization has purchased a site license. The terms of use are often listed on the software package.

Copyright and Online Publication

The ease of copying material distributed through the Internet or World Wide Web makes some people think that it is all right to do so, but material on the Web and even email messages are protected by the same copyright laws that protect print. Because publishing on the Web makes material available without purchase, you are not depriving an author of sales if you copy material from the Web. But if you distribute something you found on the Web to make money or to people who would otherwise read it on the Web and see the advertisements that accompany it, or if you copy the words and use them as your own without citation, then you are violating copyright law. Figure 21.1 identifies situations that raise the question of fair use. The judgments of fair or unfair are those of law professors.

Cyberspace law is a new area that is still uncertain. For the present it is prudent to assume that the laws for print publication apply to online text.

Trademarks, Patents, and Trade Secrets

Trademarks are brand names, phrases, graphics, or logos that identify products. The rainbow-colored apple is a trademark of Apple Corporation, as is the name Macintosh. If the marks are registered with the United States Patent and Trademark Office (PTO), no one else can use those particular marks to represent their own products. Editors look for proper representation of trademarks that may be referred to in the text. The symbol ® next to the mark means that the mark is formally registered and certified with the PTO. The symbol ™ indicates a mark registered on a state basis only or one that has not been officially placed on the Principal

1. *You get a personal message and you pass it along to one other person.*
   PROBABLY UNFAIR, though you might still have an implied license. (For instance, if the message asks for help on a nonprivate matter, there might be an implied license to pass it along to others who might be able to help.)

2. *You download an article from a newspaper's Web site and post it to a news group. The site carries advertising, and says "Do not send any copies of these articles to other people" (thus negating any implied license).*
   PROBABLY UNFAIR, since if this becomes commonplace, fewer people would access the Web site and see the advertising.

3. *You key in an article from a paper magazine that doesn't have a Web site and post it to a news group.*
   PROBABLY UNFAIR, if the magazine is available on some online service (such as DIALOG or NEXIS), or if the magazine is still on the newsstands for people to buy.

4. *You forward someone's message from one publicly accessible news group to another news group.*
   PROBABLY FAIR, because the message was published, and because the person posting it has no commercial interest in selling the message.

5. *You quote a few sentences from a news article that you downloaded.*
   PROBABLY FAIR, because it's only a few sentences.

FIGURE 21.1   Fair Use? Examples from Cyberspace

Source: Larry Lessig, David Post, and Eugene Volokh. "Cyberspace-Law for Non-Lawyers." (Alan Lewine, ed.). Available at http://www.ssrn.com/cyberlaw/. Reprinted by permission of the authors and the Cyberspace Law Institute (http://www.cli.org).

Register in the PTO. Trademarks are capitalized in print. Dictionaries identify words that are registered trademarks. Product literature is a good way to find out about whether to use the ® or ™. A typical procedure is to use the symbol on first use of each trademark or to list all trademarks used in a publication in the front matter. Constant repetition of the symbol in the text may become distracting, and the law does not require it.

Patents protect inventions in the way that copyright protects expressions. Their significance to editors is that the text records patent registration with the symbol ® just as it does for trademarks.

Law also protects trade secrets, such as the specifications for a new product or a customer list. Employees owe a "duty of trust" to current and former employers. It is illegal for a company to hire you to find out what a competitor is planning, and it is illegal for you to give trade secrets of a former employer to a new employer or of your current employer to anyone else. Documents produced under government contract projects may be classified and require secrecy. For either private or government work, you may have to use special protections to keep information in your computer files secure. The duty of trust represents one exception to the First Amendment protection of free speech.

Product Safety and Liability

According to U.S. law, companies and individuals must assume responsibility for safety of the products as they are used or even misused by consumers. The products themselves must be designed to be as safe as possible, but because design itself cannot ensure safety, instructions for use and warnings must be complete and clear. The instructions and warnings must cover use, anticipated misuse, storage, and disposal. Manufacturers cannot avoid responsibility with disclaimers (statements that the manufacturer is not responsible for misuse or accidents).

Instructions, Safety Labels, and the Duty to Warn

The first strategy of documenting safe use of a product is to write clear and complete instructions. Clarifying procedures where safety is an issue should be part of document planning. Editors may wish to request a summary of hazards so that they don't have to rely on the instructions alone to identify them. In reviewing a draft, editors rely on the principles of organization, style, visual design, and illustrations, as discussed in chapters 12–19, as well as an attitude of vigilance and care. Ambiguity must be clarified. For example, if a procedure calls for "adequate ventilation," an editor may query whether "adequate" means "fresh" air or if circulating air will suffice. Taking the perspective of the reader, the editor may note some gaps that a writer misses.

If there are hazards of using products, manufacturers and suppliers must warn of the risks unless the product is common and its hazards well known. Instructions do not constitute warnings, nor can a warning substitute for instructions. A warning calls attention to a particular procedure verbally, visually, or both. Some standards for identifying different categories of risks and for symbols and colors to identify them have been developed by the American National Standards Institute (ANSI), The Occupational Safety and Health Administration (OSHA), and the International Organization for Standardization (ISO). For example, the word "danger" and the color red are used only when serious injury or death may result. "Warning" (orange) and "caution" (yellow) warn of less serious risks.

Safety labels should be attached to products where users will see them before and as they use the product.

The Editor's Legal Responsibility

Like everyone else who is involved in product distribution, writers and editors have some responsibility in the eyes of the law for safe use of a product. They can be named in product liability lawsuits, though the usual practice is to name the company. A sense of professional responsibility as much as respect for the law encourages editors to be careful in the legal edit.

Libel, Fraud, Misrepresentation

Libel is a defamatory statement without basis in fact that shames or lowers the public reputation of an identifiable person. People who can prove libel may win damages from a publisher. The possibility of being sued for libel worries editors of fiction and periodicals more than it does technical editors because people are more likely to be discussed, referred to, or otherwise cited in works of fiction and in periodicals than in technical documents. Nevertheless, all editors should read alertly for facts to verify the accuracy of any negative statements that may be made about individuals.

Fraud and misrepresentation deceive the public. Misrepresentation may occur in labeling products or making claims about them or in claims about credentials of individuals or about data.

Ethical Issues in Editing

The requirements for ethical conduct may restrain companies and communicators more than laws. For example, one long-distance phone company trademarked the name, "I Don't Care." When consumers were asked to choose their long-distance company and responded, "I don't care," they were enrolled in the company with that name—whose rates were significantly higher than the rates of other companies. The company complied with the law in choosing its name, but consumers would probably not find the practice ethical because it takes advantage of people. Likewise the contractor who read the specification for "stainless steel nuts and bolts" and used a cheaper, less durable metal for the bolts acted within the law (arguing that "stainless steel" modified only "nuts") but did not act ethically, considering that the bolts caused the structure to fail.

Editors and others in technical communication have responsibilities that extend beyond meeting the requirements of the law. These responsibilities derive from some values that seem universal, no matter what their philosophical or religious bases, such as the responsibility not to harm others. This principle enables people to live together in groups and with some trust in others. It may reflect the wish for survival ("I won't hurt you if you won't hurt me") as well as more generous and lofty principles of mutual responsibility. Other bases for ethical choice derive from cultural values and values that are defined by the particular profession. These values anticipate the consequences of actions and help people make judgments in particular cases.

Professional Codes of Conduct

Technical communicators, like physicians, lawyers, professors, and clergy, consider themselves professionals rather than staff or laborers. Such classification brings privileges of status and material benefits—and responsibilities as well. Professionals accept responsibility for their work rather than deferring to a superior. It is unprofessional in any setting to "pass the buck." The editor's ethical responsibilities increase along with other responsibilities—a comprehensive editor is more responsible than a copyeditor. Yet all editors have a part in ensuring the quality and integrity of documents.

Figure 21.2 shows the ethical guidelines approved by the Society for Technical Communication. Like statements by other professional associations, these guidelines affirm professional competence as an ethical requirement for communicators.

STC Ethical Guidelines for Technical Communicators

As technical communicators, we observe the following ethical guidelines in our professional activities. Their purpose is to help us maintain ethical practices.

Legality
We observe the laws and regulations governing our professional activities in the workplace. We meet the terms and obligations of contracts that we undertake. We ensure that all terms of our contractual agreements are consistent with the STC Ethical Guidelines.

Honesty
We seek to promote the public good in our activities. To the best of our ability, we provide truthful and accurate communications. We dedicate ourselves to conciseness, clarity, coherence, and creativity, striving to address the needs of those who use our products. We alert our clients and employers when we believe material is ambiguous. Before using another person's work, we obtain permission. In cases where individuals are credited, we attribute authorship only to those who have made an original, substantive contribution. We do not perform work outside our job scope during hours compensated by clients or employers, except with their permission; nor do we use their facilities, equipment, or supplies without their approval. When we advertise our services, we do so truthfully.

Confidentiality
Respecting the confidentiality of our clients, employers, and professional organizations, we disclose business-sensitive information only with their consent or when legally required. We acquire releases from clients and employers before including their business-sensitive information in our portfolios or before using such material for a different client or employer or for demo purposes.

Quality
With the goal of producing high quality work, we negotiate realistic, candid agreement on the schedule, budget, and deliverables with clients and employers in the initial project planning stage. When working on the project, we fulfill our negotiated roles in a timely, responsible manner and meet the stated expectations.

Fairness
We respect cultural variety and other aspects of diversity in our clients, employers, development teams, and audiences. We serve the business interests of our clients and employers, as long as such loyalty does not require us to violate the public good. We avoid conflicts of interest in the fulfillment of our professional responsibilities and activities. If we are aware of a conflict of interest, we disclose it to those concerned and obtain their approval before proceeding.

Professionalism
We seek candid evaluations of our professional performance from clients and employers. We also provide candid evaluations of communication products and services. We advance the technical communication profession through our integrity, standards, and performance.

FIGURE 21.2   STC Ethical Guidelines

Source: Reprinted by permission of the Society for Technical Communication

This competence refers to "truthful and accurate communications"—writing and editing well—as well as to meeting expectations on time.

The guidelines also affirm adherence to laws, including laws regarding confidentiality, and to honesty and fairness. Thus, the guidelines draw on general values (honesty and fairness), law, and the expectations for professional performance in a particular field.

One point of conflict and confusion in the ethics of technical communication relates to the professional value of "objectivity." Objectivity seems to represent truth uncontaminated by the bias of opinions. The idea is that the person should get out of the way and let "facts speak for themselves." Should the communicator remain "neutral," taking no position but rather "translating technical information accurately" for the public? Furthermore, urged to "adapt" to an audience, should the communicator defer to an audience's wishes? These slippery words appear in various discussions that attempt to define the values and ethics of technical communication.

One approach to answering these questions is to consider whether facts can ever speak for themselves and whether anyone can ever be neutral. As you learned in Chapter 2, readers interpret based on their prior experiences and knowledge, and different readers will find different meanings in the same texts. Furthermore, the acts of arranging information in sentences or on a page influence interpretation. Apparent neutrality may be nothing more than avoidance of commitment.

The nature of the expertise that enables people to call themselves "professionals" also influences answers to these questions. Expertise generally requires specialized education and perhaps experience. As a consumer of services by professionals, such as your physician, you expect information about options even if you claim the right to make the ultimate choice. You respect your own limitations enough to listen to the options and welcome them.

Likewise, your client or manager may depend on your expertise. It would be unethical to withhold it because in so doing you might jeopardize their ability to make the best decision. For example, your client or manager may request the use of a particular type of software that you know is less efficient than another or may not know the environmental consequences of petroleum-based compared to soy printing inks. If you remain silent, they may never know the options. Or you may resist making recommendations in a feasibility study because the reader has higher status in the organization than you do. In all those cases, your professional expertise carries with it a responsibility to inform a client or reader, even when the reader may not want to hear your information. Your judgment represents the collected wisdom of the profession represented by your expertise, not something arbitrary and personal.

Ethics in technical communication involves more than reporting accurately what a subject matter expert says. It requires more than passively telling the truth. It also requires a proactive position in offering information and helping to guide the decisions of people who lack the specialized knowledge of the communicator and therefore do not see all the consequences of choices.

Misrepresentation of Content or Risks

Because of the serious, informative purpose of technical communication, readers expect it to be truthful. They may approach advertising with enough skepticism to question claims that are made, but because readers expect technical communication to be accurate, writers and editors have a particular responsibility for accuracy. Misrepresentations deceive readers and may cause them to make mistakes in procedures, interpretations, or decisions.

Intentional misrepresentations, such as faulty data in research or false claims about ingredients or features, are illegal, but misrepresentation is also an ethical issue because there is no certainty that instances of misrepresentation will be identified and penalized by the law.

Misrepresentation may be unintentional, as in a graph in which the segments are not mathematically defined or perspective that inadvertently emphasizes one piece of information. Editors have an ethical responsibility to understand the structure and style of visuals as well as words so that they can correct unintentional misrepresentation.

Misrepresentation includes omission as well as inaccuracy. Omissions are especially significant when there is danger of harm. Risks of purchasing or using a product should be documented. The editor may have to ask the writer directly what the risks are and not just wait to be told.

Environmental Ethics

Environmental ethics, like all other ethics, are based on the principle of responsibility for the social good as balanced against short-term prosperity. As consumers of paper, electricity, and inks, technical publications groups use products that harm the environment and, by extension, human health. The use of these products is ethical according to cultural values, but misuse is not because the danger to health and safety increases disproportionately to the benefits of use. Corporate policies can encourage product choices that are environmentally friendly. Policies as well as individual practice can minimize waste as from excessive photocopying and printing.

The paper and publishing industries spill a higher percentage of industrial toxins into the air and water than any other industry except for the chemicals and metals industries. Many of those toxins are dioxins used to bleach paper. Dioxins cause cancers and birth defects. Recycling paper and using recycled paper cuts down on waste because the manufacture of recycled paper is a less toxic process than manufacture of new paper, and it also consumes fewer trees. Uncoated papers are more readily recycled than coated papers, and white paper is more readily recycled than colored. White, uncoated paper is thus preferable except in special circumstances. Processes that minimize multiple draft copies should be encouraged.

Other choices of materials and processes can also minimize environmental and health consequences. Inks made from soybean oil are made from a replenishable agricultural product rather than from petroleum, and UV-curable inks eliminate solvents in manufacturing. Conservative use of electricity and plastic is environmentally as well as economically responsible. Electricity used to run computers, printers, and photocopiers may be generated by fossil fuels, which contribute to global warming. Plastic wrappers for manuals and other packaging will probably become landfill and may be superfluous. Thus, choices about papers, inks, and other consumables affect not just whether readers get the information they need but also the general health and welfare. As with all ethical issues, responsible corporations and individuals look beyond immediate needs and convenience to longterm and broad consequences of choices.

Bases for Ethical Decisions

Ethical decisions can be based on principles (such as honesty), law, professional standards, and assessment of consequences.

In a survey of technical communicators, Sam Dragga identified eight types of explanations for decisions about the ethics of a communication decision (see Further Reading). The most common reason was *consequences:* what happens to readers as a result of that choice. The extent to which a choice is unethical is proportional to the degree of deception or injury to the reader.

Other explanations are document *specifications* and *common practice.* Specifications refer to document descriptions, such as number of pages. Common practice establishes some expectations. If it is common practice in an employee evaluation to emphasize achievements and deemphasize limitations, then the potential deception of doing so seems less grave than misrepresentation of data in a research report. Readers may not be deceived by emphasis if practice leads them to expect it.

Technical communicators generally rejected the explanation that the burden for accurate interpretation rests on readers rather than writers.

If an important basis for ethical choice is consequences, then editors and writers must consider consequences broadly—longterm as well as immediate consequences; to secondary as well as primary readers; to the environment as well as to the corporation; and so forth.

Establishing Policies for Legal and Ethical Conduct

An editor or writer working for a company is only one person among many who are responsible for products and their documentation. Although any professional person should accept personal responsibility for the safety and welfare of others, ethical behavior in a corporate setting is also a social and cultural issue. To increase chances of the organization and its employees following high legal and ethical standards, and to minimize the potential for conflict of an individual employee against the organization as a whole, the company should have policies defining what constitutes ethical communication, procedures to include training of staff in ethical issues, and document review for adherence to the policies.

A first step for an employee is to review existing policy statements and the house style manual. If potentially useful sections are missing, or if some require revision, a proposal to research and revise the manuals could be developed. Such revisions should be the responsibility of a group of people representing different components of the organization, including technical communicators, engineers, managers, and lawyers. Developing or revising these policies as a group helps to create the sense of group responsibility for adherence to high standards of legal and ethical behavior.

A style and publications procedures manual might cover these policies and procedures:

- identification of trademarks
- registering copyrighted materials
- requesting permission to use material copyrighted by others and forms or letter templates for doing so
- visual and verbal identification of warnings and hazards
- accurate representation of verbal and visual information
- ethical use of resources and procedures to minimize waste
- review of ethical and legal issues in each document. The review should be the responsibility of various employees, including but not limited to the editor.

The employee policy manual should include commitments to

- legal and ethical conduct by the organization and its employees with examples specifically related to publications, including commitment to accurate representation of a product's value and limitations and instructions and warnings that accommodate the needs of users
- organizational as well as individual responsibility for legal and ethical conduct

Such statements affirm corporate responsibility and encourage individuals to act ethically by creating an environment in which ethical actions are expected and respected. They provide guidance for making ethical decisions on a case-by-case basis. They may prevent problems from occurring and support editors if problems do occur.

Summary

Law and ethics balance the rights of individuals in a society with the needs of the society as a whole. They aim to encourage achievement while protecting individuals or groups from harm. Different countries interpret these rights differently depending on their values. Intellectual property, product liability, and misrepresentation laws apply to technical communication. Because technical communication aims to influence the people who use documents, it is never ethically neutral. Part of the definition of what makes communication "good" is the contextual one of ethics: whether a document accomplishes its purpose of giving information or directions without harming or unduly jeopardizing people directly or indirectly. A good publications department conducts itself according to high legal and ethical standards not just because the law requires it to but also because professional responsibility and cultural values require concern for the welfare of others. Comprehensive editing requires a procedure for review of legal and ethical standards before, during, and at the conclusion of document development.

Further Reading

R. J. Brockmann and F. Rook, eds. 1989. *Technical Communication and Ethics.* Arlington, VA: Society for Technical Communication.

"Copyright Basics" (Circular 1) and "Copyright Notice" (Circular 3). Copyright Office, Publications Section, LM-455, Library of Congress, Washington, DC 20559. Also available at http://lcweb. loc.gov/copyright

Sam Dragga. 1996. "Is It Ethical? A Survey of Opinion on Principles and Practices of Document Design." *Technical Communication* 43.3: 255–265.

Richard L. Johannesen. 1996. *Ethics in Human Communication.* 4th ed. Prospect Heights, IL: Waveland.

Larry Lessig, David Post, and Eugene Volokh. 1996. "Cyberspace Law for Non-Lawyers" (Alan Lewine, ed.). http://www.ssrn.com/cyberlaw/

Jerry J. Phillips. 1993. *Products Liability in a Nutshell,* 4th ed. St., Paul: West Publishing Company. See especially "Warnings, Instructions, and Misrepresentations," pp. 210–237.

"Product Liability," 1984. *The Guide to American Law: Everyone's Legal Encyclopedia, 8.* West Publishing Company, St. Paul, pp. 318–324.

William S. Strong. 1995. *The Copyright Book: A Practical Guide.* 4th ed. rev. Cambridge, MA: MIT Press.

Christopher Velotta. 1987. "Safety Labels: What to Put in Them, How to Write Them, and Where to Place Them." *IEEE Transactions on Professional Communication.* PC 30.3: 121–126.

Westinghouse Electric Corporation. 1985. *Danger, Warning, Caution: Product Safety Label Handbook.* Westinghouse Printing Division, Trafford, PA.

Discussion and Application

1. For this activity, work in a discussion group with two or three classmates. Or, if your instructor directs, write out a response to one or more of the situations.

    For each of the situations that follow, discuss:

    - whether and on what grounds the situations might have legal or ethical dimensions
    - policies that might minimize conflict between individuals or groups within the organization regarding the document options or choices
    - options for action

    a. You are writing online help for the operating system of a new computer. As with any system, failure to follow certain instructions can result in loss of information. You try to call the user's attention to these places by writing WARNING and highlighting the word with color. Your manager objects to the emphasis with the argument that too many warnings will make the product seem deficient. The legal department confirms that there is no breach of law if the warnings are less frequent or less emphatic.

    b. U.S. law requires the manufacturers of cigarettes and beer to include a warning on each package or container about the hazards to health of using the product. Some companies comply with the law by printing the warning in all capital letters using condensed type. These typographic choices make the warning hard to read.

c. User manuals for electronic equipment frequently include a page or two of warnings at the beginning of the manual. There may be as many as 30 warnings about such hazards as shock from using the product in the bathtub, heat damage to the equipment from improper ventilation, and dangers of frayed cords. Using principles of organization from Chapter 16 and theories of reading from Chapter 2, discuss the effectiveness and ethical significance of grouping multiple warnings at the front of the manual. What would be the legal and managerial reasons for doing so?

d. A booklet for prospective investors represents the increasing value of an investment over time with a line graph that shows a sharp increase in value. Narrow segments on the horizontal axis and shading of the area beneath the line exaggerate the steepness of the line. Although the booklet is a marketing tool, its understated (not slick) style and factual information suggest the goal of informing rather than persuading. Does its marketing purpose make the exaggeration ethical?

2. Check user manuals, appliances, and power tools that you own to see how manufacturers have met their duty to instruct consumers in safe use and to warn them of dangers. Evaluate whether the instructions and warnings might be improved. Prepare a brief report of your findings.

3. Technical communication emphasizes "audience adaptation"—adjusting content, style, and design of documents to the needs, backgrounds, and uses of readers. An extension of this principle might be to give clients what they ask for. Discuss whether there are situations in which audience adaptation might go so far as to breach ethics in technical communication.

Chapter 22

Type and Production

As the link between the development and the publication of documents, editors know something about type and production. Type refers to the shape and size of the letters on the page. Production refers to the creation of camera ready originals, printing, and binding in print or the creation of hypertext ready for distribution through a network or on CD-ROM. A production editor may supervise the design, printing, and binding after development editors and copyeditors have prepared the text. However, all editors should know what happens to the document once the text is established, because choices about production influence early editorial decisions.

Knowledge of type facilitates good decisions about graphic design, whether editors work with graphic designers or do the work themselves. Collaboration of editor and graphic designer is desirable. Although graphic designers know principles of typography and layout, they do not necessarily know the content, purpose, and readers of a given document—information that will affect design decisions. Editors generally want the chance to approve design decisions for their suitability to the document content, purpose, and readers, but they need to understand design principles in order to make informed decisions.

Knowledge of the steps in production also facilitates good management. An editor in charge of production schedules editing tasks with the production schedule in mind. All scheduling is ultimately controlled by the intended document distribution date. Thus, editors will need estimates from printers about the time required for typesetting and printing. Editors also negotiate bids for printing and binding, to do this effectively they first must know the options for publication.

Desktop publishing increases the responsibilities of editors to know about type and production. Editors may assume some graphic design responsibilities rather than deferring to specialists.

This chapter provides an overview of type and production in order to familiarize editors with concepts and terms. Its purpose is to enable you, as editor, to communicate with graphic designers and printers, not to teach you how to prepare expert designs.

Type Fundamentals

Letters within the English alphabet can be recognized in a variety of shapes and sizes. The space between letters, between words, and between lines of type can also vary. Professional typesetting and desktop publishing provide multiple options for type and spacing.

Typography matters for practical as well as for aesthetic reasons. Typefaces and spacing affect the ease with which readers can read. Typefaces also express different qualities of character, such as avant garde, elegant, quiet, or masculine, that shape the way readers respond to a document. Using type well is an art, but the principles underlying the uses of type are also based on scientific research.

Typefaces

The shapes of characters identify particular designs of types. Of the hundreds of existing designs or faces, all may be classified according to whether they are serif style or sans serif. Serifs are cross strokes at the ends of the main strokes of letters. You can see examples in both the lower case and capital letters in the words "Times Roman" below. Both vertical and horizontal cross strokes appear in the capital letters. A sans serif style omits the cross strokes. The strokes for the capital letter H in "Helvetica" simply end.

Times Roman	Helvetica
TIMES ROMAN	HELVETICA

The serif styles are traditional, while sans serif styles are more recent designs. Some designers favor sans serif styles for their clean, modern look, and the style has been particularly popular in science and technology publications, advertising, and print for computer screens. However, research suggests that the serif styles may be easier to read on paper. One theory is that the horizontal lines of the serifs propel the eye across lines of type. Another theory is that the serif styles are more familiar to North American readers. Many designers prefer a serif style for long blocks of text in books and manuals.

The examples below show a few of the many typefaces available, with serif styles on the left and sans serif on the right.

Bookman	Avant Garde
Palatino	**Brandish**
New Century Schoolbook	Optima

You can see how Brandish could be difficult to read in paragraphs, but it can be attractive for display type such as headings and titles. Unless you are an expert in design, it is smart to stick with the proven favorites, such as Times Roman, Baskerville,

Bookman, Palatino, Garamond, New Century Schoolbook, and Helvetica. For computer display Arial is a good choice. Amateur designers should resist mixing typefaces in a document because a mixture can create a cluttered look.

Your choice of typeface will also depend on the faces available from the typesetter or on the laser printer. Professional typesetters can show you a type specimen book illustrating the available faces and sizes. A complete set of characters in a typeface in one size is called a *font* (though "font" in word processing programs means "typeface"). A printer may have available the 10-point Times Roman font but lack the 24-point font in the same typeface. You need to check the availability of all fonts you will need in the typeface you have chosen.

Type Size

Printers use the pica as a standard measure. The replication of a pica stick below compares picas and inches. It shows that six picas equal approximately an inch. Type is measured in points, with a point equal to one-twelfth of a pica. Thus, 12-point type will fill slightly less than a pica of vertical space, and six lines of 12-point type will fit in approximately one inch of vertical space.

The measures are not exact, however, and 12-point type in one typeface will be larger than 12-point type in another typeface. In the examples below different typefaces in the same size fill different amounts of vertical and horizontal space.

This line is set in 12-point Times Roman.

This line is set in 12-point Helvetica.

This line is set in 12-point New Century Schoolbook.

This line is set in 12-point Bookman.

The parts of the letters have different proportions in the typefaces. For example, Helvetica letters have a proportionately large x-height, the part of the letter equivalent in height to the lowercase *x*. Thus, the ascenders (the projections above the body of the letter, as on the b and h) and the descenders (the projections below the body of the letter, as on the p and y) are proportionately shorter than in Schoolbook.

The capital letter M is shown below in various sizes in the typeface Times Roman.

ᴹ ᴍ ᴍ ᴹ M M M M M M M M M

6 8 9 10 12 14 18 24 30 36 48 60 72

Type size affects readability, but it also affects document length. The choice of typeface and type size can make a difference of a number of pages in a long document such as a manual or book. It can determine whether a story fits into one column in a newsletter or spills into a second column. According to a type specimen book, 10-point Times Roman will fit 2.6 characters per pica whereas 10-point Helvetica will fit 2.4 characters per pica. A page that is 42 picas deep, with each line being 25 picas long, has 1,050 picas of space available (42 × 25). In Times, 2,730 characters would fit on the page (1,050 × 2.6), but in Helvetica, 2,620 characters would fit (about 20 fewer words). The numbers may not sound so different for one page, but a book of about 180,000 words (such as this book) would require about 20 more pages set in Helvetica than in Times.

Most blocks of text—the body copy as opposed to the headings or display copy—are set in 10-point or 12-point type. Type smaller than 9-point is difficult to read and is reserved for footnotes, indexes, and other material that readers will read selectively and in small quantities. Type 14 points or larger is reserved for children's books, large-print books for visually impaired readers, and display copy.

Headings may be set in the same size as the body copy if they are set in boldface. Boldface alone will let readers distinguish visually the headings from the body. For headings of different levels, you may distinguish the levels with space alone. For example, a level-one heading could be centered but in the same size as a level-two heading set on the left margin.

If size is the only variable in the headings, you will have to skip at least two sizes between levels if readers are to recognize the levels. Thus, if a level-one heading is left justified in 14-point type, a level-two heading can be set in 10-point type.

Leading, Letterspacing, Wordspacing, and Line Length

Various options for the treatment of individual letters, words, and lines of type—including leading, word- and letterspacing, and line length—are available.

Leading

Type would not be readable if the descenders in one line crashed into the ascenders of the line below. Thus, good typesetting includes leading (pronounced *ledding*) between the lines. The term derives from the days when type was set from metal molds. Strips of metal were placed between the lines of characters. Even if the type is set solid (without leading), there will be a tiny bit of space between the lines. Most type is set with one or two points of extra leading. Thus, 10-point type may be set on a 12-point line. The expression "set 10 on 12" and the marginal notation "10/12" both direct the typesetter to use 10-point type on a 12-point line.

The following examples show type set with the right amount of leading, with too little leading (solid), and with too much leading. The leading affects the ease of reading. The first paragraph is set in 10-point Helvetica on a 12-point line. The extra two points of leading are especially important for sans serif type in order to help readers distinguish lines and read comfortably.

Right Amount of Leading

Leading, the space between lines of type, helps the eye move horizontally. Too little leading increases reading difficulty because the type seems crowded and the letters cannot be distinguished. Too much leading makes the text look childish. The right amount of leading is comfortable to read and helps the eye find the correct line when it moves from the right back to the left margin.

The next paragraph is set in 10-point Helvetica on a 10-point line (no leading). The crowding increases reading difficulty and the chance that readers will skip lines when they look for the next line at the left margin.

Too Little Leading

Leading, the space between lines of type, helps the eye move horizontally. Too little leading increases reading difficulty because the type seems crowded and the letters cannot be distinguished. Too much leading makes the text look childish. The right amount of leading is comfortable to read and helps the eye find the correct line when it moves from the right back to the left margin.

The final paragraph is set in 10-point Helvetica on a 18-point line. The extra space, besides being wasteful and looking unprofessional, could interfere with comprehension in a long document by spatially separating related ideas.

Too Much Leading

Leading, the space between lines of type, helps the eye move horizontally.

Too little leading increases reading difficulty because the type seems

crowded and the letters cannot be distinguished. Too much leading makes

the text look childish. The right amount of leading is comfortable to read

and helps the eye find the correct line when it moves from the right back to

the left margin.

One or two points of leading are typical for body copy, assuming 10- or 12-point type. The longer the line, the more leading is needed (to help the eye move horizontally). Sans serif type generally requires more leading than serif type, and larger type sizes require more leading than smaller type sizes.

Line depth is always measured baseline to baseline. The baseline is formed by the bottom of the letters excluding descenders. The T-square feature of a pica stick makes accurate measurement easy.

Letterspacing and Wordspacing

The horizontal spacing as well as the vertical spacing affects readability. Typesetting uses proportional spacing, meaning that each letter receives space proportionate to its width. By contrast, most typewriters and older computer printers give each letter the same width so that an *i* takes as much space as an *m*. Proportional spacing makes type easier to read.

Letterspacing refers to how close together the letters are set; word spacing refers to the amount of space between words. As with leading, too much or too little letter- (or word-) spacing interferes with reading. The following examples show letterspacing that is just right, too tight, and too open. All three paragraphs are set in the same size (10/12-point) and face (Times Roman) of type.

Correct Letterspacing

Correct letterspacing helps readers recognize words because they can easily identify the shapes of the words. It also affects the number of times the eye stops in its move across a line of type. Too much letterspacing requires the eyes to make more stops than desirable. The eye has to make a wider sweep for fewer words. The crowding of condensed type also increases reading difficulty.

Too Tight Letterspacing

Correct letterspacing helps readers recognize words because they can easily identify the shapes of the words. It also affects the number of times the eye stops in its move across a line of type. Too much letterspacing requires the eyes to make more stops than desirable. The eye has to make a wider sweep for fewer words. The crowding of condensed type also increases reading difficulty.

Too Open Letterspacing

Correct letterspacing helps readers recognize words because they can easily identify the shapes of the words. It also affects the number of times the eye stops in its move across a line of type. Too much letterspacing requires the eyes to make more stops than desirable. The eye has to make a wider sweep for fewer words. The crowding of condensed type also increases reading difficulty.

You can trust a reputable typesetter to letterspace correctly. If you are using desktop publishing, use default spacing rather than playing with the computer's capacity for expanding or condensing type.

Line Length and Margins

The amount of space between words and, to some extent, between letters is partly determined by line length and justification of margins. The left margin is almost always straight in body copy; that is, the beginning letters of each line align vertically, so the text is left justified. If the lines of type align on the right margin, the text is right justified (or right aligned). If the lines align on the left but not on the right, the text is set ragged right.

Left-Justified Text

In 1993, industries in 22 categories released 2,791.4 million pounds of toxic chemicals into the air and water. The paper, printing, and publishing industries followed the chemicals and metals industries in releasing the most.

Right-Justified Text

In 1993, industries in 22 categories released 2,791.4 million pounds of toxic chemicals into the air and water. The paper, printing, and publishing industries followed the chemicals and metals industries in releasing the most.

Centered Text

In 1993, industries in 22 categories released 2,791.4 million pounds of toxic chemicals into the air and water. The paper, printing, and publishing industries followed the chemicals and metals industries in releasing the most.

Left- and Right-Justified Text

In 1993, industries in 22 categories released 2,791.4 million pounds of toxic chemicals into the air and water. The paper, printing, and publishing industries followed the chemicals and metals industries in releasing the most.

To achieve both left and right justification, extra space must be inserted between words and letters. High-quality typesetting or page layout software can achieve the spacing in such a way that the eye often won't recognize the extra space, but even with good tools, the wordspacing may be obviously exaggerated. This calls a reader's attention to the type rather than to the content. The likelihood of exaggerated spacing is greater with short lines, such as newspaper columns. Many designers, therefore, prefer ragged right text. Perhaps the page as a whole looks a little less tidy (or, from another point of view, less rigid). The choice of ragged right, however, acknowledges the goal of readable type; in this case, the reader's ease in moving across lines of type supersedes the desire for a straight right margin.

Optimal line length is relative to type size. As a general guide, lines include 1½ to 2½ alphabets, or 39 to 65 characters. The line for 12-point type will thus be longer than the line for 10-point type. Lines produced with this guideline will include nine to ten words. Another guideline is to make line length 2½ times the type size—25 picas for 10-point type. If the line length is short and words are long, end-of-line hyphenation may be necessary to prevent excessively short lines.

Paper

Paper, like typeface and layout, affects readability. Paper may be too transparent and allow shadows of the print from the reverse side to show. Paper may not accept ink well and the type may smudge, or it may be too absorbent and the type may bleed. It may have a shiny coating that glares. If you are responsible for selecting paper, you will consider size, weight, opacity, and finish as well as color. Post-consumer recycled content and potential to be recycled are also criteria for choice.

Except for quick print (photocopying) jobs in small quantities, most printing is done on large sheets of paper that are later trimmed to the proper size or on large rolls of paper. The standard size of "book" paper in the United States is 25 × 38 inches. (Sizes vary slightly in countries using the metric system.) As the diagram in Figure 22.1 shows, a sheet of paper this size will yield eight 8½ × 11 folios (a page that can be printed front and back) or sixteen 6 × 9 folios. The broken lines show additional cuts for the 16 folios. Extra space is available for trimming.

The standard size for "bond" paper is 17 × 22 inches. This paper yields four 8½ × 11 folios. "Cover" paper measures 20 × 26 inches.

Standard paper sizes determine standard page sizes. It is possible to print a book in a size other than 6 × 9, 7 × 9¼, or 8½ × 11, but the per page cost will increase because of the waste of paper. Standard paper sizes also determine desirable book length, generally a multiple of 16 pages. Each group of pages printed and bound together is called a *signature*. The most common signature size is 8-page or 16-page. Thus, a book of 97 pages rather than 96 will require a whole extra sheet of book paper for one page of text. Editing to condense slightly will be cost effective.

The basis weight of the paper is derived from the weight of a ream of the paper in its uncut size. Thus, a ream of book paper may weigh 60 pounds while a ream

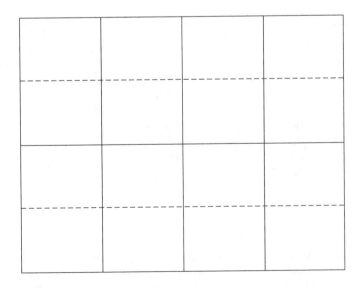

FIGURE 22.1   Book Paper, 25 × 38 Inches, Showing Cuts for Pages

of the same weight bond paper weighs 24 pounds. The following table shows equivalent weights of different kinds of paper.

Book Paper	Cover Paper	Bond Paper
basis: 25" × 38"	basis: 20" × 26"	basis: 17" × 22"
45 pounds	25 pounds	18 pounds
50	27	20
60	35	24
70	40	28
80	45	32

For most books and textbooks, 50- or 60-pound book paper is generally satisfactory. Classier publications may use 70- or 80-pound book paper.

Opacity and finish, as well as weight, affect the character of the document, success of the printing, and the extent to which the paper supports or interferes with the document's use. Opacity refers to how easy it is to see the ink through the page. Shadows from the reverse of the page distract readers. Paper weight is just one factor in opacity. Finish, or treatment to the surface of the paper, affects how well the paper accepts ink and whether the paper seems elegant or businesslike. Finishes, arranged in order of increasing smoothness, are known by the names antique, eggshell, vellum, and machine finish. Coating makes the paper smoother. Coated paper also reproduces photographs and colored inks better than does uncoated paper. The raised patterns seen on some fine stationeries are created by embossing. The choice of paper should depend on opacity and finish as well as on size and weight.

The Production Process for Print Documents

Traditional publishing requires a number of procedures and handwork. Desktop publishing may use the same procedures for printing following the creation of camera ready pages. The text is typeset, made up into pages, and proofread. Illustrations may need special treatment. If the document is a book, the pages are grouped into signatures (8 or 16 pages) for printing on book paper or rolls of paper. Metal plates are made from the signatures and then affixed to the printing press. Once the sheets are printed, they are folded, trimmed, and bound. During this process, the editor will be in contact with the printer to ensure that the job is proceeding on schedule and that the typescript is being printed correctly. More detailed information on this process follows.

Typesetting and Page Makeup

Typesetting refers to the process of using professional-quality equipment for forming the characters. If the text has been transmitted on hard copy, a typesetter may key in the characters, much like typing, but also inserts codes that direct the equipment in establishing line length, justification, letter spacing, and type style. If the text has been transmitted electronically, it should not need to be rekeyboarded.

For long or complicated documents, the text may be typeset first into galleys, long sheets of text in the chosen typeface, type size, and line length. The galleys are proofread and corrected before they are broken into pages. Corrections could change the amount of type on a given page, and the pages would have to be redefined. Galleys may be skipped if the document is uncomplicated or if pages have already been formed with desktop publishing equipment.

Once the galleys have been corrected, pages are made up, either electronically or manually, with room left for illustrations. The page proofs are then checked against the proofread galleys to ensure that no copy was omitted and to confirm the accurate correction of galleys, the correct sequence of pages, the proper placement of illustrations, meaningful page breaks, and so on.

Desktop publishing bypasses typesetting and galleys. With the use of templates from page layout or word processing software, the pages look in draft stage as they will in final form.

Illustrations

Line drawings can be printed from camera ready originals just as type can be. The originals may be produced by the computer and laser printer or drawn in ink by a skilled graphic artist. Preparation of these visuals for printing, besides their creation in camera ready form, may include enlargement or reduction.

Photographs, or continuous tone art, include shades of gray as well as black and white and require special treatment before printing. A printing press cannot print shades of a given ink color—if the ink is black, only black can be printed on the page. Thus, a photograph must first be converted to a halftone. The photograph is reproduced through a screen, which converts the grays to dots of various sizes. These dots will be printed in black. Large dots densely grouped will appear dark gray when printed, while small dots widely spaced will look like light gray. A coarse screen will produce dots that the eye can recognize, but the dots produced by a fine screen will show only if magnified. A coarse screen is used when the paper quality is poor, like newsprint. Figure 22.2 illustrates a magnified halftone to show the dot composition as well as a print of an object with a 133-line screen. This is regarded as a fine screen.

The creation of a halftone adds some expense to the preparation and printing. The cost is reasonable, but when you are getting bids for your document, you will need to specify the number of photographs to be included in order to get an accurate bid. Scanners are available for desktop publishing systems to create a halftone-like version of a photograph, but the quality of halftones prepared professionally is better.

Imposition, Stripping, and Platemaking

When the page proofs have been proofread and corrected and illustrations approved, proofs called reproduction proofs or repros are created. The proofs and illustrations are then arranged into signatures of eight or sixteen pages. The pages

FIGURE 22.2   **Halftone and Magnified Section**

must be arranged so that they will be in sequence when the signature is folded and trimmed. They must be arranged so that top and bottom margins of pages in the book or other document will be the same. Figure 22.3 shows one sequence of pages for a book. Note that the pages across the top are upsidedown. The page will be folded in half lengthwise after printing so that all the pages in the bound book will face up. The process of arrangement and alignment is called *imposition.*

A photograph is made of the signature. The negatives are arranged in a form called a flat, and negatives of illustrations are taped in. This process is called *stripping.* A thin metal plate is exposed to the negatives in a processor, and the image is transferred to the plate. A proof copy may be made of the document at this point showing not just the separate pages (as in page proofs) but their sequence and alignment as well once the signature is folded. These proofs are called blueprints or bluelines because the print is pale blue. This is the final chance to check the document before printing.

Printing: Offset Lithography

Different types of printing presses make their impressions on paper in different ways, but most large-scale commercial printing is done by a process called offset lithography. "Offset" refers to the fact that the image is transferred from a metal plate onto a rubber blanket and then onto the paper. "Lithography" means writing on stone. In art, it is a way of reproducing illustrations from plates formed by greasy crayon on stone.

A basic principle of chemistry explains printing by lithography: oil and water do not mix. The ink used in printing is oil-based. The metal printing plates are treated chemically so that the plate will accept water while the type image repels water. The plates are dampened with water. Then, when the ink is applied, it adheres to the type but not to the space behind and around the characters. The metal

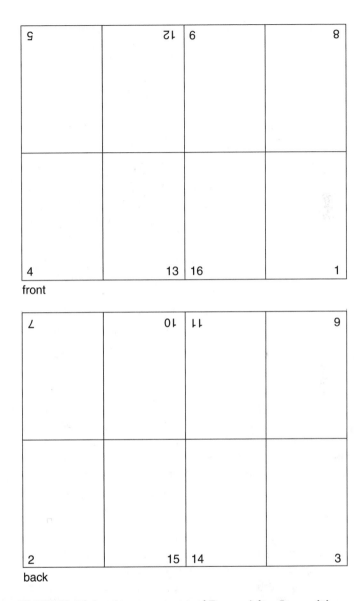

front

| 5 | 12 | 6 | 8 |

| 4 | 13 | 16 | 1 |

back

| 7 | 10 | 11 | 9 |

| 2 | 15 | 14 | 3 |

FIGURE 22.3   Arrangement of Pages After Imposition

plate is fixed to a cylinder on the printing press. It rotates while a second cylinder, covered with a rubber blanket, rotates against it in the opposite direction. A third cylinder, the impression cylinder, presses the paper against the blanket cylinder. The inked image is transferred from the metal plate onto the blanket and from the blanket onto the paper. Because the blanket is flexible, it conforms to rough surfaces on the paper. Figure 22.4 illustrates how this process works.

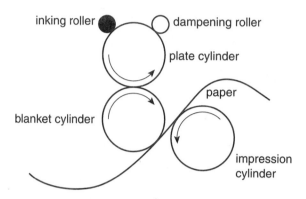

FIGURE 22.4   Offset Lithography: Schematic

When the printing is complete, the sheets are folded to create the signatures. Then, because the edges of the paper will be folded or uneven, the folded signatures are trimmed.

The production process is represented by the flowchart in Figure 22.5. The editor's responsibility during this process is proofreading—at the galley, page proof, and blueprint stages. The editor also monitors the production schedule so that the printing will be completed on time.

Color Printing

The least expensive printed documents use only one ink color. When a second color is added, the expense nearly doubles, because the paper may have to run through the press separately for each ink color. The press will have to be cleaned and set up for each new ink color, adding to the labor cost. Some presses can print more than one color at one time, but because they are more elaborate than the single-color presses, they cost more to operate.

The four-color process produces the optical illusion of all the colors that the eye can see with just four colors of ink. The four-color process works on the principle that any color can be created from the primary colors. The four colors of ink are magenta (bluish red), cyan (greenish blue), yellow, and black.

The hues of the original color photograph or painting must first be separated into the primary colors plus black. This process, called color separation, may be done by a laser scanner, or it may be done photographically. A filter on the camera lens filters out all but one hue; for example, a blue filter produces the yellow separation. A halftone image is created for each of the four ink colors.

The document is printed first with yellow ink; then it is run through the press again with magenta, cyan, and black. The black is used to correct for deficiencies in ink pigments that muddy the basic colors and to increase the overall image contrast. The press operator must be careful to align the printing images for all the colors to achieve *registration*, or the precise superimposition of one over the other. Otherwise,

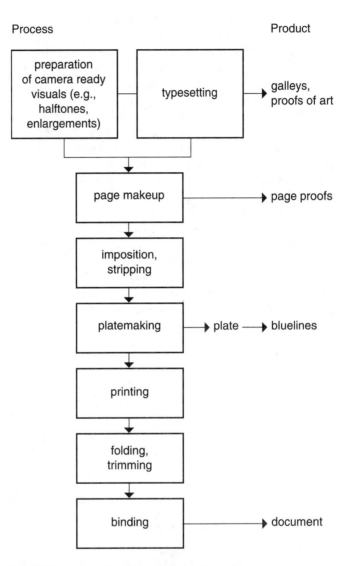

FIGURE 22.5 The Print Production Process

the final print will have streaks of yellow, magenta, and cyan rather than the desired rainbow of hues. If the printings don't align, the print is said to be out of register. The extra procedures, time, and skill required for color printing explain its cost.

Binding

Decisions about binding need to be made before pages are laid out. Binding requires extra space on the page, and different types of binding require different-sized margins. Bindings also vary in durability, and they determine how easily a book will stay open or stand on a shelf.

Manuals are typically bound in a three-ring binder or with wire spirals or plastic combs. These bindings require 5/16 inches of space along the binding edge. When you plan the page, you should allow the 5/16 inches plus the margin of space around the words.

Other types of binding are saddle stitching, side stitching, perfect binding, and library binding. Saddle stitching is appropriate for booklets. The pages are laid open, and staples are forced through the middle. A saddle stitched booklet stays open easily. Side stitching is similar to saddle stitching, but the pages are closed before staples are placed near the edge of the fold (see Figure 22.6). Side stitching is used when the bulk is too great for saddle stitching. It requires about 3/8 inches of space for binding.

Perfect binding is used for paperback books. The signatures are collected, and the spine is roughened. The cover is then affixed with an adhesive. If the adhesive dries out too much, it cracks when stressed.

The most durable binding is a library binding, a type of hardcover binding. The durability results partly from the sewing together of the signatures and from reinforcements on the spine. The rounded back allows the cover to open and close properly.

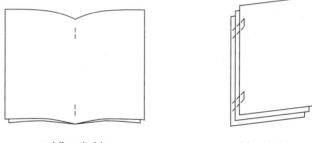

saddle stitching   side stitching

FIGURE 22.6   Common Methods of Binding

Design Tips for Beginning Designers

If you will design pages without the guidance of a professional graphic designer, you can create good pages by following a conservative approach to page design and by aiming for consistency and simplicity. One of the first steps is to use the language of typography, not of typewriters. Instead of "double spacing," you may set a heading with 12 points of space before and 3 points after. Then your headings won't float between paragraphs but will attach visually to the paragraphs that they identify. You will educate your eye to look differently at the 8½ × 11-inch piece of paper than you do as a student typing double-spaced term papers. Because the page is too wide for a readable line of type, you may create columns or side headings or use graphics on one side to create lines of the right width.

All designers, whether beginners or professionals, save time and create consistency in appearance by using a grid and by establishing templates for different document types using the styles feature of page layout and word processing programs (see Chapter 4). A grid organizes the space on the page. The simplest grid defines margins for one column of text, but even beginning designers can work with a page divided vertically in thirds. The left column becomes a place for side headings, illustrations, and comments. The two right columns may be joined to create a space twice the width of the left column. In this space, most of the text will appear. In the document template, you can embed commands for spacing, type style, face, and size as well as indentation and other features into styles and then apply all those commands simply by selecting the style.

Using characters that emulate typesetting will give your work a professional appearance—solid lines instead of two hyphens for em dashes, curly quotation marks, nonbreaking ellipses. You should probably type just one space after a period (though some companies will prefer the style of two spaces). You can discard the typewriter style of emphasis, underlining, in favor of boldface or increases in type size. You might use rules under headings or titles, but you will place them one to three points below the text so that they don't break the descenders of the type.

Indentations and gutters between columns can be narrower than you think—one em or ¼ inch will suffice. Prefer left-justification to centering. On a brochure, align tops of panels, just as you left-justify type. Leave more space at the bottom than at the top so that the text doesn't seem to fall off the page.

And, include identifying information on your documents—including a date and a way to contact the company or you if necessary.

Summary

Knowing the specialized vocabulary of graphic designers and printers enables you to communicate with them and make informed decisions about type and printing. When you are in doubt about type or have to design your own documents, choose 10- to 12-point type, in the Times Roman or Helvetica face, and set it on a line length of 2½ alphabets, or about 27 picas wide. For a book, use 50- or 60-pound paper. Schedule enough time for printing to allow for all the steps of preparation, printing, proofreading, and finishing.

Further Reading

James Craig and William Bevington. 1989. *Working With Graphic Designers.* New York: Watson-Guptill.

Judy E. Pickens. 1985. *The Copy-to-Press Handbook.* New York: Wiley.

Jan V. White. 1988. *Graphic Design for the Electronic Age: The Manual for Traditional and Desktop Publishing.* New York: Watson-Guptill.

Discussion and Application

1. Find examples of typography that encourage reading or discourage reading. Evaluate the typography in the context of the document. Do the bad examples seem to be accidental or intentional? What could be done to improve the typography? What are the positive features of the good examples?

2. Type a paragraph in Times Roman, then copy it and reformat it in two additional typefaces, including one sans serif style, such as Helvetica. Compare the amount of space each one takes. Characterize the type: Is it inviting? Quiet? Sophisticated? Masculine? Elegant?

3. To illustrate how the paper is folded on the signature, create a miniature sheet of book paper by using notebook paper. Divide it into sections, and number the sections according to the plan on page 381. Then fold the paper in half so that the two shorter sides touch and the front is outside. Fold in half again, with this fold at right angles to the first. Fold a third time at right angles to the second fold. Your pages should be numbered in sequence from 1 to 16.

4. Examine some books and other bound documents you possess. Identify the type of binding used.

Chapter 23

Management

In order to edit well, editors need to understand how texts (in print or other media) function in their contexts. They need to anticipate needs for and uses of the text by readers and users. They also need to make the right choices about organization, style, design, and visuals to enable or to persuade users. These points dominate most of the chapters in this textbook.

Although expert editorial decisions are essential to the quality of documents, they do not in themselves ensure effective documents. Quality also depends on good management and a good process of document development. Last-minute work invariably compromises quality, and unclear expectations for schedules and tasks discourage cooperative collaboration. Management must occur at both the organization and the individual levels. Editors participate in organizational management as members of project teams, planning and scheduling projects and following processes that produce the desired results. A process that includes editors from the beginning of defining the project enables developmental editing and allows time for the editing that is required. As individuals, editors must also manage their own time, often juggling several simultaneous projects, coordinating with the writers whose work they review, and meeting deadlines.

This chapter begins with a review of the impact on document quality and on editorial roles of a well-managed document development process. It then considers the key features of both organizational and individual management: planning, record keeping, estimating, scheduling, and tracking.

The Case for Managing the Document Development Process

Strategies for managing development of documents have developed because unmanaged projects compromise on quality and relegate writers and editors to peripheral positions in an organization. The unmanaged project is usually a linear one, with writing and editing occurring at the end of project development. When editors come in at the end of a project, they are often limited to basic copyediting and have little opportunity to influence structure, style, and design. Thus, their expertise is incompletely used. Writers may resent changes that are recommended when the work seems finished. The editor's role is perceived as grammar clean-up. Others who perceive editors to be peripheral may be unwilling to provide content information and support. As the last people to work on the document, editors are likely to be squeezed against impossible deadlines.

In an unmanaged project, writers also experience the squeeze of time, peripheral position in the organization, and lack of access to information. Like editors, they simply cannot accomplish all they are capable of accomplishing under such conditions, and quality suffers.

Not only do quality of documents and status of technical communicators suffer in unmanaged projects, supervisors, clients, and customers lose confidence in people who cannot deliver their products by established deadlines. If a company wishes to sell products in the European Economic Community, it must achieve ISO 9000 certification, which requires documented management procedures.

Unmanaged development is being replaced by good management that emphasizes upfront planning, creation of a team integrating all people who are involved in the parts of the project (including writers and editors), collaboration, and review. Models for management require some measurement of activities and achievements. One management model used widely in technical communication is the life-cycle model. Thinking of a project from inception to conclusion as it develops into different phases over time gives a whole project focus.

The Life-Cycle Model of Publications Development

JoAnn Hackos has developed a model for document development based on the computer software development process. She identifies five phases in this process:

1. Starting the project—information planning
2. Establishing the specifics—content specification
3. Keeping the project running—implementation
4. Managing the production phase
5. Ending the project—evaluation[1]

The planning often begins while the software to be documented is being programmed. Instead of coming in at the end of the project, the writers are integrated into development teams from the start. Hackos also classifies publications organizations according to how well they manage projects, from "ad hoc" organizations that have no standards beyond those for individual projects, to organizations with procedures that are repeated for all projects, to optimizing organizations marked by self-managed teams and close contact with users. The procedures all require planning, estimating, and scheduling.

Planning

The fundamental process of management, whatever the process or field, and whether organizational or individual, is planning. Planning means setting goals, deadlines, and schedules for meeting the goals by the deadlines. It also means setting priorities and, from the organizational perspective, assigning people to tasks. Long-term projects require identification of milestones, or points between start and finish when certain achievements should be attained. At milestones, the project team meets to review accomplishments and to revise the project, if necessary. A progress report to a supervisor, client, or external funding source may be required at intervals.

This kind of planning is best done with the team of people who will be involved in the project. Each one will bring different understanding of the tasks that need to be done. Planning for documentation will result in plans for content and for implementation. The implementation plan requires the ability to estimate and schedule based on the estimates.

Project management software can display the plan conveniently in a chart based on tasks, development periods, and dates for review and completion.

Estimating Time

If people estimate intuitively, they usually underestimate the time required for writing and editing. Even writers and editors, but certainly engineers and managers, predict that writing and editing can be completed in less time than it requires. Unless editors can estimate the time required for assigned tasks accurately and credibly, they are likely to be squeezed up against impossible deadlines and either compromise the quality of their work or work overtime to complete tasks. Time estimates help to control the expectations of superiors and train them to calculate editing time in their own product development schedules. Estimates also suggest the editor's professionalism and mastery of his or her own activities.

Estimates for document development require agreement on the type and level of editing required. Estimates also require the document specification and records from similar projects in the past. You can review sample pages from the present document, but if you have management procedures in place, you probably have records from similar projects and can use those standards. Before you can estimate how long a job will take, you need to know what responsibilities you will be expected to complete.

Classification of Editorial Tasks and Responsibilities

Good estimates begin with a clear sense of expectations. Because editing is a complex process, it does not have a single meaning. A person who thinks editing means proofreading may assume that an editor can "edit" a 200-page document in a day. Yet, if the editor understands editing to mean comprehensive editing and if the document is complex or includes numerous graphics and formulas, the editor may be able to complete only two pages in an hour or 16 pages in a day. Developmental editing that takes place as the document is being written requires meetings with the writer and others on the product team. Scheduling, supervising, and contracting for printing or media production take time beyond the editor's work with the text and are often invisible processes to people contracting for editing services. Unless you consider all the editorial tasks involved in an editing job and communicate the time required for each, you may be allotted too little time to complete a job. Or you may edit more aggressively than the writer or project supervisor wishes, creating hard feelings as well as wasting time.

An editor who manages well communicates with the writer or supervisor about expectations for editing. Time estimates follow from agreement about tasks. Table 23.1 defines editorial tasks and responsibilities and the desired outcome of each for the document and for the readers. It is the basis of the contract for editorial services in Figure 23.1.

When an editing job is assigned, you can use the classification in Table 23.1 to clarify the editing tasks that you are expected to complete, and you can estimate the required time and cost for your services on the basis of those expectations. Clients or co-workers should understand just what they will receive and understand time requirements in terms of the editing services that they request.

The copyediting and comprehensive editing tasks are typical for editors responsible for the text and visuals. The coordination, production, and desktop publishing tasks may be performed by a production editor, but in some small companies, editors assume the full range of responsibilities. Preparation of an index, product testing, and instruction are responsibilities beyond the usual expectations for editors, but editors often perform these tasks as well. Be sure to ask for clarification of such responsibilities and include them, if applicable, in your estimate.

An estimate sheet, such as the one in Figure 23.1, can prompt you to consider all the tasks that will be involved in the editing job and to clarify expectations with the person who requests editing. Many projects may be estimated informally, but it may be policy within your organization to have such an estimate signed by both the editor and the person requesting editing so that each individual will understand the expectations from the start and agree to meet them.

Although individual projects can vary in what they require, some standards have been offered in the literature on managing technical publications. Joyce Lasecke offers these formulas:[2]

Type of Topic	Guideline Hours
Step-by-step procedures	4–5 hours per procedure
Glossary terms and definitions	0.75 hours per term
Reference topics such as explanations of concepts, theories, overviews, and product information	3–4 hours per topic
Error messages (problem statement and recommended solution)	1.5–2.5 hours per message
Time to create or modify graphics other than screen captures	0.5 hours per graphic object

TABLE 23.1   Editorial Tasks and Responsibilities

Task	Responsibility (Purpose)
Basic Copyediting	
spelling, grammar, punctuation	correctness
consistency: verbal, visual, mechanical, content	consistency
match of cross-references, callouts, TOC, etc.	consistency
completeness of parts: sections, visuals, front matter, back matter, headings, etc.	completeness
accuracy of terms, numbers, quotations, etc.	accuracy
illustrations	correctness, consistency, completeness, accuracy
copymarking for graphic design	correct typesetting and page layout
Comprehensive Editing	
document planning	suitability for readers, purpose, budget
style: tone, diction, sentence structure	suitability for readers, purpose; comprehension
organization	comprehension
visual design	comprehension, usability
illustrations	comprehension, usability
cultural edit	suitability for readers according to their national and ethnic backgrounds
legal and ethical edit	comply with legal, ethical, and professional standards
Coordination, Management	
document tracking, transmission	keep production on schedule
correspondence	collaboration
scheduling editors, proofreaders, usability tests	keep production on schedule; allow time for quality control
getting permissions, applying for copyrights	meet legal requirements; protect the writer or company's rights
Production	
negotiating with typesetter, graphic designer, printer: bids, schedule	quality, economy; meet deadlines
graphic design	reader use, comprehension; document attractiveness
Desktop Publishing	
preparation and maintenance of templates	consistency of design; save time
keyboarding text, corrections	camera ready copy
preparation of illustrations	camera ready copy
HTML	
coding	consistency, readability
checking links	accuracy, ease of navigation
Indexing: preparation of an index	locate information
Product Testing	
running program, using product	usability, readability, completeness, safety
organizing and conducting usability tests	usability, readability, completeness
Instruction, Training	
in writing	writer competency
in program, product use	user competency

Client _____ Phone number _____

Address _____ Email _____ FAX _____

Document: title _____ Type _____

Date submitted _____ Date due _____

Length in pages/screens _____ Form submitted: hard copy _____

                                             disk _____

Visuals: total number _____ photos _____ tables _____ graphs _____ other _____

Other nonprose material _____

Editing required

_____	spelling, grammar, punctuation	_____	document design
_____	consistency	_____	style: tone, diction, sentence structures, globalization
_____	match of cross-references, figure numbers etc.	_____	copymarking for graphic design
_____	completeness of parts	_____	preparation of camera ready pages
_____	accuracy of terms, numbers, etc.	_____	preparation of camera ready visuals
_____	visuals	_____	HTML coding or checking
_____	organization	_____	HTML links check

Other related tasks (reference check, product testing, index, permissions requests, etc.)

_____

_____

Online editing acceptable? yes _____ no _____

Estimates of time: hours _____ working days _____

Milestone (review) dates:

Date promised:

This estimate is based on document specifications and editing tasks as shown here. It is binding only so long as the specifications and editing tasks remain constant and the document is available for editing on the date cited. Any changes will require a new estimate.

_____   _____

editor   date   client   date

FIGURE 23.1   Editing Agreement

The following more general estimates are used in technical communication as benchmarks for performance as well as for estimating large projects, but the numbers could range widely for each activity.[3] Length, technicality of the material, number of graphics, and care with which a document has been prepared by a writer will all affect the actual time it takes to complete work. As you gain experience, you will probably develop more specific guidelines comparable to Lasecke's for the type of work you do in the setting in which you work.

Writing new text	3–5 hours per page
Revising existing text	1–3 hours per page
Editing	6–8 pages per hour
Indexing (all tasks)	5 pages per hour (user guide pages, not index pages)
Production preparation	5% of all other activities
Project management	10–15% of all other activities

Thomas Duffy found that experienced editors—with a median of 11 years of experience—could edit 28.8 pages a day for comprehensive editing and 38.4 pages per day for copyediting.[4]

Record Keeping

Estimates for documents already developed should be based on records from past projects. If you know, for example, that you can copyedit ten pages in an hour of nontechnical text from a writer whose work you know, you use that figure for estimating the copyediting of an entire job. Likewise, comprehensive editing may require an hour for each two pages of typescript. Good estimating begins with good record keeping. For each job you do, you should record the job type and completion time. Soon you will have averages on which to base future estimates. Your records could be organized according to categories such as these:

Date	Job	Scope	Time
Jan. 2	proof newsletter	4 camera ready pages	30 minutes
Jan. 2–6	copyedit proceedings	200 typescript pages	25 hours
Jan. 9–13	copyedit technical manual	100 typescript pages	23 hours
Feb. 4–5	edit proposal: comprehensive	32 typescript pages	9 hours

You can begin record keeping even as a student. When you are assigned an editing project, estimate the time it will take. Then keep records and compare the estimate with the reality. Don't base the estimates just on the length of the document. Consider the nature of the document (visuals, formulas, and complex structure will all increase time). Also consider its condition (errors increase time). These records will give you rough bases for future estimates.

You should also see some increase in productivity over time. For example, if comprehensive editing requires an hour per page on your first project, your third or fourth project may require only half an hour for a comparable page.

Your records can also demonstrate productivity. Your department may be required to assess your achievements periodically, perhaps in quantitative form. To many outsiders, the work of editing is invisible, and to quantify in terms of pages or documents distributed shows productivity in a form that makes sense to other managers. Collectively, the records of individuals within a document production group form the basis of estimates for large projects. For example, you may learn that the documentation for a particular project will take 800 hours, including 250 hours of editing. This information helps project managers estimate the entire project, including documentation, rather than expecting the documents simply to appear. The collective records may also justify the hiring of new employees.

Sampling

Averages imperfectly predict the time required for a future job because of variations in writers' levels of skill, in subject matter, and in the condition of the typescript. If the document is complete when you receive it for editing, averages should be accompanied by reviews of sample pages. Sampling is especially important if you are working with a new writer or project. To estimate based on sampling, first skim the entire document to determine the number of pages of text and the number and type of illustrations, the amount of technical material, the extent of reference material, and any other features that will help you assess the scope of the editing task. Then edit sample pages, perhaps the first few pages of two chapters and pages with technical information from two parts of the document. On the basis of the time it takes you to do this work, estimate the entire editing job.

If the document has just been scheduled for development, the project manager should provide specifications for the document. These projections should be in writing in case the expectations develop to include, say, a reference manual as well as a tutorial without a proportionate increase in time.

If you estimate costs as well as time, as in freelancing and contracting as compared with editing for a salary, you will still base the estimate on a prediction of time. You may communicate your estimate to the client in terms of cost per hour, per page, or per job. However, clients deserve to know the total estimate. If they assume you can edit a 200-page typescript in three hours, your hourly fee will not accurately predict the total cost for a job that actually requires 18 hours.

Setting Priorities

In spite of all these efforts to predict and control time, editors invariably have too little time. What do you do when you estimate 100 hours for a task but only 40 hours are available? There's no easy answer to this question. Part of the solution is to establish priorities. The manual that accompanies the $600 software package on which your company stakes its reputation deserves more of your time than the inhouse employee newsletter. If you must cut corners, cut them on the lower-priority documents.

It has been said that writers and editors never finish their work; they simply abandon it. Sometimes you will have to let go of documents knowing you have not yet done your best work on them just so that you will not sacrifice more important tasks. No document is ever perfect, and in that sense writing and editing are never complete.

Document Scheduling and Tracking

Document development is usually a long-term project controlled by a specific distribution date. Projects are typically measured in terms of weeks, months, or even years rather than in hours or days. Furthermore, document development is typically collaborative, with the achievements of one collaborator determining how quickly and well another collaborator can finish his or her tasks. Editors are particularly vulnerable to inadequate scheduling or delays because the bulk of their work often comes at the end of the development cycle. They may be expected to make up for the delays of others so that the document can be distributed according to the original schedule. A thorough and realistic schedule established at the beginning of the development cycle, with periodic due dates rather than a single due date at the end, distributes the responsibility for keeping a project on schedule on all collaborators rather than disproportionately on editors. Efficient development requires a detailed management plan at the beginning of the development cycle and periodic tracking.

Management Plans

A management plan at the beginning of the project establishes the tasks and dates for completing each task. Some projects are simply document development projects. These may include newsletters, proposals, and annual reports. Other projects, particularly documentation for products, depend on the development of the product itself as well as on the document. The management plan should be formed at the beginning of the project. For example, if the project is to develop a personal computer, the plan should make allowances for the development of both the equipment and the documentation. Editors need to participate in the planning and scheduling in order to ensure sufficient time for editing.

A management plan begins with identification of the full range of tasks. For document development, these may include conceptualization of the document with consideration of readers and purpose, research and generation of a draft, technical review, comprehensive editing, copyediting, usability testing, and production. Each of these broad steps includes substeps. If document development is part of a larger project, such as development of a new software program, the management plan includes product development tasks and due dates as well. For the most efficient management, editors will participate at the outset of project development so that their work can be scheduled in advance and also so that they can begin their work with document development.

To establish due dates, managers frequently work backwards from the necessary or desired distribution date. For example, 100 copies of an annual report may be due in Washington, DC, on September 2. The document consists of 25 individual research reports to be prepared by eight staff members. Its estimated length is 300 pages of single-spaced, 8½ × 11 camera ready pages prepared by secretaries. The researchers prepare drafts of the reports on the computer. Because the reports follow a standard format, an electronic template setting margins, typeface, and headings is available for the researchers. This template results in relatively uniform reports that need minimal formatting by secretaries. The report will be printed and bound by a quick-copy shop.

Some dates must be established by phone calls to contractors, especially, in this case, the printer. Other dates may be negotiated by consulting with the contributors. The following production schedule is based on the assumption that the typist will spend no more than six hours per day on this task and the copyeditor no more than four hours per day because of other concurrent assignments.

Task	Completion Time	Beginning Date
Mail date		August 23
Printing and binding	2 days	August 20
Typing, proofreading	@ 2 hours × 25 = 50 hours	July 18
Copyediting	@ 4 hours × 25 = 100 hours	July 16

Copyediting must begin at least 25 workdays before the report is delivered for printing and binding. If concurrent projects place other demands on the copyeditor, the beginning date may be pushed backwards. To maintain a regular production pace, a copyeditor might request one report draft each day beginning five weeks before the report is scheduled to be delivered for printing and binding. However, document tracking will be easier with fewer deadlines; thus, the reports may be collected in groups of five each week.

The schedule of assignments needs to be distributed far enough in advance of the first deadline to make it possible for researchers to complete their first reports on their due dates. They will work on subsequent reports while the first reports are being edited and proofread. Researchers should be consulted for their preferences about due dates. If research projects are already completed, the reports may be due early in the schedule. The later dates should be reserved for projects that are still being actively pursued. Some researchers may wish to have all their reports due at one time to facilitate their own time management.

The schedule could be as simple as the following one, which lets each researcher see at a glance when reports are due. Researchers are identified here, for simplicity, by letter, and report topics by number.

Report Due Date

Researcher	July 16	July 23	July 30	Aug. 6	Aug. 13
A	1		2	3	
B	4	5		6	7
C	8		9	10	
D	11		12	13	14
E		15	16		17
F				18	19
G	20		21		22
H		23, 24, 25			

Without such planning, researchers are likely to wait until the last minute to write their reports, and the editor and secretaries will face a 150-hour task to be completed in just a few days. The job is not likely to be finished on time, its quality is certain to suffer, tempers will be strained—and the organization may lose funding from Washington.

More complex projects, especially those requiring comprehensive editing and technical review, those with illustrations or complex page layouts, and those being commercially typeset and printed, will require more complex scheduling.

Tracking

When there are multiple documents or sections of a document, or when the production is complex, involving typesetting and illustrations as well as editing, some systematic way to track the document through the production sequence will aid in management. You need to know both how the whole project is developing and how each individual project is progressing. You will probably maintain a file on each project, including correspondence, if any, and notes of meetings establishing due dates, plus the versions of the document as it moves through editing and production.

You may also wish to attach a tracking sheet to the document itself so that, when the document lands on your desk, you can tell at a glance exactly what needs to be done next. In the following example, the sheet will identify the report and the steps completed in production. As each step is completed, it is dated and initialed by the person who does the task.

Report #_____   Researcher_____

	date	initials
received		
copyedited		
writer proof		
editor's check		
final revisions		

In addition, you will maintain a file on the whole project, indicating its progress by showing how the individual projects are progressing. The simple form below assumes that the procedure is for the copyeditor to edit electronically. After she enters her corrections and notes queries, the report goes back to the writer for proofreading and emendations that the queries initiated. It is then returned to the copyeditor for another check. Finally, it goes to the secretary, who prepares the camera ready form.

report	received final	copyedit	writer prf	check	final
1	7/16	7/16	7/18	7/20	7/23
2					
3					
etc.					

The form records each action on the project and each time it changes hands. The form helps the editor to keep production moving and to locate any missing documents. A more elaborate project, with acquisition, graphics, and typesetting stages, would require a more elaborate tracking form.

Soliciting Bids

If you contract for services, such as printing, you will look for both economy and quality. To request bids, you will need to describe the intended document accurately and comprehensively. The specifications allow a fair comparison of bids. They also become the basis of a legal contract for services.

To request bids for typesetting and printing, you should provide the following information. If you will prepare camera ready copy, you can delete the data about typeface and fonts.

Quantity to be printed
Page size
Estimated number of pages
Composition: typeface, type style, fonts required
Halftones
Paper stock: cover, body
Ink colors
Binding
Cover (whether it requires artwork or typesetting)
Proofs required: galleys, page proofs, bluelines
Delivery site
Submission date
Delivery date (finished product)
Payment

In addition, the bid sheet may include policy statements about overruns in printing (whether you will be responsible for buying them), author's alterations (whether they will be charged per line or otherwise), and subletting (whether the bidder may sublet work to another printer).

Setting Policy

Editorial policies may govern documents, collectively and individually, and editorial procedures. In editing the text, editors will establish mechanical style. Specific documents will require other, more comprehensive policy decisions. For example, the editor of the proceedings of a conference will need to decide whether to publish printed papers or transcripts of the oral presentations.

Editors may also need to establish policies about the scope and quality of publication, including the use of illustrations in documents (such as whether to include illustrations, whether to allow photographs as well as line drawings, whether to make writers responsible for camera ready copy), the means of creating camera ready copy (desktop publishing, typesetting), the use of color in printing, and paper and binding. The policies may differ for different types of documents, such as those for inhouse distribution and those for mass public distribution. Policies can always be revised as new situations arise so they can be helpful without being rigid.

Policies may also establish editing procedures. For example, each publishing organization should have a policy about editing online, clarifying what changes may be made directly and what ones need prior approval. Other policies establish who reviews the drafts and when. Quality control may require a policy of two proofreadings by someone other than writer and editor at the page proof stage.

Just as style sheets increase editing efficiency, so can established publication policies increase efficiency by minimizing the amount of time required for decisions.

Computers in Project Management

In the course of corresponding with writers and production people, editors will depend on their word processors for form letters and individual letters and memos. Editors may also use their computers to create forms for document scheduling and tracking. Many of the files that used to be stored in cabinets can now be stored electronically.

Project management software may be useful in planning and tracking document development and production. Spreadsheets ease the task of maintaining financial records and calculating costs. Databases may maintain files of contributing writers, staff, and contract editors as well as files on printers and on different types of documents. Combined with graphics and word processing programs, spreadsheets and databases may yield reports for upper-level managers on department productivity, goals, and needs.

Business software developed to aid managers is as applicable to the management of document production as to any other type of management.

Summary

Good management means communicating clearly with others in establishing expectations for editing, scheduling adequate editing time, and keeping projects on schedule. As managers, editors also set policy and may negotiate the business of production. They guard the legal and ethical integrity of documents. All substantial editing projects require good management.

Notes

1. JoAnn T. Hackos. 1994. "A Model of the Publications-Development Life Cycle." *Managing Your Documentation Projects* (New York: Wiley), pp. 25–43.

2. Joyce Lasecke. November 1996. "Stop Guesstimating, Start Estimating!" *Intercom*, p. 6.

3. Lola Frederickson and Joyce Lasecke. 1994. "Planning Factors that Affect Project Cost." *41st*

Conference: 1994 Proceedings (Arlington, VA: Society for Technical Communication), p. 351.

4. Thomas M. Duffy. 1995. "Designing Tools to Aid Technical Editors: A Needs Analysis." *Technical Communication* 42.2: p. 267.

Further Reading

JoAnn T. Hackos. 1994. *Managing Your Documentation Projects.* New York: Wiley.

CBE Journal Procedures and Practices Committee. 1987. *Editorial Forms: A Guide to Journal Management.* Bethesda, MD: Council of Biology Editors.

Robert Van Buren and Mary Fran Buehler. 1991. *The Levels of Edit,* 2nd ed. Arlington, VA: Society for Technical Communication. Rpt. from Pasadena CA: Jet Propulsion Laboratory, publication 80-1.

Discussion and Application

1. You are the editor of an anthology of articles by different writers in different locations on the subject of computer documentation. The anthology will be published by a small commercial press. The press will print and bind the anthology from camera ready copy you provide. Make policy decisions on the following issues with the goal of creating a useful, high-quality anthology that also can be produced economically and efficiently. Identify the bases for decision making.

   a. How will the articles be acquired? How can you get quality articles and a broad coverage of topics? Will there be a general solicitation in journals that potential contributors read? Will some people be invited to contribute, and if so, will acceptance of their articles be offered in advance or will their articles be subject to peer review? Will only original articles be accepted, or will reprints also be accepted?

   b. How will camera ready copies be obtained? Will writers submit camera ready pages according to your specifications? Will they submit disks? Will they submit hard copy that will be rekeyboarded?

   c. Printing economy dictates 8½ × 11-inch pages. Will pages be laid out in two columns or in one column with wide margins? Two columns would probably be more readable and would allow more words per page, but they would be harder to produce because of controlling for column breaks as well as page breaks and because multiple writers may be doing the initial keyboarding.

   d. How consistent must the individual chapters be? If one includes a list of references for further reading, must all of them?

   e. How and where will contributor biographies be printed: at the beginning of the anthology or with each chapter?

    f. What kinds of illustrations and how many per chapter will be allowed? What will be the responsibilities of authors for providing camera ready copy?

2. For the same situation, make management decisions on the following issues with the goal of getting the anthology ready for production by the date the press has specified.

    a. When will chapters be due? Will they be due on the same date, or will you establish variable dates? If the dates vary, which ones should come first?

    b. In what order will you edit the chapters? Will you edit them in sequence, from 1 to 12, or might there be reasons to proceed in a different order? What factors will determine the order?

    c. On the basis of policies you have established for the book, what information will you have to communicate to the contributing writers about preparation of typescripts? Consider, for example, documentation style and length. Make a list of points to cover in document specifications.

    d. Suppose one contributing writer has a due date of the first of the month, and two weeks later you still have not received the chapter. What can you do?

    e. Suppose one contributing writer has submitted an improperly coded electronic file. It cannot be converted to your word processing system without extensive intervention in the text. Should you ask one of your own secretaries to do the work (which means removing that secretary from another important project and also increasing the costs of producing the anthology), or should you ask the writer to revise and resubmit the file? Your answer may not be the same for all writers involved. What will determine your choice?

    f. Outline a form for tracking the progress of the anthology overall, including acquisition, acknowledgment of the manuscript, acceptance or rejection, and other production stages.

Glossary

abstract Summary or description of document contents; part of the body of a report or book.

acknowledgments Credits to persons who have helped with the development of a publication; usually the concluding part of a preface.

adjective Part of speech whose function is to modify a noun.

adverb Part of speech whose function is to modify a verb, adjective, or other adverb.

appositive Noun or noun phrase placed with another as equivalents; e.g., Joe, *the new writer,…*

ascender Part of the letter that rises above the *x*-height, such as the top part of the letters *b, d,* and *h.*

back matter Parts of a book following the body, including the appendix, glossary, and index.

baseline Imaginary horizontal line at the base of capital letters of type. Used to measure space between lines.

basis weight Weight of a ream of paper in its uncut size; e.g., a ream of 20-pound bond paper weighs 20 pounds in its uncut size of 17 × 22 inches.

blueline or **blueprint** In offset lithography, a proof following the making of metal printing plates. Shows page sequence and folding, not just the individual pages. Printed on light-sensitive paper with blue characters.

body type Type for the text of a document as opposed to its headings and titles.

boilerplate Text that is standard for various documents and can be inserted, with minimal or no revision, in a new document.

browser Software that enables searching the World Wide Web; person who searches the Web or print pursuing particular interests rather than following the structure of the document.

bullet Heavy dot (•) used to mark items in a list. Sometimes set as a square or unfilled circle.

callout Words placed outside an illustration but referring to a part of the illustration.

camera ready   Document that may be photographed for printing without a change in type or format or special processes such as screens.

caption   Brief explanation or description of a visual printed at the bottom of the visual, beneath the title.

case   Form of a noun or pronoun that shows its relationship to other words in the sentence, whether subjective, objective, or possessive.

casting off   Process of calculating the document length according to the number of characters in a typescript and the specifications for typeface and type size, line length, and page depth of the final document. The calculation requires a character count and the number of characters per inch in a given typeface and type size.

clause   Group of words that contains a subject and verb.

> **dependent/subordinate**   Group of words with a subject and verb plus a subordinating conjunction or a relative pronoun. Cannot be punctuated alone as a sentence.
>
> **independent/main**   Group of words with a subject and verb but no subordinating conjunction or relative pronoun to make it dependent; may be punctuated as a complete sentence.
>
> **nonrestrictive**   Modifying clause beginning with a relative pronoun (*who, which*) that gives additional information about the subject but that is not essential to identify what the subject is. A comma separates this clause from the subject: "Nita Perez, *who is president this year of the users' group,...*"
>
> **restrictive**   Modifying clause beginning with a relative pronoun (*who, that*) that restricts the meaning of the subject (gives essential identifying information); no punctuation separates the modifier from the subject it modifies: "The officer *who sat at the end of the head table...*"

clip art   Predrawn pictures of common objects that can be pasted into documents. Available on paper and electronically.

cohesion   Verbal connections between sentences that create the sense that the sentences belong together thematically.

color separation   Photographic or laser process of creating the four primary printing colors from a full-color original.

complement   Words used to complete the sense of the verb. A subject complement completes a linking verb, while an **object** completes a transitive verb.

complex sentence   Sentence structured with a dependent as well as an independent clause.

compositor   Person who uses typesetting equipment to key in text for a typeset document; a typesetter. Person who assembles type and illustrations into pages.

compound sentence   Sentence composed of at least two independent clauses.

compound-complex sentence   Sentence consisting of two independent clauses plus a dependent clause.

comprehensive editing   Editing for the full range of document qualities, including content, organization, and design as well as grammar and punctuation, with the goal of making a document more usable, suitable for its purpose and readers, and comprehensible. Sometimes distinguished from basic copyediting.

conjunction   Word that joins words (nouns, verbs, modifiers), phrases, or clauses in a series.

    coordinating   Joins items of equal value, including two independent clauses: *and, but, or, for, yet, nor, so.*

    subordinating   Joins items of unequal value, especially a dependent to an independent clause. Makes a clause dependent: *although, because, since, while,* etc.

contract   Agreement for a person or company to perform work for another company. Instead of being employed by the company for which they write and edit, technical communicators may work independently on contract or work for a contract company, one that places its employees in short-term assignments for other companies.

copy   Typescript or graphics used in preparing a document for publication.

copyediting   Editing a document to ensure its correctness, consistency, accuracy, and completeness.

copyfitting   Process of fitting copy into a prescribed space. May be done mathematically, before typesetting, with the character count and document specifications from casting off, or electronically, with a page layout program.

copymarking   Placing notations on a typescript to inform the keyboard operator of corrections and directions for spacing, typeface, and type style.

copyright   Protection provided by law to authors of literary, dramatic, musical, artistic, and certain other intellectual property, both published and unpublished. The law gives the owner of copyright the exclusive right to reproduce, distribute, perform, or display the copyrighted work and to prepare derivative works.

copyright holder   Person or organization that owns the rights to the copyrighted work.

copyright page   Page in a document identifying the copyright; usually the back of the title page.

cropping   Cutting one or more edges from an illustration to remove irrelevant material and to center and emphasize the essential material.

dead copy   Version of a document during production that has been superseded by a later version. The typescript becomes dead copy once typesetting has produced a galley.

descender   The part of a letter that descends below the baseline, such as the tail on the letters *g* and *y.*

design, document   The plan for a document and all its features (content, organization, format, style, typography, paper, binding) to make it useful and readable.

design, graphic   The use of typography, space, color, paper, and binding to create a functional and attractive document.

design, visual   The visual features of the document, including columns, headings, and lists, that reflect its structural parts and enable user access to information.

desktop publishing   The preparation of final (camera ready) copy using office equipment and thus bypassing the professional typesetter; usually done with a computer, laser printer, and graphics and page layout programs that emulate the quality of typesetting. Printing and binding could then be done professionally or inhouse.

display type   The titles and headings of a document as opposed to the body copy.

document set   A group of related documents, such as all the manuals for a piece of equipment or all the manuals published by a particular organization.

dummy   A graphic designer's sketch of pages as they are to be printed, showing line length, margins, and placement of headings and illustrations.

editing, substantive   Evaluation of a document's substance—including content, organization, format, style, and use of visuals—in context of intended readers and use, and recommendations for revision. Used as a synonym for *comprehensive editing*.

em   Linear measure about equal to the width of a capital letter *M* in any given typeface and size; the square of a typesize. The normal paragraph indent in typeset copy.

en   Half an em. Used in ranges, as in ranges of numbers.

figure   Illustration that is not tabular; e.g., line drawing, photograph, bar graph, line graph.

folio   (a) Page number. (b) Leaf of a manuscript, including front and back sides.

font   Collection of characters for a typeface in one size, including roman and italic characters, capital and lowercase characters, and sometimes small caps. Any one typeface may be available in different fonts, such as Times 10 and Times 12.

footnote   Explanatory information or publication data at the bottom of an illustration or page of text.

foreword   Part of the front matter of a book; introductory remarks written by someone other than the writer or editor.

format   Placement of text and graphics on a page. Relates to the arrangement of information on the page, including the number and width of columns; the dimensions of margins, spacing, and type; also relates to whether the text is in prose paragraphs or some other form.

four-color printing   Printing of full color reproductions from four ink colors, the three primary colors plus black.

front matter   Parts of a document that precede the body; e.g., title page, table of contents, preface.

galley proof   Copy of text after typesetting but before page breaks have been established. Printed on long shiny paper. Used to check accuracy of typesetting.

gerund   Noun substitute formed from a verb plus the suffix *-ing; see* **verbal.**

GIF   Graphics Interchange Format used to create images for the World Wide Web.

globalization   Process of making a document suitable for international audiences by minimizing the number of adaptations needed for localization. Includes minimization of cultural metaphors and illustrations and use of language that can be translated readily.

glossary   Short dictionary, including definitions, of key terms used in the document.

grammar   System of rules describing the relationships of words in sentences.

graphics   (a) Text with a strong visual component consisting of more than words arranged in paragraphs; e.g., tables, line drawings, graphs. (b) Visuals with mathematical content (as would be drawn on graph paper); e.g., graphs, architectural drawings.

grid   A graphic designer's illustration of the sections of a printed page, showing margins and width of columns.

gutter   Inner margin of a book, next to the binding.

half title   Page at the beginning of a book or division that names only the main title, not the subtitle or other identifying information.

halftone   Picture with shading (different tones of light and dark) created by dots of different density. Produced by photographing the subject through a screen.

hanging indent   Paragraph or list form in which the first line protrudes farther into the left margin than subsequent lines. Also called an *outdent*.

hard copy   Document or draft printed on paper as opposed to the electronic version.

home page   The first screen of a World Wide Web site that functions like a book cover and table of contents to identify contents and structure of the site. The site may consist only of the home page if the information is limited.

house style   Style choices for mechanics preferred by a publishing organization. Derives from the designation of any publishing organization as a *house*.

HTML   Hypertext Markup Language. Code to create various type styles in hypertext publications on the World Wide Web.

hypermedia   Multimedia, such as text, sound, and video, with electronic links.

hypertext   Text with electronic connections (links) to other documents or to sections within the text.

icon   Visual representation of a process or concept; a visual symbol.

imperative mood   *See* **mood.**

imposition   The arrangement of page proofs in a form before platemaking so that they will appear in correct order when the printed sheet is folded.

index   List at the end of a book or manual of key terms used in the document and the page number(s) where they are used.

indicative mood   *See* **mood.**

infinitive   Verbal consisting of the word *to* plus the verb; e.g., *to edit*. Used primarily as a noun.

inhouse   Adverb or adjective indicating that work is completed or applies within the organization rather than without; e.g., an inhouse style manual designates style choices preferred by that organization but not necessarily by other organizations.

inflection   Change in the form of a word to show a specific meaning or a grammatical relationship to another word. The verb *edit* would be inflected as follows: *edit, edits, edited*. The adjective *good* inflected is *good, better, best*.

inspection Formal review of a document by a group that may include writers, editors, managers, subject matter experts, and legal experts. An inspection meeting brings them together to determine revisions needed.

Internet Global network of regional networks; the cables and computers that form the network. The World Wide Web uses the Internet to transmit information.

intransitive verb Verb that does not "carry over" to a complement. The predicate ends with an intransitive verb.

introduction Substantive beginning section for a document; in a book, generally the first chapter rather than part of the front matter.

italics Style of printing type with the letters slanted to the right.

iteration Version of the typescript as it moves through various editorial passes. A new iteration incorporates some editorial emendations.

justification Adjustment of lines of text to align margins; alignment.

keyboard operator Person who types copy for a new version of a document. This person may be a word processing specialist or typesetter (compositor), depending on the equipment he or she uses.

landscape orientation Position of lines of type on pages parallel with the long side of the page to create pages wider than they are tall. *See* **portrait orientation.**

layout (a) Spread and juxtaposition of printed matter. (b) Dummy or sketch for matter to be printed.

leading (pronounced *ledding*) Space between lines of type.

legacy document Existing document that must be adapted and converted for a new publication medium, new software, or new version of a product or organization.

legend Explanation of symbols, shading, or type styles used in a graph.

legibility The ease with which type or illustrations can be read. Legibility refers to recognition, while readability refers to comprehension.

line drawing Drawing created with black lines, without shading. It may be photographed or printed without a halftone.

linking verb Verb that connects a subject with a predicate adjective or predicate nominative rather than with a direct object.

list of references List of works cited in a document, with publication data.

localization Adapting a document for specific area. Includes translation but also use of cultural values of the readers' country.

manuscript Unpublished version of a document. Because the term literally suggests handwriting, it is often replaced by *typescript.*

markup Process of marking a typescript for its final form; can be done on paper or electronically.

mood Verb form indicating the writer's attitude toward the factuality of action or condition expressed.

    imperative Verb form used to express commands: "Turn on the computer."

    indicative Verb form used for factual statements: "The computer is turned on."

subjunctive   Verb form used to indicate doubt or a hypothetical situation: "If the server were turned off every night, browsers could not access our Web site."

multimedia   Information expressed in more than one medium, such as print, sound, video, CD-ROM, and graphics.

navigation   In hypertext, the process of locating desired destinations.

noise   Distracting material in a document, such as errors, excess words, or an inappropriate voice, that interferes with the reader's attention to the content.

nominalization   Noun formed from a verb root, usually by the addition of a suffix; e.g., consideration, agreement.

noun   Part of speech representing a person, place, thing, or idea.

object   Noun or noun substitute that is governed by a transitive active verb, a nonfinite verb, or a preposition. A direct object tells what or who. An indirect object tells to whom or what or for whom or what.

offset lithography   Common printing method. The design or print is photographically reproduced on a plate, which is placed on a revolving cylinder of the printing press; the print is transferred to, or *offset* on, a rubber blanket that runs over another cylinder, and from the blanket onto the paper.

online   Electronic or digital information. *Online* documentation consists of instructions that appear on the computer screen rather than in a printed manual.

page proof   Copy of typeset text that follows correction of the galley proofs and page breaks. Used to check the accuracy of corrections and the logic of page breaks.

parallel structure; parallelism   Use of the same form (e.g., noun, participle) to express related ideas in a series.

participle   Modifier formed from a verb with the addition of the suffix *-ing* or *-ed*; e.g., *dripping* pipe, *misplaced* cap.

perfect binding   Binding for paperback books in which the cover is attached to the pages with adhesive.

persona   Character or personality of the writer as projected in a document by his or her style.

phrase   Group of related words that function as a grammatical unit; does not contain both a subject and verb.

  infinitive phrase   Includes the infinitive form of a verb plus modifiers; e.g., "to write well."

  noun phrase   consists of a noun and its modifiers; e.g., "stainless steel."

  participial phrase   includes a participle plus modifiers; e.g., "diffusing quickly."

  prepositional phrase   begins with a preposition; e.g., "above the switch."

pica   Unit of linear measure used by graphic designers and printers; roughly one-sixth of an inch. Used to describe both vertical and horizontal measures.

pica stick, pica ruler   Measuring device marked with increments of both picas and inches.

plate   Light-sensitive sheet of metal upon which a photographic image can be recorded. When inked, will produce printed matter in offset lithography.

point   Unit of linear measure used by graphic designers and printers, especially in describing type size; one-twelfth of a pica.

portrait orientation   Position of lines of type on pages parallel with the short side of the page to create pages taller than they are wide. See **landscape orientation.**

predicate   Division of a sentence that tells what is said about the subject. Always includes a verb; may also include a complement of the verb and modifiers.

preface   Part of the front matter of a document stating the purposes, readers, scope, and assumptions about the document. Often includes acknowledgments as well.

preposition   Part of speech that links a noun with another part of the sentence.

printer   Person who reproduces or who supervises the reproduction in multiple copies of a document.

production   Process of developing a document from manuscript to distribution. Requires scheduling and coordination of services such as editing, graphic design, typesetting, printing, and binding.

pronoun   Part of speech that takes the position and function of a noun. May be personal (*I, we, you, they*), relative (*who, whose, which, that*), indefinite (*each, someone, all*), intensive (*myself*), reflexive (*myself, yourself, herself*), demonstrative (*this, that, these*), or interrogative (*who?*).

prose   Words in sentence form, as opposed to verse.

publisher   Person or organization that funds the publication and its distribution. Usually separate from the printer.

query   Question to the writer posed by the editor requesting information that is necessary for completing the editing correctly.

query slip   Piece of paper on which a query is written; attached to the typescript.

ragged right   Irregular right margin. Characters are not spaced to create lines of equal length.

readability   As applied to formulas, a quantifiable measure of the ease with which a text can be read. Based on counts of sentence and word features such as number of syllables and number of words per sentence. More broadly, the ease with which a reader can read and understand a document, based on content, level of technicality, organization, style, and visual design.

recto   In a book, the righthand page, numbered with an odd number. The *verso* is on the back.

redundant   Duplicate information; unnecessary repetition.

register   Alignment of printing plates one on top of the other to reproduce colored prints accurately.

relative pronoun   Pronoun that introduces a relative clause and has reference to an antecedent; *who, which, that.*

resolution   Number of dots of ink per inch (dpi) used to form characters. Low resolution (e.g., 300 dpi) can produce coarse or bumpy strokes.

RFP   Request for proposals; document that identifies a need for research, a service, or a product and invites competitive proposals to provide it.

river   White space running through a paragraph that forms a distracting diagonal or vertical line.

roman type   Type style characterized by straight vertical lines in characters rather than the slanted lines that characterize italic type.

root screen   The first page of online help. It identifies the main content divisions and establishes the layout to familiarize readers with the design of the other screens. Also called *main menu.*

running head   Title repeated at the top of each page of a book. May be the book title, chapter title, or author's name. May vary on recto and verso pages.

saddle stitching   Binding for a booklet with staples through the fold in the middle.

sans serif   Type style characterized by absence of *serifs* or short horizontal or vertical lines at the ends of the strokes in letters. Also called gothic. Common sans serif typefaces are Helvetica and Arial.

schema   (plural *schemata*) Structured representation of a concept in memory.

screen   Glass plate marked with crossing lines through which continuous-tone art is photographed for halftone reproduction.

script   Outline of visual and verbal elements of a document or module.

semantics   Study of meanings.

serif   Small horizontal or vertical line at the end of a stroke in a letter; also a category of typefaces characterized by the use of serifs, such as Times Roman and Bookman.

server   Computer and software that send electronic information, such as email and World Wide Web pages, to client computer programs requesting the information.

SGML   Standard Generalized Markup Language. Code for marking structural parts of documents to facilitate transfer among computer platforms, archiving, and databases.

side stitching   Binding in which staples are forced through the edge of the book.

signature   Group of pages in a book folded from a single sheet of paper. Typically includes 16 pages, but may include 8, 32, or even 64 pages depending on the size of the pages and the number of folds.

simple sentence   Sentence consisting of one independent clause.

solidus   Slanted line (/) used in math to show division and in prose to mean *per* or to separate lines of poetry that are run together.

specifications   Written, detailed description of a product's functions, capabilities, and physical characteristics. Establish a plan for product development and enable documentation before the product is complete.

standard American English   Widely accepted practice in North America with regard to spelling, grammar, and pronunciation; the speech and writing patterns of educated persons in America; edited American English.

storyboard   Poster-sized visual and verbal outline of the structure and contents of a document; used in document planning.

stripping   In printing, the arrangement and taping of negatives from text and illustrations in a flat before platemaking; also the cutting and pasting of a corrected version in place of an incorrect one.

style   Choices about diction and sentence structure that affect comprehension and emphasis as well as projecting a *voice* or *persona.*

style, electronic   Typography and spacing specifications for a document part such as a heading; once defined, a style enables application of all the specifications with one command.

style, mechanical   Choices about capitalization, spelling, punctuation, abbreviations, numbers, and so forth, when more than one option exists.

style, typographic   Choices about appearance of type, such as bold, italic, or condensed.

style manual   Collection identifying preferred choices on matters of mechanical style including capitalization, abbreviations, and documentation.

style sheet   List of choices for spelling, mechanics, and documentation for a specific document. Helps the editor make consistent choices throughout the document.

subjunctive mood   *See* **mood.**

syllable   Unit of a word spoken as a single uninterrupted sound. Includes a vowel or a syllabic consonant.

syntax   Structure of phrases, clauses, and sentences.

table   Text or numbers arranged in rows and columns.

table of contents   List in the front matter of a document of the major divisions, such as chapters, and the page numbers on which the divisions begin.

template   In word processing or publishing software, a collection of styles (formatting directions) that define a document. *See* **style, electronic.**

tense   Form of the verb that indicates time of the action as well as continuance or completion. Indicated by inflection

    past tense   Inflection of the verb that indicates time in the past. With regular verbs, formed with the addition of the suffix *-ed.*

    present tense   Verb form that indicates current time.

    future tense   Verb form that indicates action that will occur in the future. Usually formed with a helping verb, *shall* or *will.*

title page   Page in the front matter of a document identifying the title. May include other information, such as the name of the writer or editor, date of publication, and publisher.

tone   Sound that the voice of the writer projects—serious, angry, flippant, concerned, silly, and so forth.

type size   Height (and proportionate width) of a letter, expressed in points. For example, 12-point type will almost fill a 12-point (1-pica) line and is good for body copy, while 72-point type will almost fill a 6-pica line and is so large that its use would be restricted to banners, announcements, and book titles.

type style   Shape of letters as determined by the slant and thickness of the lines and the presence or absence of serifs. *Roman* style uses vertical lines while *italic* style uses slanted ones; *serif* style uses serifs while *gothic* or *sans serif* style does not. *Bold* or *regular weight* and *condensed* or *expanded* may also define type styles. Also denotes classes of type such as *body type,* or the body of the text, as compared with *display type,* or titles and headings.

typeface   Type design produced as a complete font and named; e.g., Times Roman, Helvetica.

typescript   Draft of a document, before it is typed in final form or typeset; the copy on which an editor works; the parallel of *manuscript*, when documents were written first in longhand.

typeset   Adjective describing text that has been prepared by photoelectronic typesetting equipment rather than by a typewriter or desktop publishing equipment. Typeset copy is generally of higher quality than copy produced by desktop publishing because the letters are more finely shaped and the options for spacing and type style are greater.

typesetter   Person whose job is to prepare typeset copy. May refer to the owner of a typesetting business or to the compositor, the person who keyboards the documents.

typesetting   Process of keying text into photoelectronic typesetting equipment in order to produce typeset galleys. Formerly done manually or mechanically, with lead characters.

usability   Ease and accuracy with which a document, such as a manual, can be used.

usability test   Test with representative users to determine whether they can readily use the document for its intended purposes, such as completing a task accurately or finding reference material.

usage   Accepted practice in the use of words and phrases.

verb   Part of speech that denotes action, occurrence, or existence. Characterized by *tense, mood,* and *voice.*
   intransitive   Verb that does not carry over to a complement.
   linking   Verb that links the subject to a subject complement.
   to be   "Is" or a variant of *is;* links the subject to a subject complement.
   transitive   Verb that requires a direct object to complete its meaning.

verbal   Verb used as a noun, adjective, or adverb. Verbals may be *participles* (modifiers formed from a verb plus the suffix *-ing* or *-ed*), *gerunds* (noun substitutes formed from a verb plus the suffix *-ing*), and *infinitives* (verbs plus *to,* used chiefly as nouns).

version control   Process of ensuring that all people who work on a document use the latest version.

verso   In a book, the page on the left side as the book lies open, numbered with an even number; the back side of a *recto* page.

voice   Form of a verb that indicates the relation between the subject and the action expressed by the verb.
   active voice   Verb form indicating that the subject performs the action expressed by the verb.
   passive voice   Verb form indicating that the subject of the sentence receives the action expressed by the verb; always identified by a *to be* verb plus a past participle.

URL   Uniform Resource Locator; the address for a World Wide Web site.

VRML   Virtual Reality Modeling Language; code that describes 3-D images and animation for display on the Web.

white space   Graphic design concept: blank space on the page that functions to draw attention to certain parts of the page, to provide eye relief, to signal a new section, or to provide aesthetic balance.

wordspacing   Amount of space between words. Manipulated in order to achieve right justification and to eliminate rivers.

World Wide Web; the Web; WWW   Multimedia, graphical international database of information stored in multiple sites that are linked by the Internet. Accessed electronically through a browser.

x-height   Size of a letter without its descender or ascender, or the equivalent to the x in the alphabet.

Index

Abbreviations, 106–109
  apostrophes in, 147–148
  consistency of, 86
  copymarking for, 35
  identifying, 107
  of Latin terms, 107–108
  of measurements, 108–109, 160
  periods with, 107
  plural of, 148, 160
  of scientific symbols, 108, 160
  spelled-out words vs., 109
  of states, 108
Abstract, 70
Accents, 87
Accessing information, 24
  via headings, 292
  in tables, 167, 276
  typographic devices for, 277
  via visual design, 279
Accuracy
  content, 67–68
  ethics and, 365
  of illustrations, 72, 170, 365
Acknowledgments, 70
Action verbs, 228–229
Active listening, 349–350
Active voice, 231–233
Adjectives, 117
  comma with series of, 144
  compound, 148–149
  verbs as, 231
Adverbs, 117
Agreement
  subject-verb, 119
  pronoun-antecedent, 125
Alignment
  of equations, 163
  of table columns, 171
  of paragraphs, 276. *See also*
    Justification
Analogy, 255

Analysis
  editorial, 199–200, 214–216,
    244–246, 265–269
  as means of understanding,
    256–257
  in project planning, 9, 269
Anti-aliasing, 76
APA Publication Manual, 87, 90,
    92, 163
Apostrophes, 147–148
  proofreading marks for, 179
  uses of, 147
Appendix, 71
Appositive phrase, 114, 118, 147
Arrangement, document. *See*
    Organization
Arrangement, sentence, 220–222
Ascender, 372
Author–editor relationship
    See Editor–writer relationship

Back matter, 71
Baseline, 375
Basic copyediting. *See* Copyediting
Basis weight, 377–378
Bids, 398
Binding, 278, 383–384
Block quotation, 151, 276
Blueline, 177, 380
Blueprint, 378, 380
BMC Software, 4–7, 11–12, 93, 94
Boilerplate, 344
Boldface
  copymarking for, 35
  for headings, 284
  highlighting with, 106
  proofreading marks for, 180
Bond paper, 377
Book, parts of, 68–71
  back matter, 71
  body, 70–71

  front matter, 69
Book paper, 377
Bottom-up editing, 201
Brackets
  copymarking for, 36
  in fractions, 161
  proofreading marks for, 179
British Imperial System, 108, 159
Browser, 291, 329–330, 334
Budgeting, 363
Bullets, 277

Callouts, on illustrations, 73–74
Camera ready copy, 68
  typing punctuation for, 152–153
  See also Desktop publishing;
    Illustrations
Capitalization, 105–106
  consistency of, 86
  copymarking for, 35
  distracting, 106
  for emphasis, 277
  principles of, 105
  proofreading marks for, 180
  in statistics, 166
  in symbols of measurement, 160
Caption, 74
Case, pronoun, 126
Categories and organization, 256,
    261, 327
Cause-effect order, 260
CD-ROM, 321, 322
Charts. *See* Illustrations
Chicago Manual of Style, 69, 90, 91, 291
Chronological order, 259–260
Chunks of information, 263
Clause
  definitions of, 114, 134
  dependent vs. independent,
    134–136
  punctuating, 138–142

Clause *(continued)*
restrictive vs. nonrestrictive, 141–142
Clip art, 314
Cohesion, 220–222, 233, 263–264
Collaboration, 7–11
See also Editor–writer conference; Editor–writer relationship; Management
Colon
in compound sentence, 139
copymarking for, 36, 41
with introductions, 163
with lists, 152
proofreading marks for, 179
quotation marks with, 151
when to use, 152
Color
and international communication, 278
as design option, 278
in hypertext, 329–330, 334
Color printing, 382–383
Color separation, 382
Columns
in page design, 273, 276, 385
in screen design, 330
in tables, 167–168
in text, 276
Comma
with adjectives in series, 143–144
in complex sentence, 140–141
in compound sentence, 139
in compound-complex sentence, 142
with coordinate adjectives, 144
copymarking for, 36, 41
with introductory and internal phrases, 146–147
proofreading marks for, 179
with quotation marks, 151
in series, 143–144
in simple sentence, 138
Comma fault, 139
Comparison-contrast order, 260
Complement, 113, 116, 120
Completeness, document, 68–71
back matter, 71
body, 70–71
of hypertext, 72, 77
of illustrations, 72, 170
preliminary pages, 69–70
Complex sentence, 140
Compositor, 34
instructions for in copyediting, 36, 38
instructions for in proofreading, 184
Compound adjectives, 144
Compound sentence, 139
Compound–complex sentence, 142
Compound nouns, 148

Compound verbs, 138
Comprehension
multimedia and, 331
organization and, 253, 254
reading and, 25–27
signals and, 25–29
style and, 212, 213–214
templates and, 255
visual design and, 278–279
Comprehensive editing
computers and, 208–209
context and, 198, 202
copyediting vs. (ex.), 194–198
definition, 193
example of, 203–206
ethics and, 206–207, 362
of illustrations, 298–307
overview of, 13–14, 193–202
plan for, 10, 199, 201, 206
process of, 198–202
analyzing readers, purposes, and uses, 199–200, 203
completing the editing, 201–202
evaluating outcome of, 202
establishing editing objectives, 201
evaluating documents, 200, 203–204
for visual design, 275–285
when to perform, 206–208
working with writer in, 201, 202
Computers
comparison function, 238–239
comprehensive editing with, 208–209
consistency and, 95–96
copymarking and, 49
creating illustrations via, 76
editing illustrations via, 314–316
management and, 389, 399
proofreading via, 184–185
sample style sheet prepared on, 96
Confidentiality, 359, 363, 364
Conjunctions
in compound sentences, 138–139
definitions, 115, 134
in series, 143
types of, 135–136
Conjunctive adverbs
definition of, 115, 139
vs. subordinating conjunctions, 139–140
Consistency, document, 82–96
content, 83, 88
copymarking and, 38
excessive, 88
in hypertext documents, 10
of illustrations, 71, 83
mechanical, 85–88
noise and, 88
of screen design, 85, 279–282, 330–331
of style, 84

of tables, 170
types of (table), 83
of typography, 84
verbal, 83–84
visual, 84–85, 279–282
of tables and figures, 170
Content
completeness of, 26, 54, 68–71, 170
complexity of, and visual design, 292
consistency in, 83, 88
linking familiar and new, 26, 221, 258–259, 263–264
of online documents, 325–326
visual design and, 291–292
See also Comprehension; Organization
Context of documents, 19–23
attributes of (table), 21
comprehensive editing and, 198
constraints as, 23
culture and, 22–23, 243–251
origins and impact, 20–21
relationship to text, 20, 23
style and, 216, 221, 236, 243
Contract company, 7
Contract for editorial services, 392
Contracting for printing, 398
Contractions, 10
Conventions
of number usage, 158–159
of organization, 257–258
of passive voice, 232–233
of visual design, 279–280, 291
See also Reader expectations
Coordinate adjectives, 144
Coordinating conjunctions, 135–136
in compound sentences, 138–139
definitions, 115, 134
in series, 143
Copyediting, 65–79
for accuracy, 67, 72
for completeness, 68–71
comprehensive editing vs. (ex.), 194–198
for consistency, 66. See also Consistency
for correctness, 66, 71, 170. See also Abbreviations; Capitalization; Grammar; Punctuation; Spelling
description of, 14
ethics in, 206–207
goals of, 65–66
hypertext, 77
of illustrations, 71–76
method of, 77–79
of nonprose text, 78
proofreading vs., 176
style manuals for, 89–95
of tables, 167–171
guidelines, 167–168
example, 167–171

Copymarking, 34–46
  ambiguous letters and symbols, 40, 165
  consistency in, 38
  descriptive versus procedural, 53
  desktop publishing and, 49
  electronic, 49–61
  in the editing process, 6, 12, 65
  equations, 162–164
  fractions, 161–162
  for graphic design, 43–44
  illustrations, 43, 168–170
  marginal notes vs. text emendations, 38
  mathematical material, 161–165
  nonprose text, 42–43
  placement of symbols in, 36–40
  for punctuation, 41–42
  symbols for (tables), 35–37
Copyright, 357–360
  international protection, 358
  notice, 326, 358
  ownership, 358
  page, 69
  registration and fee, 358
  of software, 359
Correctness, document, 66, 71, 170.
    See also Abbreviations;
      Capitalization; Grammar;
      Punctuation;
Spelling
Correspondence with writers, 352–353
Cover, 69
Cover paper, 378
Cropping, 314, 315
Cross-references
  consistency of, 83, 87
  illustrations and, 73
  parenthetical, 151
Cultural editing
  categories for (table), 250
  See also International communication
Cyberspace law, 359, 360

Dangling modifiers, 123–124
Dash, 151–152
  copymarking for, 36, 41–42
  em vs. en, 41–42
  proofreading marks for, 179
  when to use, 151–152
Databases and SGML, 54
Dates, 86
Dead copy, 177, 185
Decimal fractions, 159
Deletions
  copymarking for, 35
  proofreading marks for, 179
Delivery. *See* Visual design
Dependent clause, 134–135
  in complex sentence, 137, 140–142

  in compound-complex sentence, 137, 142
Descender, 372
Descriptive markup, vs. procedural, 53
  See also SGML
Design
  document, 12, 19, 29, 274
  graphic, 12, 43, 274
  information, 3, 11, 29, 61, 326
  of online documents, 325–332
  page (ex.), 279–281
  screen, 8, 10, 11, 83, 85, 279, 291, 329–331
  specifications, 274, 291
  tips for beginners, 384–385
  usability and, 29, 279–282
  visual, 11, 83, 273–293
  See also specific types
Design literacy, 322
Desktop publishing
  illustrations and, 314–316, 379
  markup and, 49
  page design and, 12, 49
  printing and, 378
  responsibilities of editors in, 370
  templates and, 49
  typesetting vs., 379, 384–385
Development of publications, phases of, 12, 388
Developmental editing, 193, 390. *See also* Comprehensive editing
Discriminatory language, 243–247.
    See also Nondiscriminatory language
Display copy, 284
  equations as, 162
  type size of, 373
  See also Headings
Documentation of sources, 83, 87
Document design, 12, 19
  definition and functions of, 29, 274, 278
  visual design vs., 274
  See also Design
Document genres, 16, 274
Document set, 257, 280
"Doublespeak," 234
DTD (Document Type Definition), 53
Duty of trust, 360

Editing agreement, 392
Editor
  basic functions of, 3–13, 343–347
  as manager, 344–346. *See also* Management
  in production, 12, 370
  tasks and responsibilities of (table), 391
Editorial review, 6, 10, 202
  ethics and, 367
  vs. instruction, 348
  See also Editor–writer conference

Editorial tasks and responsibilities (table), 391
Editorial trespass, 58–59
Editor–writer conference, 347–351
  effective language in, 349–351
  furniture arrangement for, 351
  organization of, 348
  purpose of, 347–348
  reviewing draft in, 348–349
Editor–writer relationship, 341–354
  collaborative vs. competitive, 7, 342–343
  in comprehensive editing, 199–202
  in conference, 347–351
  correspondence in, 352–353
  effective editing and, 343–344
  efficient managing and, 344–346
  establishing, 341–343
  professionalism and, 346–347
  strategies for effective, 343–347
Electronic copymarking, 60
Electronic transmission, 34, 60, 79
Ellipsis points, 152
Email, 353
Em dash, 41–42, 152
Emphasis
  in illustrations, 302–304
  with visual design, 279
En dash, 41–42, 152
End focus in sentences, 220–221
Environmental ethics, 365–366
Equal sign, 36
Equations, 162–164
  breaking, 163
  displaying and numbering, 162
  grammar and, 164
  punctuating, 163–164
Estimating time, 344, 389
Estimating costs, 394
Ethics in editing, 206–208, 362–368
  bases for ethical decisions, 362, 366
  consequences and, 366
  environmental ethics, 365–366
  establishing policies for, 366–367
  expertise and, 364
  misrepresentation and, 332, 365
  online documents and, 332
  policies and, 366–367
  professional codes of conduct and, 362–364
  safety of users and, 206–207, 259, 361
  style manuals and, 366–367
  See also Professionalism; Safety; Warnings

Fair use, 332, 358–360
Familiar-new sequence, 26
  in document organization, 255, 258–259, 263–264
  in paragraph organization, 221–222
Faulty predication, 119–120

Feasibility, 260, 265
Figure. *See* Illustrations
Finish, paper, 378
Flat, 380
Flow, 263. *See also* Cohesion
Flowchart, 300
    as planning tool, 324, 329
Folio, 377
Font, 372. *See also* Typeface
Footnotes
    in tables, 168
    in illustrations, 74
Forecasting statement, 27, 88
Foreword, 70
Format
    genre vs., 274
    styles and, 52
    visual design vs., 274
    See also Visual design
FOSI, 53
Four-color process, 382
FrameMaker, 5
Fractions, 161–162
Fragment, sentence, 118
Front matter, 69

Galley proof, 176, 379
Gender and language, 243–244;
    246–247. *See also*
    Nondiscriminatory language
General-to-specific order, 258–259
Genres, 16, 274
Gerund, 115, 123, 135
Globalization, 248. *See also*
    International communication
Glossary, 71
GPO Style Manual, 90
Grammar, 112–129
    checking via computer, 60
    common errors (table), 127
    equations and, 164
    guidelines for editing for, 128–129
    meaning and, 112, 118
    parts of speech, 113
    sentence patterns, 113–118
    sentence parts (table), 114
Graphic design
    copymarking for, 43–44
    editing and, 370
    tips for beginners, 384–385
    visual design vs., 274
    See also Design; Layout;
        Typography
Graphics. *See* Illustrations
Graphs, 299
Grid, 274–275, 385
Guidelines for writers, 344–345
Gutter, 76, 385

Half title page, 69
Halftone, 379, 380
Handbooks of grammar, 112
Hanging indent, 276

Hard copy
    in comprehensive editing, 208
    copymarking, 34–46
    online editing and, 209, 349
Headers. *See* Running heads
Headings, 283–285
    as access devices, 292
    in business letters, 279
    checking, 279
    frequency of, 284–285, 292
    functions of, 276
    levels of, 283–284
    as structural signal, 283–284
    in tables, 74, 168
    visual distinctions of, 284, 373
    wording of, 283
Hierarchy of information
    in hypertext, 326–327
    organization and, 28, 255, 263
    signals and, 27–29
    visual design and, 278
Home page, 324, 325, 326, 327
HTML 54–55, 322, 323
    editing and, 55
    example of, 56
    hypertext and, 49
    SGML and, 54
Hypertext, 54–58
    accuracy of, 72
    color and, 329–330, 334
    copyediting, 77, 324
    editing process for, 332–336
    flowchart and organizing of,
        324, 329
    grammar and, 331
    hierarchy of information in,
        326–327
    identifying information in,
        326, 327
    illustrations and, 331, 334
    links, 55, 72, 279, 322, 323–324
    organizing, 256, 258, 263, 324–325,
        326–329
    proofreading, 10, 11
    reading and, 323, 324, 326, 329
    structural signals to, 27
    style and, 329
    See also Online documents
Hyphenation
    of compounds, 148–150
    copymarking for, 36, 41–42
    dashes vs., 41–42
    of measures, 160
    proofreading for, 186
    proofreading marks for, 179
    trends in, 150

Icons, 295, 325
Idioms, 128
Illustrations, 72–77, 295–316
    comprehensive editing of, 298–307
    appropriateness and number of,
        301

arrangement of, 302
    emphasis and detail in, 302–304
    matching form, content, and
        purpose of, 301
    perspective size, and scale in,
        304–305
computers and, 76
copyediting of
    for accuracy, 72, 170
    for completeness, 170
    for consistency, 71, 83, 170
    for correctness, 170
    for readability, 72, 170–171
copyediting of, guidelines, 71–72
copymarking of, 43, 76
desktop publishing and, 314–316,
    379
discriminatory language in, 307
evolution of, 295
in international communication,
    296
in online documents, 76, 331, 334
organization of, 302
parts of, 72–75
    callouts, legends, captions, and
        footnotes, 73–74
    example of, 75
    labels, numbers, and titles,
        72–73
placement of, 75–76, 302
preparing for print, 312
reasons to use
    convey information, 296–297
    support text, 297
    enable action, 297–298
relationship of text to, 297
reproduction quality of, 76
schemata and, 297
taste in, 307
types of, 298
types of (table), 299–300
Image map, 323
Imperative mood, 135
Imposition, 379–380
Indentation
    of block quotes, 276
    copymarking for, 37
    proofreading marks for, 181
    uses of, 276
Independent clause
    in complex sentence, 140
    in compound sentence, 138–139
    in compound-complex sentence,
        142
Index
    as access device, 279
    definition of, 71
    as editorial responsibility, 391
Information design, 3, 11, 29, 61, 325,
    326, 333
Information management, 54
Infinitive phrase, 118
Infinitives, 123, 124

Ink
  data ink (maximizing), 307
  environment and, 365
  excess, 72
  in four-color printing, 382
  in offset lithography, 380
  paper and, 376, 378
Inline fraction, 161–162
Inspection meeting, 347
Intellectual property, 357–360
International communication
  color and, 278
  copyright for, 358
  correspondence and, 247
  cultural sensitivity and, 247
  illustrations in, 296
  SGML and, 52, 53
  spelling and, 105
  standards for markup, 52, 53
  translation and, 248
  visual design and, 276, 291
International Organization for
    Standardization
  ISO 9000 and, 172, 388
  SGML and, 52, 172
International System. *See* Metric
    system
International System of Units. *See* SI
    units
Internet, 321, 322, 330
Intranet, 7, 321
Intransitive verb, 116
Introduction, 259
ISBN, 69
ISO 9000, 172, 388
Italics
  copymarking for, 35
  for headings, 284
  highlighting with, 277
  in mathematical copy, 164–165
  proofreading marks for, 180
"It is" openers, 218

Justification
  copymarking for, 36, 37
  of headings, 284
  left vs. right, 276, 375–376

Known-new contract. *See* Familiar-
    new sequence

Labels, in illustrations, 72–73
Landscape orientation, 76, 277
Latin terms, 107–108
Layout
  consistency in, 83, 85
  in desktop publishing, 15
  visual design vs., 274
Leading, 373–374
Learning
  document structure and, 253,
    254–257
  reading for, 25, 26

schema theory of, 254–257, 259
  See also Comprehension
Legacy documents, 9, 10
Legal issues in editing, 356–362
  copyright, 357–359
  fair use, 332, 358–360
  intellectual property, 357–360
  libel, fraud, misrepresentation,
    332, 361–362, 365
  privacy, 332
  product safety and liability,
    361–362
Legend, for illustrations, 74
Legibility, 277
  of online documents, 330
Letter of transmittal, 352–353
Letterspacing, 375
Levels of edit. *See* Editorial tasks
    and responsibilities
Libel, 361–362
Library binding, 384
Life-cycle model of publications
    development, 388
Line drawing, 379
Line length
  copymarking for, 37
  justification, spacing, and, 375–376
  readability and, 276, 376
  screen design and, 330
Links (hypertext), 55, 72, 279,
    323–324
  editorial considerations regarding,
    77, 324
  graphic, 322
  See also Hypertext
Linking verbs, 116
List of contributors, 70
List of references, 70
List of tables and figures, 69
Lists
  consistency in, 10, 87, 88
  grouping related material in,
    260–263
  numbers, letters, bullets in, 276
  punctuation of, 10
  as visual design option, 276, 289
Localization, 249; *See also*
    International communication
Lowercasing
  copymarking for, 35
  proofreading marks for, 180
  See also Capitalization

Macro editing. *See* Comprehensive
    editing
Macrostructure, 255, 257
Main clause. *See* Independent clause
Management, 387–400
  basic responsibilities of, 12, 388
  classification of editorial
    responsibilities, 389–393
  collaboration and, 388
  computers and, 389, 399

contract for editorial services, 393
  of editor–writer relationship,
    344–346
  estimating time, 344, 389–393
  of information and SGML, 54
  life-cycle model and, 388
  planning, 389, 395–397
  productivity and, 394
  record keeping, 393–394
  sampling, 394
  scheduling and tracking, 344,
    395–398
  setting policy, 398–399
  setting priorities, 394–395
  soliciting bids, 12, 398
  tracking documents, 397–398
Management plan, 395–398
Margins, 273, 274, 276–277, 279, 282,
    284, 384
  justification and, 375–376
Markup. *See* Copymarking
Mathematical material
  computer-assisted typesetting of,
    161
  copyediting principles for,
    158, 161
  copymarking of, 161–165
  equations, 162–164
  fractions, 161–162
Meaning
  grammar and, 112
  illustrations and, 298
  organization and, 253, 256,
    259–260
  punctuation and, 133
  sentence structure and, 118, 121
  style and, 212, 213–214, 216–220,
    240
  See also Comprehension;
    Semantics
Measurement
  abbreviations for, 160
  systems of, 108, 159
Mechanical style, 85–88
Memory and organization, 262
Menu bars, 10, 275, 276
Metric system, 92, 108, 159
Microsoft Word, 31, 33
Microstructure, 255, 257
Milestones in long-term
    development, 389
Misrepresentation, 332, 365
Mnemonics, 101, 147
Modifiers
  compound, 148–150, 160
  dangling, 123–124
  definition of, 114, 117
  meaning and, 117
  misplaced, 122–123
  phrases as, 117
Motivation of readers, 282
Multimedia, 7, 321, 322, 331
Multisyllabic words, 235

Navigation, 9, 25, 275, 279, 324, 325
  organization and, 327
Negative constructions, 224–225
Noise, 25
  in hypertext, 29
  mispunctuation as, 133
  in tables, 167
  types of, 28–29
Nominalizations, 230–231
Nondiscriminatory language,
    243–247
  editing to achieve (ex.), 244–245
  in illustrations, 307
  nonsexist language, 246–247
Nonfinite verb. *See* Verbal
Nonrestrictive clause, 141–142
Nonverbal communication, 351
Nouns, 115
  collective, 119
  compound, 148
  concrete vs. abstract, 233–234
  verbs as, 230
Noun phrase, 118
Numbers
  as access device, 277
  in compound adjectives, 149–150
  consistency of, 86
  conventions of use, 159
  for cross-reference, 292
  for illustrations, 72, 73
  in lists, 277
  spelling of, 86, 159
  in technical vs. nontechnical
      material, 159
  in translation, 159

Object
  direct, 114, 116, 117
  of preposition, 152
Objectivity, limits of, 364
Offset lithography, 380–382
Ongoing document use, designing
    for, 282
Online editing, 49–61
  comparison programs and, 60
  for comprehensive goals, 208–209
  copymarking in, 60
  hard copy and, 60–61, 209, 349
  vs. markup, 56–58
  policies and methods, 55, 58, 59
  reasons for, 10, 11
  tracking changes in, 59–60
  trespass and, 59
  version control and, 10, 58–59
Online documents
  color and, 329–330, 334
  comparisons with print, 322–323,
      325
  content of, 325–326
  copyright of, 332, 359
  design of, 325–332
  editing process for, 332–336
  fair use of, 332

grammar and, 331
hardware and software
    considerations in, 329–330
identifying information on, 326,
    327
illustrations and, 331, 334
navigation in, 9, 25, 275, 279,
    324–325
organization of, 256, 258, 263,
    324–329
screen design and, 329
style and, 329
uses by readers of, 25, 334
visual design of, 291, 329–331
  See also Hypertext
Online help, 322, 327–328
Opacity, 378
Operational signs, 165
Order of importance, 260
Organization, 11, 25, 27
  comprehension and, 253, 254–257
  content-based, 254–257
  and content completeness, 253,
      260, 262
  conventional patterns of, 259–260,
      327
  editing for (ex.), 264–269
  of editor–writer conference, 348
  general-to-specific, 258–259
  grouping related material, 260–263
  hierarchy in, 27, 28, 255, 263
  of hypertext, 256, 258, 263,
      324–325, 326–329
  of illustrations, 302
  learning and, 254–257
  of lists, 260–263
  meaning and, 253, 256, 259–260
  of paragraphs, 263–264
  for performance, 253–254
  principles of, 257–263
  reader expectations and, 258
  signals to, 27
  task-based, 253–254
Outdent, 276

Page proof, 176, 379, 382
Paper
  basis weight, 377
  coating, 365, 376, 378
  opacity, 378
  readability and, 376
  recycled, 365
  size, 377
Paragraph
  copymarking for, 37
  organization of, 263–264
  proofreading marks for, 181
  short vs. long, 276
Parallelism, 143
  faulty, 144, 146, 343
  in headings, 283
  in paragraphs, 263
  punctuation and, 145

sentence structure and, 144–146,
    219–220
Parentheses
  copymarking for, 36
  in equations, 162
  in fractions, 161
  proofreading marks for, 179
  punctuation with, 151
Participial phrase, 118
Participle, 115, 123, 135
  dangling, 123–124
  in passive voice, 231–233
Parts of speech (table), 115
Pass, editorial, 77, 198, 201–202
Parts of a book, 68–71
Passive voice
  dangling modifiers and, 124
  as style choice, 231–233
Patents, 357, 359–360
Patterns of information. *See*
    Organization
Perfect binding, 384
Periods
  abbreviations with, 107
  copymarking for, 36, 41
  as ellipses, 152
  with parentheses, 151
  proofreading marks for, 179
  with quotation marks, 151
  typing spaces after, 153
Permissions, 332, 358–359
Persona
  appropriateness of, 29
  style and, 213
Perspective in illustrations, 304–305
Photographs, 379
Phrase
  clause vs., 135
  introductory and internal,
      punctuating, 146–147
  single word vs., 234–235
  types of, 114, 117–118
Pica, 372
Pica stick, 375
Planning, 4–5, 9, 254, 346
  in comprehensive editing, 10, 199,
      201, 206
  in management, 388–389
Plate, 381
Plurals, 148, 160
Point, 372
Policies
  for editing, 398–399
  for ethics, 366–367
  for online editing, 55, 58, 59
Portrait orientation, 277
Positive constructions, 224–225
Possessives, 87, 147–148
Predicate, 113, 118
Preface, 70
Prefixes
  hyphenation with, 150
  spelling words with, 101

Prepositional phrase, 117–118
Prepositions, 115
Printing, 380–383, 398
Procedural markup, vs. descriptive, 53
    See also Copymarking
Production
    basic process of, 14–15
    binding, 383–384
    color printing, 382–383
    of illustrations, 379
    imposition, stripping, and platemaking, 379–380
    printing, 380–383
    role of editor in, 12, 370, 378
    typesetting and page makeup, 378–379
Productivity, 394
Product safety and liability, 361–362
Product team
    composition of, 4, 8
    in editorial review, 202
    in project management, 388
Professionalism
    codes of conduct and, 362–364
    editor–writer relationship and, 346–347
    language use and, 113, 134
    responsibilities associated with, 364
    management and, 389
    See also Ethics
Project management. *See* Management
Pronoun
    agreement with antecedent, 125
    case, 126
    definition of, 115, 125
    number, 125
    possessive, 125
    reference, 126, 127
    relative, 125, 128, 136
Proof, 176, 177
Proofreading, 176–187
    basic procedures in, 176–177
    computer programs for, 184–185
    copyediting vs., 176
    goals of, 177
    marks for (tables), 179, 180, 181
    online publication, 185
    strategies for, 185–187
Publication, 7
Punctuation, 133–153
    of complex sentences, 140–141
    of compound sentences, 138–140
    of compound-complex sentences, 142
    copymarking of, 36, 41
    copymarking symbols for (table), 36
    of equations, 163–164
    marks, 150–153
    method of editing for, 153

of phrases, 143–147
proofreading marks for (table), 179
of sentence types (table), 137
in series, 143–144
of simple sentences, 138
value of, 133–134
within words, 147–150
    See also specific types
Punctuation marks, typing, 152–153

Quality assurance
    editing and, 4, 14, 362, 363
    ISO 9000 and, 172
    management and, 387–388
    proofreading and, 399
    in World Wide Web documents, 332
Quantities and amounts, 127–128
Quantitative material. *See* Mathematical material; Statistics
Query, 44–46, 59, 261, 352
Query slips, 44
Question mark, 179
Quotation marks
    with other punctuation, 151
    proofreading marks for, 179

Ragged right, 376. *See also* Justification
Readability
    comprehension and, 22
    of illustrations, 72
    justification and, 376
    leading and, 374
    letter- and wordspacing and, 375
    line length and, 376
    paper and, 376
    serif vs. sans serif type and, 371
    sentence length and, 223
    of tables, 72, 169–171
Reader expectations
    culture and, 22–23, 247, 250
    organization based on, 258
    for passive voice, 231–233
    style and, 213, 221, 232–233
    visual design and, 279–280
Readers, 19–30
    creating meaning and, 24
    editorial focus on, 16–17
    editors as, 16, 258–259
    interaction with text and, 22, 24–25
    motivating via visual design, 282
    needs of, and illustrations, 295–298
    online documents and, 25, 323, 324, 326
    purposes for reading and, 200
    questions asked by, 26, 258, 267
Reading
    for comprehension, 25–26
    hypertext and, 323, 326, 324, 329
    interactive nature of, 24–25

to learn vs. to do, 25–26
    noise in, 28–29
    purposes for, 22
    selective, 24–25, 279, 324
    signals to aid in, 25, 27–29
Record keeping, 393–394
Recycling paper, 365
Redundant categories, 235–236
Redundant pairs, 235
References, 70, 83, 87
Registration, 382–383
Relationships of words in sentences, 118–122
Relative pronouns, 125, 128, 136
Repetition for cohesion, 264
Representational illustrations, 300
Repros, 379
Request for proposals. *See* RFP
Restrictive clause, 141–142
RFP, 257, 264–266, 291
Risks and warnings, standards for, 361
Root screen, 327, 328
Root words, 101
Rules in tables, 167
Running heads, 71

Saddle stitching, 384
Safety
    editor's responsibility and, 206–207, 248, 361
    labels, 361
    misrepresentation and, 365
    product, 361
    warnings and, 206–207, 277, 278, 282, 361
    See also Liability
SAIC, 7–11
Sampling, 394
Sans serif, 371
Scale of illustrations, 304–305
Scanner, 379
Scheduling, 5, 344, 395–397
Schemata
    illustrations and, 297
    organization and, 27, 254–257
    visual design and, 288
Schema theory of learning, 254–257, 259
    reading and, 27
    visual design and, 288
Screens for halftones, 379
Screen design
    contrast and, 330
    effects of hardware and software on, 329
    instructional goals and, 8, 10, 11
    legibility and, 330
    tables and frames in, 330
    visual consistency in, 83, 85, 279, 330–331
Scripts, 9
Selective reading, 24–25, 279

Semantics, 83, 84, 117, 121
Semi-block format, 279
Semicolon
    in complex sentence, 137
    in compound sentence, 140
    in compound-complex sentence, 142
    copymarking for, 36, 41
    proofreading marks for, 179
    quotation marks with, 151
    in series, 144
Sentence core, 129
    action verbs and, 228–229
    independent clause as, 134, 216
    main idea in, 217–218
    meaning and, 216
    style and, 210
    in "there is" and "it is" sentences, 218
Sentences
    complex, 140
    compound, 138–140
    compound-complex, 142
    elements of, 113
    length of, 222–223
    modifiers in, 117
    objects and complements in, 113–117, 119–121
    parallelism in, 144–146, 219–220
    parts of (table), 114
    patterns of, 113
    pronouns in, 125
    punctuation of, 138–142
    relationships of words in, 118–122
    simple, 138
    subjects and verbs in, 112–117
Sentence structure
    arrangement of parts, 220–222
    cohesion and, 220–221, 233
    end focus in, 220–221
    main idea and, 217–218
    meaning and, 118, 216–220, 226
    parallelism and, 144–146, 219–220
    positive vs. negative constructions, 224–225
    subordination and, 218–219
    S-V-O and S-V-C word order, 222
    "there are" and "it is" openers, 218
    verbs and, 113
Sentence types and punctuation (table), 137
Series
    adjectives in, punctuating, 144
    commas in, 87, 144–146
    semicolons in, 144
Series comma, 144–146
Serif, 371
Setting Priorities, 394–395
Sexist language, 246–247
SGML, 52–54
    completeness check and, 54
    for databases, 54
    as descriptive markup, 53

DTD, 53
    editing and, 53, 54
    electronic markup and, 49, 52–54
    FOSI, 53
    information management and, 54
    as ISO standard, 172
    tags, 52, 53
Shading
    in illustrations, 302
    in tables, 167
Side stitching, 384
Signals
    comprehension and, 26, 27
    headings as, 279
    structural, 27
    verbal, 27
    visual, 28
    See also Noise
Signature, 377, 378, 379–380
Signs of aggregation, 161
Signs of relation, 165
Simple sentence, 138
SI units, 108, 159
Size of illustrations, 304–305
Small caps
    copymarking for, 35
    proofreading marks for, 180
Society for Technical
        Communication
    Ethical Guidelines, 363
    Guidelines for Authors, 345
Solid fraction, 161
Solidus, 161
Spacing
    copymarking symbols for (table), 37
    letter- and word-, 373–375
    proofreading symbols for (table), 181
    See also White space
Spatial organization, 259–260
Specifications
    design, 274, 291
    for documents, 78, 172, 344, 394, 398
    estimates and, 394, 398
    ethics and, 362, 366
    for products, 172–173
Spelling, 98–105
    consistency in, 85–86
    guidelines in editing for, 99–102
    international variations of, 105
    frequently misspelled words (table), 100
    frequently misused words, 102–105
    with prefixes and suffixes, 101–102, 150
Spelling checker, 60, 78, 99–100, 184
Spiral binding, 384
Stacked fraction, 161–162
Standard American English, 66

Standards
    estimates and, 389
    ethical and legal, 366–368
    for publications, 5, 172
    for risks and warnings, 361
    SGML as, 52, 53
State names, abbreviations for, 109
Statistics, 165–166
"Stet," 35, 179
Storyboard, 9, 329
Stripping, 380
Structural signals, 27
Structural illustrations, 300
Style, 212–251
    analysis of (ex.), 214–216, 236–238, 244–246
    comprehension and, 212, 213–214, 216–220
    computers and editing for, 238–239
    context and, 216, 221, 236
    consistency of, 84
    cultural expectations for, 213, 247, 250
    definition of, 212–214
    editing for (ex.), 236–240
    global issues and, 247–251
    guidelines for effective, 216–225, 228–236
    human agents and, 223–224
    localization and, 249–250
    meaning and, 240
    method of editing for, 240
    nominalizations and, 230–231
    nondiscriminatory language and, 244–247
    persona and, 213
    positive and negative constructions and, 224–225
    sentence structure and, 216–220
    tone and, 213
    translation and, 248–249
    types of (box), 89
    verb choices and, 228–233
    word choices and, 233–236
Style manuals, 89–94
    comprehensive, 90
    discipline, 90–92
    house or organization
        contents of, 5, 92–94,
        editor–writer relationship and, 344
        ethics and, 357, 366–367
Style sheet, 94–95
    electronic, 94–95
    examples of, 95, 96
    function of, 10, 94, 98
    hyphens and dashes on, 150
    illustrations and, 73
    what to record on, 5, 9, 94
Styles, electronic, 50–51, 78, 95, 185, 208, 385
    See also Templates

Subject, 113, 114, 118
Subject complement, 120–121
Subject matter expert, 259, 260, 342
Subject-verb agreement, 119
Subordinate clause. *See* Dependent
    clause
Subordinating conjunctions,
    135–136, 140
Subordination, as clue to meaning,
    218–219
Subscript, 35
Substantive editing. *See*
    Comprehensive editing
Suffixes
    in nominalizations, 230
    spelling words with, 101–102, 150
Superscript, 35
S-V-C, S-V-O word order, 222
Syntax, 83, 84

Table of contents, 69, 279
Tables, 167–171
    copyediting (ex.), 168–171
    copyediting guidelines for,
        167–168
    functions of, 167
    headings for, 170
    placement of, 302
    when to use, 276
Tables and figures, list of, 69
Tables and frames in screen
    design, 330
Tags (SGML, HTML), 52, 53
Task analysis, 254
Task-based order, 253–254
Team, product, 4, 8, 11
Technical editing
    ethics of, 17, 206–208, 362–368
    genres of, 16
    goals of, 11
    production functions of, 12,
        370, 378
    qualifications for, 16–17
    "technical" aspect of, 15
    setting for, 15
    subject matter and method of, 15
    text editing function of, 12–14
    See also Copyediting;
        Comprehensive editing
Technical editor. *See* Editor
Technical material. *See* Mathematical
    material; Statistics
Technical review, 6, 14, 58, 202,
    346–347
Templates, 5, 59, 60, 78, 385, 396
    desktop publishing and, 49, 50–52
    learning and, 255
    limits of, 22
Tense, verb, 121–122
"There are" openers, 119, 218
Title page, 69

Titles, in illustrations, 72, 73
Tone, 213
Top-down editing, 129, 153, 201, 266
Tracking, 59, 395–398
Trademarks, 357, 359–360
Trade secrets, 357, 360
Transitive verb, 116
Translation, 248–249
Transmission, electronic, 34, 60, 79
Transposition
    copymarking for, 35
    proofreading marks for, 179
Trespass, 58–59
Turnover, 46
Tutorial, development of, 7–11
Typeface, 371–372
Typesetting
    desktop publishing vs., 379,
        384–385
    distinguishing headings in, 284,
        373
    of mathematical material, 161–164
Type size, 372–373
Type style
    consistency in, 85, 88
    of headings, 284, 373
    serif vs. sans serif, 371
    as visual signal, 28
    when to vary, 277
Typography, 84, 370–376
    consistency in, 85
    leading, 373–374
    letter- and wordspacing, 375, 376
    line length, 375–376
    proofreading marks for (table), 180
    typefaces, 371–372
    type sizes, 372–373
    type styles, 85, 88, 273, 277,
        378, 385
Underlining, 277
Unit modifiers, 150, 160
URL, 322, 323, 326
Usability
    definitions of, 5, 11
    editorial responsibilities and, 256,
        391
    visual design and, 29, 279–282
Usability testing, 14, 166, 336, 395
Usage, 127–128
    idioms, 128
    quantities and amounts, 127
    relative pronouns, 128
    in statistics, 166
Usage panel, 112
U.S. Customary System, 108, 159

Verbal, 115, 123–124
Verbal signals, 27
Verbs, 113–117, 228–233
    action, 228–229
    active vs. passive, 231–233

faulty predication, 120
    intransitive, 116
    linking, 116
    nominalization of, 230–231
    pronoun agreement with, 125
    sentence arrangement and, 220
    sentence patterns and, 113
    strong vs. weak, 229–230
    subject agreement with, 119
    tense of, 121–122
    to be, 116
    transitive, 116
    types of, 113–117
Variations for cohesion, 264
Version control, 10, 58–59
Visual design, 273–293
    aiding comprehension via (ex.),
        285–290
    content and, 291–292
    for ongoing use, 282–283
    functions of, 278–283
    guidelines in editing for, 291–292
    instructions and, 292
    motivating readers via, 282
    options for, 275–278
    production method and, 291
    providing access to information
        via, 279
    reader expectations and,
        279–280
    reading complexity and, 292
    visual identity and, 280
    warnings and, 282
    See also Headings; Screen design
Visual signals, 28
    See also Visual design
Voice. *See* Passive voice, Active
    voice.

Warnings, 277, 278, 361
White space, 277–280, 292, 302
World Wide Web
    illustrations and, 295, 307,
        331, 334
    political, ethical, and legal issues
        of, 332
    quality assurance and, 332
    See also Hypertext; Online
        documents
Words
    frequently misspelled, 100
    frequently misused, 102–105
    multisyllabic, 235
    phrases vs., 234
    spacing between, 375
    using concrete, accurate,
        233–234
Wordspacing, 375, 376
Work for hire, 357

x-height, 372